Sonic Interaction Design

Sonic Interaction Design

edited by Karmen Franinović and Stefania Serafin

The MIT Press
Cambridge, Massachusetts
London, England

MIT Press books may be purchased at special quantity discounts for business or sales promotional use. For information, please email special_sales@mitpress.mit.edu or write to Special Sales Department, The MIT Press, 55 Hayward Street, Cambridge, MA 02142.

This book was set in Stone Sans and Stone Serif by Toppan Best-set Premedia Limited, Hong Kong. Printed and bound in the United States of America.

Library of Congress Cataloging-in-Publication Data

Sonic interaction design / edited by Karmen Franinović and Stefania Serafin.
  pages cm
Includes bibliographical references and index.
ISBN 978-0-262-01868-5 (hardcover : alk. paper)
1. Sonic interaction design. 2. Product design. 3. Sound in design. 4. Human-computer interaction. I. Franinović, Karmen, 1975—editor of compilation. II. Serafin, Stefania, 1973—editor of compilation.
TS171.S624   2013
004.01'9—dc23
2012028482

10 9 8 7 6 5 4 3 2 1

Contents

Introduction

Sound is an integral part of every performative and aesthetic experience with an artifact. Yet, in design disciplines, sound has been a neglected medium, with designers rarely aware of the extent to which sound can change the overall user experience. The shaping of the auditory character of products has been dominated by functional priorities such as noise reduction, the representational uses of sounds as auditory icons, and sound as emergency signal or alarm. These roles that sound has played as functional, iconic, and signaling element during the interaction with an artifact have arguably limited our everyday experiences with sound. *Sonic interaction design* (SID) emerged from the desire to challenge these prevalent design approaches by considering sound as an active medium that can enable novel phenomenological and social experiences with and through interactive technology.

By using sound in an embodied and performative way, this burgeoning new field works with emergent research topics related to multisensory, performative, and tactile aspects of sonic experience. Yet, it builds on existing knowledge and themes that have been at the heart of sound discourses for decades, such as the relationship between sound and location, issues of privacy and power, and the ability of a sounding artifact to communicate information. A deep examination of such topics requires the involvement of diverse perspectives and methods belonging to a range of different disciplines including interactive arts, electronic music, cultural studies, psychology, cognitive sciences, acoustics, interaction design, and communication studies. Thus, SID can be defined as an interdisciplinary field of research and practice that explores ways in which sound can be used to convey information, meaning, and aesthetic and emotional qualities in interactive contexts. As a design practice, SID is the creative activity of shaping the relationships between artifacts, services, or environments and their users by means of interactive sound. In this context, the goal of SID is not only to extend the existing research on interactive sound but also to foster new application areas and new domains of practice for sound designers, architects, interaction designers, media artists, product designers, and urban planners.

Product Design: Reduction versus Quality

Since its origins, sound design for products has focused on the elimination of undesirable noise that is produced through physical interaction. In the early 1950s, several companies begun to apply evaluation to their products through precise measurements of the level of sound produced during their use. Such tests allowed designers and engineers to accordingly modify the mechanical design of their products in order to reduce sonic annoyance for the user. Today, similar measurements are combined with psychophysical methods well suited to characterize the acoustic annoyance of the user for products such as vacuum cleaners, coffee machines, and air conditioners [14]. However, this type of evaluation often fails to account for the users' emotional and cognitive responses to the functional and aesthetic aspects of a product [3]. Thus, more recently, psychologists have begun to explore emotional responses to sound in products beyond so-called user preference tests [15], reflecting a more general trend in design research of exploring the emotional impact of products [11].

Although in practice the trend toward reducing annoying sound, rather than designing its quality, still dominates in product design, there are a few cases in which sound is used to affect the overall perceived quality of a product. For example, in order to provide strong product identities, sonic branding professionals have begun to use sound to enhance the use of a product rather than to deploy sound solely as an addition to the visual advertisements. Various luxury goods are designed in a way that action performed with the product generates desirable sounds, as for example in opening of a perfume bottle, starting a motorcycle, or closing the door of a car [12]. Meanwhile, scientific researchers have demonstrated that other senses such as the sense of taste can be affected through sound, leading them to conclude that

the auditory cues elicited by our contact or interaction with different surfaces (such as abrasive sandpapers or even our own skin) and products (including electric toothbrushes, aerosol sprays, food mixers, and cars) can dramatically change the way in which they are perceived, despite the fact that we are often unaware of the influence of such auditory cues on our perception. [13]

Although designers could benefit from applying such perceptual phenomena to products, the ethical issues related to the use of multisensory effects without the awareness of the user still remain to be discussed.

Human-Computer Interaction: Metaphors and Sonification

In digital products, sound has been predominantly used to accompany screen-based interaction or to present information in the form of sound. Auditory displays can help extract meaning from complex data, alert to a particular event, or signal the accomplishment of an action performed with a graphical icon on the screen [8]. Such representational uses of sound have defined a number of functional roles for computer sound as well as contributed to development of new design tools. Extensive research was conducted on *earcons*—abstract sounds that accompany different computer events [2]—and *auditory icons*—everyday sounds associated with interactions on the computer desktop [5]. In this context, everyday sounds have proved to be efficient interactive metaphors because users could easily associate such sounds with physical events that cause them. A typical example is the sound of crushing paper when moving a file to a trash can on a computer desktop [6]. This research also presented a designerly way of thinking about sound, suggesting that its qualities should be exaggerated in order to improve the perception of the sound event and thus facilitate the interaction.

More recently, the auditory display community has begun to address the role of human action in exploring sonified data sets, introducing the notion of interactive sonification [7]. The auditory signals provide feedback about the data under analysis and can be used to continuously refine a targeted activity. Interactive sonification has proved to be a useful approach to monitoring activities related to sport, rehabilitation, and everyday tasks. However, limited to screen icons and the computer mouse, the user is engaged only in reduced hand movements, and the performative aspects of sound cannot be exploited to their full potential. In order to foster an embodied experience, both the interface and its sonic behavior must be carefully designed. This is an important issue in interaction design, where the availability of physical computing resources has enabled the development of tangible interfaces.

Musical Instruments: Physicality and Gesture

New musical instruments research explores the importance of tangible computing in creating expressive interaction. New interfaces for musical expression are designed to engage the user's gestures, most often through physical manipulation of the interface. Such research and practice are complementary not only to traditional computer software interaction limited to the mouse and the keyboard but also to an empty-handed gesture, one that seems to dominate video art installations and HCI research in virtual environments [4, 10]. These developments in music technology research led scholars in perception and cognition to expand their focus from studying the human as a receiver of auditory stimuli to gaining deeper understanding of the perception-action loops that are mediated by acoustic signals [9].

Although the consideration of sound as a temporal medium that can engage performative uses remains largely in the field of music technology, the sound and music computing community started to bridge the gap between musical instrument development and product design. They were among the first to argue for the importance of design principles and methods for the development of sonic interactive systems rather than addressing solely engineering and computing problem-solving approaches [1]. Because both musical and sound artifacts designed for gestural interaction share numerous engineering, perceptual, sound synthesis, and design problems, there is much to learn from research on new interfaces for musical expression, especially on the issues of mapping and continuous feedback. However, the nonmusical interaction poses social and cultural questions that differ largely from those linked to musical context and stage performance.

Sonic Interaction Design: From Representation to Performance

Stemming from existing approaches to designing sound, SID aims to combine knowledge from a number of fields such as interactive arts, electronic music, cultural studies, psychology, cognitive sciences, and acoustics. It shifts the focus from reception-based auditory studies and representation-based design strategies toward more active, embodied, and emotionally engaging uses of sound. Multimodality, especially the tight connection among audition, touch, and movement, is examined in an ecological framework in order to develop new design principles and apply these to novel interfaces. Confronting the challenge of creating adaptive systems that sonically respond to the physical actions of the user, SID explores questions of embodiment and performance and investigates how sonic gestures can convey emotions and promote expressive and creative experiences. In this sense, SID follows the trends of the so-called third wave of human–computer interaction, where culture, emotion, and experience, rather than solely function and efficiency, are the scope of interaction between human and machines [11].

From the methodological point of view, this requires novel perspectives that move away from the rigid guidelines and techniques that have been traditionally adopted in the sound and music computing communities. Strictly technical recommendations and formal listening tests are moved aside in SID and replaced by design and evaluation principles that are more exploratory in nature. The design challenges are no longer of a technical nature because nowadays the wide availability of physical computing resources allows practitioners who are not technically trained to easily produce interactive sound artifacts. Instead, the challenges rely on ways in which designers may successfully create socially meaningful, physically engaging, and aesthetically pleasing sonic interactions. Thus, new pedagogical and design methods are needed in order to bring SID to professionals and students in a creative and designerly manner. For example, situated and explorative studies of the connections between actions and sound in the real world may help designers learn how to transfer the natural and ecological relationships to the development of intuitive sonic interfaces.

The multisensory aspects of interactive sonic experience that SID is concerned with must be designed with consideration of the orchestration of the auditory, tactile, visual, and kinesthetic senses within real-world context. To come closer to reaching such a complex goal, the field of SID, which is in its infancy, must engage with a wide range of research topics including the perceptual, cognitive, and emotional study of sonic interactions, improved models for the reception of sound and its role in performance of actions, adapted design methodologies, sound synthesis technologies and their use, and finally, design and evaluation methods addressing the individual and social experience with sounding objects. For a new generation of sound designers to be capable of addressing the interdisciplinary problems the field raises, a more solid foundation needs to be developed that can draw on such bodies of knowledge. It is our hope that this book partially accomplishes this task.

Overview

Sonic Interaction Design collects contributions from historians, scientists, artists, and designers working in the fields of sound culture studies, interaction design, sound and music computing, interface engineering, auditory cognition, and interactive arts. Through an overview of emerging topics, theories, methods, tools, and practices in sonic interaction design research, the book aims to establish the basis for further development of this new design area. The book is divided in two parts: a theoretical part composed of long chapters presenting key themes in SID and an application-oriented part with a number of case studies.

Chapter 1 reflects on everyday practices with sound-producing objects in the past. Three auditory aspects are examined: the significance of sounding objects as sources of information, as objects of noise control, and as artifacts facilitating privacy through acoustic cocooning. Chapter 2 presents a critical reflection on the uses of interactive sound in art and design research addressing related social and cultural aspects. The discussion of various explorations into multisensory experience traces a context for novel application of interactive sound. Challenges encountered when designing for continuous and adaptive sonic response are discussed in chapter 3, which stresses the importance of the use of tactile and kinesthetic aspects of sound in interaction. Chapter 4 introduces novel pedagogical methods focused on teaching new design approaches to interactive sound by actively engaging students with sonic material. The methods and problems in evaluating the self-produced sound are discussed in chapter 5, and ways of extending classical auditory cognition listening tests to include physical performance are proposed.

The second part of the book presents various examples of products and systems that are created for the context of mobile music, enactive learning, rehabilitation, gaming, participatory media, and others. Thirteen case studies are presented, which cover practical applications and theoretical considerations of the use of sound in interactive artifacts. The first set of case studies presents examples outlining the tight connection between sound and touch, a connection that was already investigated in chapter 3. Examples of interfaces exploring such connections are the *Pebblebox*, a box filled with pebbles whose motion is detected by a microphone and used to control a granular synthesizer, the *Sound of Touch*, an installation where sound is produced by interaction with different physical gestures, and the *Gamelunch*, an exploration of a sonically augmented dining scenario. *Zizi* uses sound to allow the user to develop a more affective relationship with his or her furniture. Interaction between sound and touch in mobile devices is described in the *SonicTexting* application. An interface for interacting with musical content called *A20* exemplifies a user-centered design process applied to SID. Interactive sound has also been used for rehabilitation purposes and to sonify EEG signals, as two case studies describe. A novel approach to sonifying data is the high-density sonification technique, described in one case study. The use of sound in virtual reality and mixed reality is also described, with particular attention given to impact sounds and surround-sound synthesis. Rhythmic interaction is important in creating a deep relationship between people and technology using sound. As a case study outlines, rhythmic interaction is important to provide feedback in repetitive tasks. As a last contribution, workshop-based methods to explore the use of sound in interactive commodities are proposed.

Combined together, the chapters and the case studies present an overview of the different applications and theoretical approaches that are currently being investigated in the sonic interaction design field. While presenting the existing discourses and providing the state of the art in sonic interaction design, this book proposes a fresh perspective on interactive sound as a situated and multisensory experience.

Acknowledgments

This book would not have been possible without the support of researchers involved in the Sonic Interaction Design Action funded by COST (COST-SID IC601), an intergovernmental framework for European Cooperation in Science and Technology. Its members contributed much of the content presented here. The SID COST community was also crucial for reviewing the individual chapters of the book. Karmen would like to thank her colleagues at the Interaction Design program and the Institute for Cultural Studies in the Arts at Zurich University of the Arts. Stefania would like to thank colleagues at the Department of Architecture, Design, and Media Technology at Aalborg University. Finally, we would like to thank Doug Sery of MIT Press and the anonymous reviewers for MIT Press who provided us with invaluable feedback that led to improvement of the book.

References

1. Bernardini, N., Serra, X., Leman, M., Widmer, G., & De Poli, G. (2007). The roadmap for sound and music computing. http://www.soundandmusiccomputing.org/roadmap.

2. Blattner, M., Sumikawa, D., & Greenberg, R. (1989). Earcons and icons: Their structure and common design principles. *Human-Computer Interaction*, 4(1), 11–44.

3. Blauert, J., & Jekosch, U. (1997). Sound-quality evaluation—A multi-layered problem. *Acustica—Acta Acustica*, *83*(5), 747–753.

4. Cadoz, C. (1988). Instrumental gesture and musical composition. In *Proceedings of the international computer music conference* (pp. 1–12). International Computer Music Association.

5. Gaver, W. (1989). The sonic finder, a prototype interface that uses auditory icons. *Human-Computer Interaction*, 4, 67–94.

6. Gaver, W. W. (1993). How do we hear in the world? Explorations in ecological acoustics. *Ecological Psychology*, 5(4), 285–313.

7. Hermann, T., & Hunt, A. (2005). An introduction to interactive sonification. *Multimedia, IEEE*, *12*(2), 20–24.

8. Kramer, G. (Ed.). (1994). *Auditory display: Sonification, audification, and auditory interfaces*. Reading, MA: Addison-Wesley.

9. Leman, M. (2007). *Embodied music cognition and mediation technology*. Cambridge, MA: MIT Press.

10. Miranda, E., & Wanderley, M. (2006). *New digital musical instruments: Control and interaction beyond the keyboard*. Middleton, WI: A-R Editions.

11. Norman, D. A. (2004). *Emotional design: Why we love (or hate) everyday things*. New York: Basic Books.

12. Sottek, R., Krebber, W., & Stanley, G. (2005). Tools and methods for product sound design of vehicles. In *SAE 2005 noise and vibration conference and exhibition*. Grand Traverse, MI: SAE.

13. Spence, C., & Zampini, M. (2006). Auditory contributions to multisensory product perception. *Acta Acustica united with Acustica, 92*(17), 1009–1025.

14. Susini, P., McAdams, S., & Winsbe, S. (1999). A multidimensional technique for sound quality assessment. *Acustica—Acta Acustica, 85,* 650–656.

15. Vastfjall, D., & Kleiner, M. (2002). Emotion in product sound design. In *Proceedings of the international symposium les journées du design sonore*. Paris, France.

I Emergent Topics

1 Listening to the Sounding Objects of the Past: The Case of the Car

Karin Bijsterveld and Stefan Krebs

In 1929, the German engineering professor Wilhelm Hort published an article [57] on street noise in *Der Motorwagen*, a journal for the automotive industry (figure 1.1).[1] The article was a characteristic product of the 1920s. At that time, papers and pamphlets discussing the "nerve-racking" noise of big cities such as New York, London, Paris, and Berlin appeared quite frequently. What makes Hort's article particularly noteworthy in the context of this book, however, is his alphabetical list of ninety-one words that describe a wide variety of sounds. The list starts with words such as *ächzen, balzen, bellen, blöken, brausen, brodeln,* and *brüllen*. It further includes sounds such as *donnern, dröhnen, gellen, girren, hämmern, heulen, klatschen, klingeln, knallen, krähen, lachen, meckern, pfeifen, prasseln, prusten, rollen, schellen, schlagen, schreien,* and *schmettern*. And it ends with *ticken, trampeln, trappeln, trommeln, tropfen, weinen, wiehern, wimmern, zirpen, zischen,* and *zwitschern*.[2] In between brackets, Hort indicated which sounds were directly related to street life: the barking of dogs, the trampling of horses' feet, the whiplash of coachmen, the yells of street vendors, the clang of coal being poured out, the pulling down of trash cans, the rumble of street cars, the bang of automobiles, and the shrieks of automobile signals [57].

Hort's list expresses the enormous variety in the sounds he experienced as well as his special interest in the abatement of street noise. Today, we do not instantly recognize all the sounds he identified. Nor do we automatically understand why they were a source of such concern to him. Because some of the sounds mentioned—the horses' trampling, the coachmen's whiplashes, and the coal's clanging—are no longer everyday phenomena of street life, the words for sounds may not speak to us in the same way as they did to early twentieth-century people. Most of us, for instance, will not immediately understand the sonic subtleties connected to the *trampeln* or stamping of ungulates versus their *trappeln* or trampling. Car noise is still omnipresent, but automobiles do not *bang* anymore. If concerns about vehicle noise survive to this day, they certainly do not include Hort's obsession with street noise in all its various manifestations.

Figure 1.1
Frontispiece of *Der Motorwagen*.
Source: *Der Motorwagen* 7 no. 1 (1904).

Our brief consideration of William Hort's early twentieth-century list of sounds serves to introduce what we wish to explore in this chapter. This book is about designing objects—ranging from domestic appliances to interfaces and data sets—in such a way that their sounds help users to interact with the objects in a purposeful, engaging, and expressive manner. As a field of research and practice, sonic interaction design thus "explores ways in which sound can be used to convey information, meaning, and aesthetic and emotional qualities in interactive contexts" (Franinović and Serafin, introduction in this volume). This implies that not only is the construction of the sounds themselves relevant to sonic interaction designers but also the way in which users evaluate these sounds, give meaning to them, or are affected by them (Giordano, Susini, and Bresin, chapter 5 in this volume).

Still, the ways in which people listen to and interpret sounds are highly dependent on context and vary across time and space, even though the historiography of sound has also shown some remarkable continuity in the cultural connotations of particular sounds. This chapter aims to raise the awareness among sonic interaction designers of such historical continuities and discontinuities between the meanings of sounds from the past and the present. Our argument underscores that sonic interaction design is never a static endeavor: with the passage of time, sounds are bound to acquire new meanings. If sound designers remain concerned with the design of sounds for more and more new objects, it is our hope that they also feel challenged to do so for objects we have already lived with for some time.

We show the historical shifts and stabilities in the meaning of sound by focusing on the history of the car as sounding object in particular and by largely limiting our scope to the twentieth- and twentieth-first-century Western world. Unraveling the shifting sounds of one sounding object over a limited time span allows us to track the specific continuities and discontinuities in the auditory cultures featuring cars. Because the automobile was one of the earliest artifacts to be subject to both noise regulation and sound design, it is a highly relevant sounding object in the context of a sonic interaction design book. And although our chapter is not conceived to be a history of sonic interaction design as a field, we included many early examples of engineers who aimed to design artifacts, notably cars, that would speak to their users in meaningful or agreeable ways.

Our historical focus on sounding objects implies that we leave out many of the other sounds associated with the past. For one thing, the sounds produced by the human body, such as the laughing and yelling listed by Wilhelm Hort, are ignored in our story. The same holds for the many animal sounds he wrote about. One should not forget, however, that human and animal sounds comprised much of the everyday sonic environment in the era before the Industrial Revolution, and even today such sounds spark new debates every now and then. The putative nuisance of childrens' cries at playgrounds, for instance, was put on the political agenda in both Germany and the Netherlands very recently.

We develop our argument in four steps. First, we discuss what the rapidly expanding historiography of sound has to say about sounding objects in the era prior to the twentieth century—prior to the age of the automobile. This not only widens the scope of this chapter but also clarifies how the twentieth-century sound design of cars has its roots in pre-twentieth-century cultural connotations of sound. Second, we focus on the early days of car sound design, notably the rise of the ideal to create a luxuriously "silent" interior car acoustics. Third, we explain how such aims acquired an even higher significance in the automotive industry once listening to the car engine as a source of information about its functioning had been reassigned from the ears of the driver to those of the mechanical engineer. Finally, in our fourth section, we explain how the rise of international noise abatement regulation, sensory marketing, and digital technologies fostered an enhanced focus of the automotive industry on interior *and* exterior sound in the 1990s.

As we will show, however, these new contexts and options in the 1990s did not mean that the automotive industry could afford to neglect the continuities in the cultural meaning of sound. On the contrary: sound design happened to work best when taking culture into account. So, let us first introduce the roots of this culture of sound by exploring how pre-twentieth-century everyday life and its material objects sounded—and what it meant to those who heard these sounds.

1.1 Sounding Objects in Pre-Twentieth-Century Everyday Life

What do we know about the objects people heard before 1900 and about what these sounds meant to them? Less than we would like, actually. This is partly because today's generation of historians seems more interested in writing about the late nineteenth century and after than about previous ages. But it is also a result of the prevailing approaches and perspectives in the historiography of sound since its establishment in the 1970s. How can we understand their focus on the long twentieth century?

In many ways, sound historians have been obsessed with *before and after* concerns, with what happened to our sonic environment before and after the first and second Industrial Revolutions, or before and after the introduction of sound recording. In 1977, the Canadian composer and environmentalist Raymond Murray Schafer set the stage with *The Soundscape: Our Sonic Environment and the Tuning of the World* [94]. In this seminal study, Schafer coined notions such as "soundscape" and "schizophonia." The word "soundscape" referred to our sonic environment, conceptualized as both our everyday environment of sounds and the musical compositions designed to improve this sonic environment. The term "schizophonia" stood for "the split between an original sound" and its "reproduction" [94, p. 273].

Schafer argued that the Industrial Revolution, with its many machines and intensified mobilization, had negatively affected the "keynote" of the Western soundscape. Before industrialization evolved into an all-pervasive phenomenon, Westerners lived in a "hi-fi" sonic environment: an environment with a "favorable signal-to-noise ratio," in which many single sounds could be heard clearly. The spread of the Industrial Revolution, however, has basically reduced this hi-fi sonic environment to a "lo-fi" one, in "which signals are overcrowded, resulting in masking or lack of clarity" [94, p. 272]. This situation even deteriorated after the split between original sounds and their reproduction, itself the result of the introduction of the phonograph in the last quarter of the nineteenth century. The "aberrational effect" of these technologies had contributed to the nervous *condition humaine* reflected in the notion of "schizophonia."

To help the world recreate its sonic environment for the better, Schafer put forward two solutions: ear cleaning, or training our ears to listen more carefully, and a better acoustic design of our sonic environment. In order to advance this objective, Schafer and his colleages Hildegard Westerkamp and Barry Truax [112–114] established and developed the World Soundscape Project in Vancouver. In their view, the new, improved sonic environment should be inhabited by "sound objects" that please the ear. Such sound objects—defined as "acoustical objects for human perception" after composer Pierre Schaefer, or as "the smallest self-contained particle of a soundscape" in Schafer's redefinition—can be "referential": they may refer to concrete objects such as bells or drums, just like "sounding objects." At the same time, however, "sound objects" should be treated as "phenomenological sound formation[s]" [94, p. 274]. In Schafer's understanding, then, "sound objects" differ conceptually from "sounding objects" [see also 9].

Schafer's focus on how industrialization and new sound technologies changed the Western soundscape has inspired many environmental historians, historians of urbanity, and historians of science and technology to focus on the history of noise in the twentieth century. Hence the dominance of the twentieth century and after in the historiography of sound. Examples are the studies by Lawrence Baron [14], Richard Birkefeld and Martina Jung [20], Peter Bailey [11, 12], Hans-Joachim Braun [24], Emily Thompson [109], and Peter Coates [33]. We return to their work later.

Yet completely independent from Schafer and his group in Vancouver, historians working in the tradition of the French *Annales* began to pay attention to the history of sound as an interesting aspect of everyday life, including people's everyday sensory experiences. Fortunately, they developed an ear for the historiography of sound in the nineteenth century.

Initially, these historians of the senses started out in a rather descriptive vein: they cataloged all the sounds citizens could hear at particular moments in time. Guy Thuillier, for instance, described the sounds villagers in the Nivernais could hear in the middle of the nineteenth century. In this inventory, he mentioned sounds varying from the lament of the death's harbinger to the sounds of forges, clocks, and steam machines [110, pp. 231–234]. Alain Corbin, however, stressed that such work wrongly implied that the past habitus of citizens did not condition their listening. He argued that modalities of attention, thresholds of perception, and configurations of the tolerable and intolerable ought to be taken into account [35]. His own work on bells in the nineteenth-century French countryside showed how these bells not only structured the villagers' days and mediated news in ways we would generally not be able to understand today but also contributed to the villagers' spatial orientation and expressed the symbolic power of towns. The more extensive and louder the bells, the bigger the power of the towns in which these bells rang [36].

Such relationships among sound, rituals, and power have been the subject of many studies by cultural historians and anthropologists [28, 30, 31, 53, 60, 101, 102]. Taken together, their work makes clear that what a particular society defines as wanted "sound" or "unwanted noise" expresses deeply rooted cultural hierarchies. The right to make oneself heard and to decide which sounds are allowed or forbidden has long been the privilege of the powerful, whereas those lower in rank—such as women, children, and servants—were supposed to keep silent. And if the lower-ranked did not behave in the proper docile manner and, for instance, used their drums and horns, they were soon under suspicion of intentionally disturbing societal order by making noise. And as we have underlined elsewhere, "positively evaluated loud and rhythmic sounds have had connotations of strength, significance, and being in control, whereas noise has often been associated with social disruption" [17, p. 40].

Yet between the sixteenth and nineteenth centuries, the elite became increasingly obsessed with mastering its own sound. Being able to keep silent became a sign of social distinction and civilization, whereas making noise was seen as vulgar, uncivilized behavior. Even in the opera houses and concert halls of the nineteenth century, the restless social talk by the aristocracy gave way to the concentrated silence of the rising bourgeoisie. This new class bolstered its self-confidence by claiming its ability to pay respect to the genius of the composer and to be touched by pure and authentic emotions [63, 104].

Whereas Schafer's claims on the West's deteriorating soundscape inspired the historiography of twentieth-century noise abatement, scholars in music, media, and technology studies increasingly addressed his statements on "schizophonia." One strand of studies on recorded music lamented the rise of muzak—originally a brand name—as the excess of late capitalism. This music was designed to seduce factory workers to withstand their tiring and boring work or to lure consumers into buying commodities they did not really need [8, 13, 59, 74].

A series of studies on the crossroads of media studies and technology studies responded more critically to Schafer's ideas, however. Jonathan Sterne, for instance, questioned Schafer's hidden assumption that the quality of communication with help of "original" sound is higher than communication via "recorded" sound [106]. Moreover, he and many others pointed out the complex relationships between the recording and the recorded sound. In their view, one cannot simply speak of the establishment of a split between original and copy; instead, one should make it historically understandable how people started accepting recordings as "copies" of original sound events.

These scholars unraveled the cultural appropriation of recordings by explicating the many cultural practices that had to be put into place to accomplish this process. These included the rise of "audile techniques," such as in the use of headphones necessary for individual and concentrated listening, and the organization of public "tone tests," meant to convince the audience of the idea that a recording of a singer referred to and was as good as the singer himself. Of similar significance were the introduction of new musical instruments such as the Stroh violin (figure 1.2) in order to enhance the sound quality of studio recording and the establishment of a new aesthetics of violin playing—the use of vibrato—to experience recorded classical music as lively as music presented during a stage performance with visually expressive musicians. Even the shift from the explorative listening in two-way amateur radio to the more passive listening to professional broadcasting was relevant in this respect [40, 41, 53–55, 64, 65, 99, 106, 108].

Although such studies again centered on the late nineteenth century and after, more recent work has shifted this focus now that the paradise-lost theme is losing its force. A mix of studies from the areas of literary studies, cultural history, and history of the senses has opened up the soundscapes of earlier pasts [49, 58].[3] Literary scholar Bruce Smith has, for instance, claimed that the acoustic world of early modern England never reached above 60 decibels "apart from barking dogs and the occasional gunshot,"

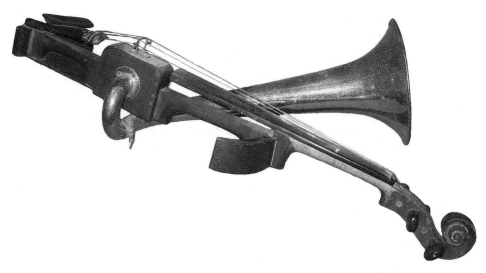

Figure 1.2
Stroh Violin, 1934.
Source: Christophe Monié, CC BY 2.5.

whereas the sound level produced by average city traffic today is 80 decibels and that by jets 140 decibels [100, pp. 50, 76]. Examples of the pre-twentieth-century sounding objects he and his colleagues have discussed most extensively relate to bells, war noises, the hammering trades, and street musicians. Because we have already referred to Corbin's study of the use and meaning of bells in the nineteenth century, and because the sound of war dominated people's sonic environments only in particular episodes,[4] we focus on the sounds of the trades and street musicians in the remainder of this section.

Indeed, the sounds produced by the various skilled trades were ubiquitous. These and other sounds were subject to many forms of local ordinances, even long before industrialization. Local bylaws targeting singing and shouting on Sundays, barking dogs, crying vendors, nightly whistling, street music, and making noise in the vicinity of churches, hospitals, and other institutions were commonplace. Moreover, in many nineteenth-century European towns, citizens put straw or sand on the pavement in front of hospitals and the homes of the sick to reduce the roar of traffic [11, p. 59; 37, p. 132; 94, p. 190; 116, p. 21]. Most regulations prohibited making noise on particular days, at particular hours, in particular places, or without a license. Local regulations that were drawn up to police the noise made by blacksmiths, coppersmiths, mills, and other trades were based on a similar approach: they placed restrictions on both when and where these industrial activities could take place.

The oldest proscriptions on the trades date back to ancient Rome. Roman law prohibited coppersmiths from establishing their trade in any street where a professor already resided. In 1617, the law faculties at the universities of Leipzig and Jena, consciously or not, imitated Rome by proclaiming that "no noisemaking handworker" was permitted to set up his business at any place inhabited by "Doktores." Both legitimated their actions by claiming that the handworkers' noise forced "the scholars to give up their research." More than a century later, in 1725, the university town of Turin banned the work of all noisy handworkers from the town's borders [122, p. 121]. By that time, all the coppersmiths' workplaces in Venice had already been centered in one quarter [92, p. 514]. This spatial intervention in the noise of one particular trade was not a Venetian invention, however. By the end of the fourteenth century, an Antwerp ordinance ordered that citizens were not allowed to put up a grease or flatting mill within its inner walls [90, p. 102].

The noise of mills was, in fact, a highly common nuisance to city dwellers. Millers literally hammered no less than smiths did. The mills themselves were sounding objects that are now almost forgotten because we are used to linking mills with a tranquil pastoral life rather than with a noisy urban life. Yet a city like Antwerp had grease mills for pressing oil from crushed seed, snuff or powder mills for grinding tobacco, and madder mills for milling and mashing madder, a seed plant used for the production of the color Turkish Red. In 1691, the municipal authorities decided that owners of grease mills must obtain a license before starting their business. This was, again, a regulation with spatial consequences because granting or withdrawing licenses had a direct effect on the density of trades in particular areas within the city's walls. Five years later, the council began regulating the hours of production: grease and soap makers were required to restrict their seed crushing activities to the hours between 5:00 am and 10:00 pm. Authorities in Brussels instituted the same regulation, and Bergen officials adopted a similar one for iron forges in 1737. Apparently the rules were difficult to uphold because Antwerp citizens were still complaining about "the violent noise and big roar" in the vicinity of mills in the mid-eighteenth century, which led to the doubling of fines in 1765 [90, pp. 80, 107, 114, 160–163]. Yet it was only around 1875 that many European countries issued nuisance legislation covering a wide variety of industries. These laws protected the properties of the owning classes, churches, hospitals, and schools against the damage noise could create to their possessions.

It is highly important to note, however, that the sounds of cities and town life were not always unwanted. That we know much more about the negatively evaluated sounds of the past than about the positively evaluated ones is partially because governmental and municipal archives and the records about regulation are so well preserved. Yet on the basis of other sources, historians have been able to show that the sounds of town life also often referred to work and wealth, and that the buzz of towns could be an enormous reassurance to early nineteenth-century American travelers coming from the fields where they could still hear the howling of wolves [103]. In contrast, the wolf's howl became a treasured sound, the sign of an unspoiled environment, in the 1960s [33]. Such examples make the often contested nature of sounds, as well as shifts in their evaluation, understandable.

Historian Emily Cockayne [34] has recently documented how urban dwellers in seventeenth- and eighteenth-century England coped with a wide variety of street noises—the sounds made by coaches and carts on cobblestones, pigs and dogs, sounds emanating from taverns, inns, and alehouses, horns and bells, mills and smitheries, and street hawkers and street musicians. Cockayne claims that attitudes shifted toward a heightened sensitivity to noise at the end of the seventeenth century due to a swelling population and increased trade. According to her, the noises that were restricted, for instance through licensing, were those made by the poorest classes: the popular entertainers and low-profit traders whose sounds did not fit into the lifestyle of a growing professional urban class longing for peaceful reading and studying. Such does not hold for the West in general, however, as the examples of restrictions for mainstream trades illustrate.

Several historians and literary scholars situate a change in attitude to street noise in the nineteenth rather than in the seventeenth century. Some connect it to the rise of the idea that the bourgeois interior had to be a safe haven from a harsh world [37]. Others emphasize that campaigns against street music, such as the one in London in the 1860s, had strong undertones of contempt toward persons from the lower classes and of a foreign origin. These historians see the campaign against street noise, in which Charles Dickens, the historian Thomas Carlyle, and the illustrator Richard Doyle joined forces with the mathematician Babbage, as the attempt by a rising elite of "authors, artists, actors, and academics" to secure professional standing and the right to work at home without being distracted. A London bill of 1864 restricted, at least on paper, street music "on account of the interruption of the ordinary occupations or pursuits," whereas a previous statute had only allowed for restrictions on account of illness [86, p. 53]. In practice, the bill helped the pursuits of the bourgeois because successful enforcement happened only in middle-class residential areas [7,

p. 194]. Cockayne acknowledges this nineteenth-century trend but adds that there is "evidence of the beginnings of this drive from the mid-seventeenth century" [34, p. 128].

We return to noise abatement campaigns below because the twentieth-century versions of such campaigns affected the design of the car as a sounding object immensely. First, however, we explain why and how the earliest forms of car sound design did not focus on the noise emission of cars, or their *exterior* sound, but on *interior* car sound—the sound in the ears of those who inhabited these cars. As we show, such early ideals of interior car sound were closely related to the cultural hierarchies of sound we have just discussed on the basis of the historiography of sound.

1.2   A Luxurious Silence: The Early Days of Car Sound Design

It took a long way for the carriage-like bodies of early automobiles to become the "acoustic cocoons" [19] of today's limousines, a shift that can be traced back to the development of the utilitarian vehicle in the interwar period [81]. This change in the car's function was accompanied by the transformation of the open sports car into the closed owner-driven sedan. Until the early 1920s, open touring bodies or runabouts with folding tops dominated the American and European markets. Better protection from bad weather conditions and more comfort for everyday use were decisive motives to enclose the body [48]: "No longer would wind, dust, and rain blow into the passenger compartment, nor would hair be 'blown to bits'; the solid top prevented all of that" [121, p. 58]. In the United States, where car ownership was already widespread in the late 1910s, the enclosed car had long remained the privilege of the wealthy nonetheless. Yet manufacturers hoped that the comfort of the permanently closed owner-driven sedan would attract new consumer groups such as self-driving women and families. As advertisements from the 1920s said: "If you really want your wife to drive," choose a closed sedan, and buy "year round comfort" for the family [29, 62]. In Europe, still behind in mass motorization [79, p. 40], enclosing the body was highlighted as the key to success in stimulating sales: "The new direction [in body design] seems to be right: The automobile must lose the eggshell of its luxury-childhood to become what it should be: a downright traffic-machine" [10].

The "closed car revolution" quickly transformed the closed car body from an elitist symbol to a basic feature of everyone's car. In 1926, the *Automobile Digest* reported that "the open car is losing ground each year in popularity with the closed type setting a fast lead" [cited in 82]. As a result, 82 percent of the cars built in the United States were closed in 1927 versus a mere 10 percent in 1919 [48, p. 164]. In

1924, about 90 percent of the German cars built had an open body, but by the end of the decade the relation between open and closed bodies had reversed completely [42, p. 98].

For automotive engineers, this change in technical properties came with additional implications. The traditional separation of chassis and body was now considered as technically and aesthetically insufficient and obsolete: the old mix of materials, the machine-like appearance of the metal chassis and the painted wooden body were conceived as unwelcome. In fact, the closed body symbolized the emancipation of automotive engineering from carriage building. People denounced the traditional carriage-type body as the "old dodderer" who feels bad weather in his wooden "bones" and expresses his feelings with creaks and squeaks. In contrast, the new closed bodies were expected to be much less noisy [83].

In 1921, the French engineer Charles Weymann copied airplane constructions of thin wooden chevrons covered with oilcloth for his closed-body design (figure 1.3). In France and Germany, Weymann's elastic timber and fabric sedan soon became the ideal "body of tomorrow" [66, 85]. In contrast, the American body manufacturer Fisher Body favored the composite design of wooden frames sheathed in steel.[5] Another competing technological concept was offered by Edward G. Budd: the all-steel body.[6] This concept had been introduced in 1915, and the key years for its diffusion in the automotive industry were the 1920s. Budd's technology facilitated the production process and is now seen as a "cornerstone to the foundations of the mass production car industry" [82].

The manufacturers of composite and all-steel bodies equally stressed the noiselessness of their constructions. To acquire this degree of quiet, insulation with soundproofing materials and better body mounting were crucial. "Composite body silence" was to be achieved through the extensive use of pads and bumpers "to eliminate squeaks at metal to metal and wood to metal contacts" [43]. Weymann pointed at the elasticity of the body that improved riding comfort and reduced body noise [120]. Fisher Body engineers also believed in the "superior strength and quietness" of wood as construction material; this belief is the main reason why Fisher Body, despite the greater labor intensity, stood by the composite body design until 1937. The Fisher brothers made a virtue of necessity and developed craftsmanship and elegance into their trademark, symbolized through the company's logo, which showed the outline of a royal ceremonial coach [121]. The accompanying philosophy stated that the user was more conscious of the body than of any other part of his car.

The Edward G. Budd Manufacturing Corporation, however, reclaimed the "greatest progress" in noise elimination for the design principle of the all-steel body [3]:[7] "For

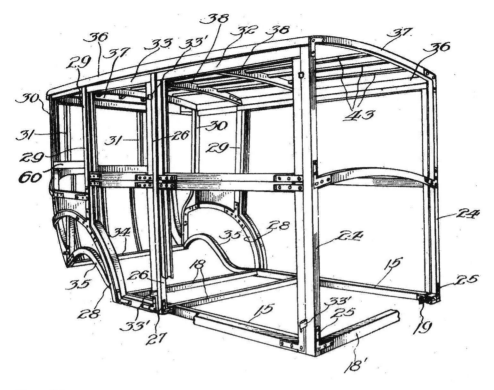

Figure 1.3
Engineering drawing of Weymann's flexible body frame (US Patent 1694572, 1925).

body noises are caused by joints where parts rub together or work loose. Where there are no joints, there can be no squeaks or rattles. Budd has eliminated joints by eliminating parts and by the extensive use of flash welding" [25]. As the advertising campaign highlighted, the one-piece body did not only reduce service costs on "squawks about squeaks"; "its quietness is an important factor in selling cars—and helping to keep them sold" [26] as well.

The Budd advertisements read like a direct response to a study on the correlation of subjective feelings and riding quality, published in the September 1930 issue of the *SAE Journal*. Psychologists from Purdue University had interviewed 125 persons about their attitudes to the sensory qualities of automobiles. The researchers "believed that these feelings have much to do with the apparent comfortable qualities of a machine" [23, p. 355]. To investigate this subjective side of automobile comfort, the survey analyzed 24 items ranging from vertical motion to the interior decoration. One of the most apparent clear-cut results was the unanimous complaint over body noise, which was experienced as highly distracting: "Body squeaks are particularly objectionable" [23, p. 358]. In addition, the assessment showed that noises arising from unknown sources caused anxiety. The authors concluded: "This item and the one on squeaks suggest radical changes in chassis and body construction" [23, p. 358]. Altogether, the study underpinned the crucial role of subjective perception in the evaluation of riding comfort.

Moreover, silence became incorporated in a set of luxurious attributes, such as fine fabrics and advanced styling features. Pontiac's 1935 "Silver Streak" advertisement announced:

If your ideas of luxury run to fine fabrics exquisitely tailored, sparkling chromium fittings, and an adroit use of color, the jewel-box perfection of Pontiac interiors is sure to win your heart. And if you measure luxury by *comfort* Pontiac scores again! Deep cushions receive you . . . ample weight and special springs make your silent progress smooth and steady. [89]

In the same year, Fisher Body pointed at "the good solid thud of a [Fisher] door swinging shut" [44]. And Plymouth proudly claimed for its models that they were "scientifically insulated against engine vibrations and road noises." Only in a Plymouth can "you *really* relax" [87]. Sociologist Mimi Sheller has stated that "different emotional registers are produced through the variations in the embodied driving experience" and that a smooth and silent ride is historically aligned with ideas of luxury, privilege, and wealth [98, p. 228]. Indeed, the advertisements of American and European companies show that from the 1930s onward, quietness became a sign of the car owner's sense of good taste. By stressing this meaning of quietness, the manufacturers both used and expressed the age-old connotation of silence with standing, distinction, and civilization—a connotation that, as we have seen, acquired increasing significance for the societal elite in the sixteenth century and after and became an obsession for some of the professional elites in the nineteenth century.

After World War II, Fisher Body continued to foster the company's image of high-quality craftsmanship and luxury, gradually transforming the connotation of the car interior from a noisy space into a silent sanctuary. A 1953 magazine ad used the picture of a deer looking in a car interior to stress the already achieved soundproofing—here visually associated with the peaceful quiet in the woods. The accompanying text elucidated the decisive role of Fisher Body's engineers:

Yet their search of silence continues—at the special Fisher Body sound laboratory. Right now, for instance, they are even making use of the latest binaural sound equipment in their work. This consists of double microphones, attached to a tape recorder on the rear seat of a test car, which register sound exactly the way a person with normal hearing catches it. So—you can add scientific noise protection to the other luxuries offered you in Body by Fisher. [45]

As a result of this scientific quest, Fisher Body introduced the new "sound barrier" in 1958, which "shrugs off the shakes and shuts out road noise" [46]. Another 8 years later, automobilists were literally invited to "escape" into the quiet world of Fisher Body: "It starts with the closing of a Fisher Body door. A solid 'thunk'—and sound is snuffed out like a candle. Road sound: soaked up by rubber body mounts. Traffic noise: muffled by thick layers of insulation" [47]. According to Fisher Body, by then the automobile had become a place to unwind. The individual car began to function as a private space where drivers could calm down and keep the ever-growing traffic intensity, acoustically and emotionally, at a distance. It is no coincidence that in the very same years, the car radio increasingly became advertised as mood regulator and radio pilot. Listening to one's favorite radio program could help the motorist to move through traffic politely and prevent a bad temper in case fellow drivers behaved less politely [19]. And by providing traffic information, the radio pilot could guide the driver safely through heavy traffic [117].

Today, the car is widely configured as an ideal listening booth—equipped with CD players, digital radios, mp3 players, multiple speakers, and mobile phones. It seems to be the natural place to enjoy hi-fi listening: the proximity of the aural enhances the kinesthetic experience of the car's acceleration and movement, and vice versa. Michael Bull's ethnography of the car has shown its particular character as a technology-enabled aural habitat. People feel uncomfortable without the mediated sound of car radio or other sound technologies. They habitually switch on their radio when they enter the car: "Automobile users thus appear to prefer inhabiting an accompanied soundscape" [27, p. 192; 38].

1.3 Listening to the Engine: Car Sound as Source of Information

Drivers who listen to their car radio or other audio systems will probably listen less carefully to the sounds of their car proper. Yet in the early days of the automobile, listening to the car engine for information was crucial to drivers. To sonic interaction designers it will come as no surprise that people listen to machines for information. With respect to the gasoline car, engineers, car mechanics, and car drivers tried hard to make sense of what a car's engine would tell them. But they also argued intensively about who was and who was not allowed to speak about or interpret those sounds.[8] And if those sounds themselves would be subject to change and variation, the meanings attached to car engine sounds would also shift over time.

Before further developing this argument, we first briefly return to the early history of automobility. The historiography of technology conceives the interwar era as a turning point in the culture of the automobile [67]. During these years, the automobile's image transformed from an adventure machine to a utilitarian means of transportation [81]. This shift in function accompanied not only a rise in the overall number of automobiles but also an increase in the number of self-driving motorists. Even for Germany, relatively late in mass motorization, the magazine *Allgemeine Automobil-Zeitung* predicted twice as many self-drivers than chauffeurs by 1928 [42, p. 98].

Until around 1930, it was common for chauffeurs to be also in charge of maintaining and repairing the cars of their wealthy upper-class employers [21, pp. 13–30]. Handbooks and driving manuals considered it necessary for chauffeurs to possess a good sense of hearing, their ears serving as the best set of diagnostic tools available [72]. In 1905, the *New York Times* suggested that a would-be chauffeur should "learn to study his machine by his ears as much as by his hands. By listening to the car and noting the different noises, one may be able to detect immediately any flaw in the perfect running of the mechanisms" [Birdsall, cited in 21, p. 111].

As this quotation suggests, listening for trouble was key to gathering information about the location and cause of technical problems. Other authors of drivers' manuals stressed that a fine sense for the rhythm and pitch of the engine could be learned only by experience. Theoretical understanding of automotive technology was seen as a prerequisite for the auditory diagnosis of engine trouble yet insufficient for identifying audible malfunctions. Until one actually heard piston slap, no written description could communicate what it sounded like [21, p. 112]. Following the philosopher Michael Polanyi [88], the skill of listening carefully to machines can be characterized as "tacit knowledge"—implicit knowledge that is hard to communicate verbally or write down in a manual. The French sociologist Marcel Mauss [76] pointed at the bodily dimension of this kind of knowledge. He called it a "technique of the body" that is engraved into the body through practice and manual repetition. Listening, as a technique and skill, is thus structured by practical experience and, concomitantly, oriented toward practical reasoning. This made the mode of "diagnostic listening," as we coined it [18, 70], into a crucial aspect of repair knowledge. It is a specific case of the sound-information relationship.

The authors of the drivers' handbook underlined that for noticing this information, for being able to hear what the engine said, one had to avoid any unnecessary noise while driving. Nothing should distract the motorist's sense of hearing from carefully observing the engine sound. The authors highlighted, for example, the skill to change gears silently as a way to recognize a good driver. Only when the engine ran as quietly as possible did it become possible to notice deviances from normal motor functioning

Figure 1.4
Cartoon depicting the unpleasant concert of a noisy engine.
Source: A.G. von Loewe. "Um die Auspuffklappe." *Allgemeine Automobil-Zeitung* 27 no. 7 (1926): 21.

plainly and early [70] (see figure 1.4). At this point, automobile practice converged with a relatively new discourse of mechanical engineers about noise as a sign of mechanical inefficiency. After the turn of the century, noisy machinery had become a sign of mechanical friction, and mechanical friction was seen as something that could easily result in lower productivity and increased cost. In automobilism, the design of noiseless components and the quieting of the entire car construction became ends in themselves. To achieve these goals, automotive engineers focused on oiling moving parts, improving gearing, balancing the engine, and reducing vibration and resonance.

With this new meaning, noise gradually lost its older connotation of power, which highlights an important societal reconfiguration of sound and noise in the twentieth century [16, pp. 325, 328]. Not only did engineers redefine noise from a sign of power to a sign of wear and tear, the 1910s, 1920s, and 1930s also saw a rise of noise-abatement societies, pamphlets, and exhibitions all over Europe and the United States. In this context, the engineers' perspective on noise became more dominant, strengthened both by the introduction of the decibel and other sound level measuring units in the 1920s and by the introduction of legislation, in the second half of the 1930s, that limited the maximum noise emission of cars and car horns. In addition, antinoise activists did everything to educate the public in order to have them behave more "civilized" and less "rowdy" on the road. This implied that pedestrians and bikers should use the streets in a more orderly and predictable manner, that car drivers should use their horns much less than they were used to, and that they should take care of maintaining their cars and oiling the gears so that these would make less noise [17]. This does not imply, however, that the connotation of loud sound with power lost its cultural force completely—our next section attests to this phenomenon.

In line with a new ear for noise as the expression of mechanical problems, the 1920s automobile literature distinguished between two kinds of sound. The first group of sounds signified the normal running of the engine, the gears, and the axles. Regular sounds indicated that the car was running well. Drivers were taught it was the rhythmic and silent run of the engine that required their full attention: "The regular humming of the gear box or chain drive indicates that everything is in best order. He [the driver] will notice soon that every engine and every car has its own tempo and that even the slightest technical problem alters this lovely rhythm" [73, p. 304]. Familiar car sounds were thus a comfort to both drivers and mechanics: such sounds indicated that the engine was behaving as it was supposed to.

Similar reassuring functions of sound are known in other technical domains. Cyrus Mody, sociologist of science, has investigated how sound helps to define ways of performing laboratory work. He claims that sound structures experimental experience: "When things run smoothly, these sounds unfold regularly, marking out the running of a clean experiment. Learning these sounds, and the experimental rhythm they indicate, is part of learning the proper use of the instrument" [80, p. 186]. In addition, Mody has underlined the epistemological relevance of hearing, as data expressing periodicity can, in some cases, be better processed with one's ears than with the eyes [80, pp. 186, 188]. Historian of science Gerard Alberts has referred to a related story. He describes how in the 1950s computer operators in the Philips Physics Laboratory amplified the sounds of their new electronic equipment because they missed the comforting rattling sounds of their mechanical calculators [1, 2].

The second group of sounds can be associated with symptoms of trouble. Motorists were advised to listen carefully to deviations in tempo and rhythm; once they noticed a discord they were warned to drive cautiously and "prick up their ears" [56, p. 217]. Back home, motorists had to examine thoroughly the symptomatic sounds they registered. They had to learn to distinguish between, for example, the knocking of preignition or the puffs and pops of carburetor problems. In doing so, they had to match sonic evidence and theoretical knowledge of automotive technology into the already mentioned technique of diagnostic listening, which can be linked to yet another notion: that of "sound mapping." This notion of sound mapping stems from the field of psychoacoustics, understood as the assignment of sounds to the information they represent [52, pp. 55–56; 69]. It seems a useful metaphor for grasping the social practice of diagnostic listening. However, this technique could only be acquired through practical experience, as the authors of drivers' manuals claimed.

To distinguish the various relevant sounds of car engines, these authors also suggested particular methods and tools, including simple hearing rods (see figure 1.5), stethoscopes, and the Tektoskop. The latter was a stethoscope with two probes. The idea behind it was to enable mechanics to examine two engine spots at once, helping them to compare two sounds at the very same time. The method was related to what Jonathan Sterne has called "audile technique," a technique that isolates the listening experience from other sensory input in order to intensify and focus the listening activity [106, pp. 93–98]. In the world of auto mechanics, however, the usefulness of listening devices was contested. Whereas some manufacturers advertised their tools as the ideal troubleshooters, others kept stressing the significance of listening skills as such:

For diagnosing engine sounds a very fine sense of hearing is required, especially when the sounds are very faint, and when several sounds from different sources have to be distinguished simultaneously, as is often the case. [. . .] Those who are unaccustomed to the use of a wooden rod or a stethoscope, which are both put against the engine from the outside, will easily be misled by the effect of resonance. [4, p. 81]

In this line of reasoning, listening tools could not compensate for lack of sonic expertise.

One key issue chauffeurs, self-drivers, and even car mechanics had to deal with in the process of diagnostic listening was the verbalization of sonic experience. Tables, listings, and fault trees could help to investigate audible defects. Less clear, however, was how mechanics should know what handbook authors exactly meant when they referred to, for instance, knocking sounds. One author considered this sound to start

Figure 1.5
An engineer examines a series of automobile generators, listening for unusual sounds.
Source: Krafthand 14 no. 48 (1941): front cover.

with a "light pinking" and to rise until "diabolic detonations" pressed themselves on one's ears. Another author distinguished no less than seven audible malfunctions of the engine, varying from "metallic knocking," "high knocking," and "damped knocking," to a "muffled clang" [70].

Obviously, no shared or standardized vocabulary existed to express the audible characteristics of engine knocks. At the very same time, it was crucial to distinguish between these sounds, as the "metallic" one was just a nuisance, but the "high" one was a serious threat. Knocking was the most prominent problem, but the verbalization of other audible malfunctions was just as complex. At times, the communication failed completely. In a letter to the *Allgemeine Automobil-Zeitung*, one reader asked what his car's "hot noise" might mean, a description that proved incomprehensible to the newspaper's technical advisers. Eugen Mayer-Sidd, author of many articles in the trade journal *Krafthand*, even questioned the use of written repair accounts by definition: "it is exceptionally difficult to give someone else a detailed and graspable description of a technical work or method that enables him to do it himself later" [77, p. 57]. He was notably pessimistic about communicating malfunctions: "it is even more difficult to give a comprehensible description of a malfunction in such a way that someone else, without physical inspection [of the engine], can give appropriate advice on how to repair it with certainty" [77, p. 57]. Still, he and other authors regarded practical experience as the only way to learn diagnostic skills.

During the 1920s, chauffeurs and self-driving motorists cultivated two intertwined, yet distinct, listening techniques: they relied on auditory skills to monitor their car's sounds while driving, and in this way they acquired the sonic skill of diagnostic listening—a technique also applied by car mechanics. We distinguish between *auditory* skills and *sonic* skills. Listening for information implies not only skills for understanding what exactly one is listening to but often also technical skills linked to the listening devices used, such as positioning the stethoscope properly on the car or car engine. Even musical skills may be required, for instance, when engineers single out the different "melodies" and "rhythms" of humming machines. This explains why we speak of *sonic* skills rather than *auditory* skills [18]. At the end of the decade, however, the automotive world was hit by a severe crisis: the number of complaints about the reliability and qualification of car mechanics escalated. Car owners articulated their increasing distrust and dissatisfaction in letters to trade journals and consumer magazines. In such letters, they questioned the diagnoses of their car mechanics, asked for the evaluation of repair bills—excessive in their eyes—and requested legal advice on refusing to pay these bills. Because of this repair misery, car dealers and manufacturers feared waning car enthusiasm and declining sales numbers. Even worse, experienced mechanics were in short supply [21, pp. 76–98; 70].

In the United States, the Ford Motor Company responded to the repair crisis by reforming the company's own repair branch: "Ford engineers began to adapt production methods they had introduced in the factory—including time-and-motion studies, standardized procedures, modern machinery, and an extensive division of labor—to automotive repair" [78, p. 278]. But even the establishment of a flat-rate system did not entirely solve the problem of distrust.

In Germany, the response was rather different. Here, a lengthy struggle for expertise began in the car repair business, culminating in the question *who* the genuine car sound expert was. In the 1920s, motorists had embodied automobile listening as a technique of the body, as we have seen. On a more fundamental level, they had even engraved these techniques into their habitus. Displaying sonic skills became a means of social orientation: motorists considered car repair as matching their upper middle-class social standing. It had become a bourgeois cultural technique.[9] The German car mechanics' exit from the repair crisis was both an answer to this situation and the result of a long-standing German craft system. Their way out was to professionalize their business. In 1933, new legislation introduced an obligatory 4-year apprenticeship in the car mechanics trade. A subsequent adjustment restricted the right to open up a repair shop to certified master craftsmen [70].

In this process, mechanics started to question the position of car owners as knowledgeable amateur mechanics. Editorials, testimonials, and cartoons in trade journals claimed that lay motorists simply lacked the necessary skills to repair their cars themselves. What was more, drivers should learn to ignore minor noises that were nothing but a nuisance. Only in case they heard a threatening noise, should they bring their car to a specialized repair shop. In the 1930s, then, the craft profession came up with a discourse that clearly demarcated the professional expertise of the car mechanics from the lay listening practices of the motorists. Increasingly, car owners complaining of noises were denounced as noise fanatics: "He [the driver] is often bothered by noises that only exist in his fantasy" [6]. With the topos of the noise fanatic, auto mechanics claimed the practice of diagnostic listening as their exclusive domain:

The expert's trained ear knows the sound of the engine. He distinguishes between healthy sounds and noises that indicate an upcoming problem. In contrast, the layman is often anxious about harmless noises. It is nonetheless better, however, that he consults an expert for nothing than that he disregards every noise until the engine has a serious problem. [5]

Our genealogy of automobile listening practices shows that the car mechanic's claim to diagnostic listening expertise was substantiated only by their reference to traditions of "high-quality craftsmanship" and the "master craftsman's honor." Not the listening techniques themselves but their cultural and societal underpinning empowered the mechanics' knowledge [50, p. 120]. Because of this situation, drivers gradually restricted their listening practices to mere "monitory listening" [18]. In addition, the apprenticeship system can be interpreted as the institutionalized response to the problem of how to transfer tacit knowledge: new car mechanics now learned their skills during their 4-year apprenticeship on the job.

After the Second World War, drivers kept listening to their cars to monitor the engine. In case of trouble, however, they left the auditory diagnosis to professional car mechanics or even bought a new and quieter car [19]. Car mechanics themselves partially lost their listening expertise whenever more diagnostic routines were delegated to new testing equipment, shifting the focus from sonic to visual information. A similar sensorial shift occurred with the introduction of on-board diagnostics and the malfunction indicator lamp [21, pp. 138–169]. Yet sonic information did not become completely obsolete. Even today, car manual authors advise motorists to listen for any "unusual engine noises" that might indicate a technical problem [111, p. 249].

1.4   A Smart Sound: Car Sound Design Today

Although listening to the engine in order to detect mechanical defects has more recently lost its predominance in automotive engineering due to diagnostic tools that visualize rather than amplify data about the car engine, the design of car sound proper—both interior and exterior sound—became increasingly important in the 1980s and notably the 1990s. Engineers are still listening, yet with a new focus on creating a sound that would attract the right sort of customer for the right brand of car.[10]

In the 1960s and 1970s, national governments and European bodies imposed increasingly stringent limits on the maximum noise emission of cars [71, 93]. Between 1970 and 2000, the European maxima for passenger cars shifted from 86 to 74 decibels [32]. As a result, automobile makers were forced to further limit car noise. Engineers had their hands full when it came to reducing the low-frequency sound in cars [107]. There were many such sounds: the rustle of the wind, the vibrations of the engine, the sound of tires touching the road surface, the hydraulic system of power steering, and the whining of the fuel pump. Acoustic engineers tried to tackle such sounds by, for instance, applying acoustic glass or by introducing an innovative design of car tires [39, 68, 105]. Time and again, however, they would find that removing one noise rendered some other one audible [51, 95]. So, even though car manufacturers had been interested in interior car sound from the early 1920s onward, it was the obligatory reduction of exterior noise that brought formerly masked interior sound into the limelight again after the 1970s [32].

Reducing the interior noise that is audible now that other sources are silenced is not the only reason for the automotive industry to work on sound design, however. Manufacturers have invested considerable time and money in sound design to make sure that switches, warning signals, direction indicators, windshield wipers, the opening of car windows, the locking of car doors, and the crackle of the leather uphol-stery come with the *right* sound. Such sounds, or "target sounds," should express their make's identity and attract particular groups of consumers. Two simple figures may demonstrate just how much today's automotive industry invests in their product's sonic characteristics: BMW employs over 150 acoustic engineers, and Ford has an acoustical department of 200 employees [61, p. 106; 115, p. 9].[11]

Both people from the automotive industry and sociologists have explained this increasing investment in finding the exact match between the sonic experience of a particular make of car and specific groups of consumers by referring to structural societal changes. The German sociologist Gerhard Schulze argues that sensory experi-ence plays a much larger role in the selling of consumer goods today than in the past because many products have been perfected to such a degree that differences in tech-nical specifications between automotive brands have grown smaller over time [97]. Moreover, most consumers today choose from an enormous array of consumer goods on account of the booming postwar economy. The many options available, however, also give rise to uncertainty. To compensate for the absence of differences in technical quality and make selection easier for consumers, products are increasingly sold by cashing in on the emotional meaning and inner experience products evoke in their buyers.

In the marketing of new products, therefore, sensory experience has become crucial. How appliances feel, smell, or sound and how these experiences fit the buyer's identity have thus become as relevant as how the cars look [75, p. 114; 118]. All this has led to a strong "aestheticization" of everyday life, captured by Schulze under the heading of the *Erlebnisgesellschaft* (experience or event society). In such a society, experience value trumps functional value, both in selling strategies and in the motivation of consumers [97, p. 59].

Many stakeholders in the automotive industry present a similar analysis of today's society [32]. Now that there are many wealthy people, one interviewee claimed, their concern for quality of life has come into play. And because these people do not have to worry any longer about whether or not their cars actually run, they may start com-plaining about their sound.[12] What is more, the automotive industry considers sound more easily observable by consumers than other aspects of automotive quality, such as safety [123, p. 1]. Depending on the specific preferences of customers, cars should have "decent," "luxurious," "dynamic," or "sporty" sounds [15]. Renault, for one, made its Clio 3 "silent" and gave its Megan Sport a "dynamical sound."[13]

Given these design shifts, it is not surprising that automobile manufacturers started to advertise the sonic qualities and interior sonic tranquility of their vehicles with increasing fervor in the 1990s. The new owner of a Chrysler Voyager would find a "whispering stillness" waiting for him. The Jaguar S-type was said to be propelled by a "silent force," whereas in a Toyota Avensis "even the silence comes standard." Amid the bombardment of noises our ears have to put up with, drivers would experience "tranquility" in a BMW diesel, Mercedes-Benz, and Volkswagen Passat. Better still, in the Ford Focus, everything we touched would come with "the sound it is supposed to make."[14] An ad for the Chrevrolet Epica Business Edition even claimed that the "smartest kid in the classroom" is "all too often also the most quiet."[15] By doing so, Chevrolet cleverly reused the age-old association of learnedness with quietude—recalling the privileges granted to the professors in the seventeenth and eighteenth centuries. At the same time, these marketing campaigns remind us of automobile advertisements from the late 1920s and 1930s (see section 1.2).

In this context of sensorial marketing, an extensive research industry has evolved that fully concentrates on developing testing methods for tracing the feelings, emotions, and experiences when one is driving a car. The German-based company HEAD acoustics GmbH is one of the players within this new research industry. It provides the automotive industry—and increasingly the domestic appliances and personal computer industries—with testing methods and also actual testing of sound. In turn, companies such as HEAD acoustics collaborate with universities, whose staff members equally address the shifts associated with today's "experience society" [96, p. 12]. The notion that sound "is well known to enhance or detract from our pleasure in possessing or using a product" [22, p. 1] has thus been reinforced by an emerging and growing network of manufacturers, designers, testing companies, marketers, and academics who reciprocally spread the word of sensorial branding and design.

It is also in the research and testing labs focusing on the subjective experience of car sound that it became clear, however, that the association between loud sound and power was still alive and kicking and could not be ignored. Of course, particular sporty makes, such as Porsche, have treasured their loud and impressive sound as an indispensable characteristic of their make. Yet adding any sound to any car, which became technically possible as a result of digital sound technology, and which HEAD acoustics tested with help of a moving Sound Simulation Vehicle, happened to have its limits. In one such test the sound of a glamorous 500 PS Aston Martin was added to a relatively ordinary make, a Ford Focus. Test drivers opening the throttle of the Ford Focus would hear the sonic feedback of an Aston Martin: "and acoustically you get such feedback, that you think 'wow, now we're really driving,' but then not much really happens because the car altogether lacks performance." This could be quite dangerous in fact when one is trying to pass another car.[16] In this way, the testing situation itself created shifting conceptions of what listening while driving was "really" about. In addition, it showed that sound designers cannot simply ignore our history of listening and our history of how particular sounds, or levels of sound, are connected to particular meanings, such as the connection between loud and powerful.

Car manufacturer Ford similarly discovered that its engineers could not leave the historically developed cultural conventions of sound behind in any random way. In the ideal world of a Ford sound engineer, the future Ford customer would be able to upload a series of car sounds (for the blinker, seat belt warning signal, windshield wiper, and so on) of his or her choice, just as he or she uploads a ring tone for a cell phone. It was decided this option should only be available to customers after the Ford sound engineers had created a full "sound composition" in which all sounds would be both typically Ford and go together perfectly well.[17] Yet when Ford designed a new fancy blinker sound ("pock-pock-pock") for its Ford Focus, many test subjects and automotive journalists initially rejected it because of their nostalgia for the age-old "click-clack" originally linked to the relay. Only after the Ford Focus had been advertised as a first-rate modern car, and a story had been linked to its futuristic blinker sound, did the new blinker sound come to be accepted.[18] It thus became clear that design needs to be accompanied by marketing techniques and that the history of sound cannot simply be treated as a relic from the past.

1.5   Conclusions

By considering the history of sounding objects both before and after the introduction of the automobile, we have sought to clarify some of the continuities and discontinuities involved in sound and its meanings. That we consider this focus important largely follows from old themes and new trends in the historiography of sound. Like many of today's historians we have listened more extensively to the sounds *after* the second Industrial Revolution than before because the quantity and quality of sounding objects drastically changed after this revolution.

Unlike Murray Schafer, however, and more in line with cultural and media historians, we do not start from the assumption that the possibility to record and replay sounds merely created schizophonia. In contrast, sound-recording technologies not only created noise pollution—a continuous theme in the historiography of sound— but also, in tandem with new acoustic isolation technologies, enhanced people's options for acoustic privacy, that is, control over which sounds enter and do not enter their private space, notably in the car. Moreover, it is not the mere description of what could be heard that has been our focus but what these sounds meant to those who listened and how these connect to cultural connotations and societal hierarchies.

In the course of the twentieth century, the car transformed from a loudly sounding and hooting contraption into a place in which to unwind and find tranquility, privacy, and control over one's sonic environment. In this era, mechanics would rely heavily on their ears when it came to repairing cars, while engineers constantly tinkered with and endlessly redesigned the sounds of automobiles.

At the wider societal level, loud car sounds have largely transformed from a sign of power to a sign of uncivilized behavior, and after the 1970s, European and national legislation forced the automotive industry to reduce exterior noise. However, as we have just seen, particular groups of car owners still perceive the sportive sound as a sign of engine strength. In this respect there is also much continuity in the meanings associated with sound over time. Furthermore, listening to machines gained respect as a skill in its own right, one that was secured by professional demarcation until the rise of visual monitoring tools reduced its significance.

Our exploration of several episodes from the history of sounding objects provides perhaps a threefold source of inspiration to sound designers in particular. First our account elucidates the many skills involved in listening to sounding objects in order to acquire information. Not only do sounds speak to users in different ways; to understand what objects do by listening to their sounds requires, so it seems, intricate and ambiguous processes of cultural learning. Although the German car mechanics in our account may have overstated the exclusiveness of their expertise, even car drivers had to learn to listen to their vehicle from the manuals they read and then by driving their vehicles day after day.

The second kind of inspiration might be derived from reading more about the history of the automobile's sound design. Apart from the efforts of architectural acousticians and their work on churches, schools, and concert halls [109], automobiles have figured prominently in sound design—and they did so from early on. From its inception, the car has been subject to redesign as a sounding object to create a better sound. Of course, we are fully aware of the irony of this because traffic is still listed as number one in many noise annoyance overviews. But this also explains perhaps why the history of the automobile's sound design, including that of its interior, involves an ongoing, open-ended narrative. Finally, and more generally, sound designers and others active in the same field will hope to find inspiration in a greater understanding of the cultural variation and historical shifts in the meaning and interpretation of sounds. This may contribute to their ability to listen carefully to future changes in the world of sound and its meanings.

Notes

1. In 1914, Wilhelm Hort was chief engineer of the AEG turbine factory, as well as *Privatdozent* at the Polytechnic University (*Technische Hochschule*) of Berlin. All translations of quotes from German, French, and Dutch sources into English are made by the authors. We would like to thank Ton Brouwers for checking our English.

2. Providing translations that accurately express the meaning of these words is difficult, yet the following listing gives a first impression: groaning, creating mating cries, barking, bleating, roaring, bubbling, yelling, thundering, booming, reverberating, cooing, hammering, howling, clapping, ringing, banging, crowing, laughing, moaning, whistling, crackling, snorting, rolling, ringing, beating, crying, hurling, ticking, stamping, pattering, drumming, weeping, whinnying, whimpering, chirping (insect), hissing, chirping (bird).

3. The remainder of this section is largely drawn from chapters 2 and 3 of *Mechanical Sound* [17], courtesy MIT Press.

4. For interesting accounts of the sounds of wars, see Rath [91] and Ross [in 103].

5. The lightweight construction of composite bodies like Weymann's could be mounted to cars with small engines thereby introducing enclosed comfort to the bottom quarter of the automobile market.

6. Budd did not invent the core technologies to produce the all-steel body but introduced the set of production methods leading to mass production.

7. Ambi-Budd, a German-American joint venture founded in 1926, introduced the all-steel body in Germany [84].

8. This section is a shortened and revised version of Krebs [70]. We are grateful to Oxford University Press for granting us permission to do this.

9. Andrea Wetterauer argues that self-driving and maintaining cars became a legitimate part of bourgeois culture in this era. A feuilleton of the *Frankfurter Zeitung*, for instance, compared the embodied expertise of car steering and repairing with traditional cultural techniques such as piano playing or dancing [119, pp. 155–166].

10. This section draws from and is largely based on Cleophas and Bijsterveld [32]. We are grateful to Oxford University Press for granting us permission to do this.

11. Transcript of interview by Kristin Vetter with Ralf Heinrichs (engineer and Ford employee), June 8, 2004, Cologne, Germany (originally in German), p. 20. We would like to thank Kristin Vetter and Fleur Fragola for the interviews they did in the context of the Selling Sound project and NWO (the Dutch NSF) for funding this project.

12. Transcript of interview by Kristin Vetter with Nicholas Chouard (Acoustic Engineer, former Renault employee), June 4, 2004, Aachen, Germany (originally in German), p. 11.

13. Transcript of interview of Eric Landel and Virginie Maillard (Renault employees), October 16, 2007, Guyancourt, France (originally in French), p. 2.

14. "Let your car do the talking" (magazine ad Chrysler Voyager, published between 2002 and 2007); "U rijdt geen diesel, u rijdt Jaguar" [Jaguar brochure, The Netherlands, 2006]; *Zelfs de stilte is standaard*, Commercial Toyota Avensis, Videotape [Talmon: AV Communicatie, 1999]; "Een mens krijgt 469.082 geluidsprikkels per minuut," [*Elsevier*, October 23, 1999, pp. 12–14]; *Business Travel*, Commercial Mercedes-Benz, Videotape [Stuttgart: DaimlerChrysler AG, 1990]; *Blind Brothers Three*, Commercial Volkswagen Passat, Videotape [Almere: TeamPlayers, 1999]; "De nieuwe Ford Focus: Voel de verfijning," [*Dagblad de Limburger*, January 24, 2005, p. B11].

15. "Het slimste jongetje van de klas," [*Elsevier*, July 27, August 24, and September 21, 2006 (publication dates based on oral communication Chevrolet-General Motors, Main Office The Netherlands, August 21, 2007)].

16. Transcript of interview by Kristin Vetter of Ralf Heinrichs (engineer and Ford employee), June 8, 2004, Cologne, Germany (originally in German), p. 7.

17. Transcript of interview by Kristin Vetter of Ralf Heinrichs (engineer and Ford employee), June 8, 2004, Cologne, Germany (originally in German), pp. 8, 12.

18. Transcript of interview by Kristin Vetter of Ralf Heinrichs (engineer and Ford employee), June 8, 2004, Cologne, Germany (originally in German), pp. 9–10.

References

1. Alberts, G. (2000). Computergeluiden. In G. Alberts & R. van Dael (Eds.), *Informatica & Samenleving* (pp. 7–9). Nijmegen: Katholieke Universiteit Nijmegen.

2. Alberts, G. (2003). Een halve eeuw computers in Nederland. *Nieuwe Wiskrant, 22*, 17–23.

3. Ambi-Budd. (1926). Advertisement: Der größte Fortschritt. *Motor, 14*(10), 127.

4. Anonymous. (1932). Abnorme Fahrgeräusche und ihre Ursachen. *Das Kraftfahrzeug-Handwerk, 5*(6), 81–82.

5. Anonymous. (1936). Der Motor hat einen "Ton." *Das Kraftfahrzeug-Handwerk, 9*(11), 330.

6. Anonymous. (1939). Es rattert und quietscht. *Krafthand, 12*(26), 779–780.

7. Assael, B. (2003). Music in the air: Noise, performers and the contest over the streets of the mid-century metropolis. In T. Hitchcock & H. Shore (Eds.), *The streets of London: From the Great Fire to the Great Exhibition* (pp. 183–197). London: Rivers Oram Press.

8. Attali, J. (1985). *Noise. The political economy of music*. Manchester: Manchester University Press.

9. Augoyard, J.-F., & Torgue, H. (Eds.). (2005). *Sonic experience: A guide to everyday sounds*. Montreal: McGill-Queen's University Press.

10. B.R. [initials only] (1924). Neue Wege im Karosseriebau. *Auto-Technik, 13*(4), 20.

11. Bailey, P. (1996). Breaking the sound barrier: A historian listens to noise. *Body & Society, 2*(2), 49–66.

12. Bailey, P. (1998). *Popular culture and performance in the Victorian city*. Cambridge: Cambridge University Press.

13. Barnes, S. J. (1988). *Muzak, the hidden messages in music. A social psychology of culture*. Lewiston/Queenston: The Edwin Mellen Press.

14. Baron, L. (1982). Noise and degeneration: Theodor Lessing's crusade for quiet. *Journal of Contemporary History, 17*, 165–178.

15. Bernhard, U. (2002). Specific development of a brand sound (pp. 103–115). *AVL Engine and Environment* (Graz, September 5–6).

16. Bijsterveld, K. (2006). Listening to machines: Industrial noise, hearing loss and the cultural meaning of sound. *Interdisciplinary Science Reviews, 31*(4), 323–337.

17. Bijsterveld, K. (2008). *Mechanical sound. Technology, culture and public problems of noise in the twentieth century*. Cambridge, MA: MIT Press.

18. Bijsterveld, K. (2009). *Sonic skills: Sound and listening in the development of science, engineering and medicine, 1920s-now*. Unpublished Proposal for the NWO-VICI competition in the Netherlands.

19. Bijsterveld, K. (2010). Acoustic cocooning: How the car became a place to unwind. *The Senses & Society, 5*(2), 189–211.

20. Birkefeld, R., & Jung, M. (1994). *Die Stadt, der Lärm und das Licht. Die Veränderung des öffentlichen Raumes durch Motorisierung und Elektrifizierung*. Seelze (Velber): Kallmeyer.

21. Borg, K. L. (2007). *Auto mechanics. Technology and expertise in twentieth-century America*. Baltimore: The Johns Hopkins University Press.

22. Boulandet, R., Lissek, H., Monney, P., Robert, J., & Sauvage, S. (2008). How to move from perception to design: Application to keystroke sound. *NOISE-CON 2008* (Dearborn, MI, July 28–30).

23. Brandenburg, G. C., and Swope, A. (1930). Preliminary study of riding-qualities. *SAE Journal, 27*(3), 355–359.

24. Braun, H.-J. (Ed.). (2002). *Music and technology in the 20th century*. Baltimore: Johns Hopkins University Press. (Originally published by Hofheim: Wolke, 2000.)

25. Budd. (1932). Advertisement: The squeak. *Automotive Industries, 66*, 109.

26. Budd. (1932). Advertisement: To the man. *Automotive Industries, 66*, 45.

27. Bull, M. (2001). Soundscapes of the car: A critical ethnography of automobile habitation. In D. Miller (Ed.), *Car cultures* (pp. 185–202). Oxford, New York: Berg.

28. Burke, P. (1993). Notes for a social history of silence in early modern Europe. In *The art of conversation* (pp. 123–141). Ithaca, NY: Cornell University Press.

29. Chandler. (1924). Advertisement: If you really want. *National Geographic*, April, page unknown.

30. Classen, C. (1993). *Worlds of sense. Exploring the senses in history and across cultures*. London, New York: Routledge.

31. Classen, C. (1997). Foundations for an anthropology of the senses. *International Social Science Journal, 151*, 401–412.

32. Cleophas, E., & Bijsterveld, K. (2012). Selling sound: Testing, designing and marketing sound in the European car industry. In T. Pinch & K. Bijsterveld (Eds.), *The Oxford handbook of sound studies* (pp. 102–124). Oxford: Oxford University Press.

33. Coates, P. A. (2005). The strange stillness of the past: Toward an environmental history of sound and noise. *Environmental History, 10*(4), 636–665.

34. Cockayne, E. (2007). *Hubbub—Filth, noise & stench in England, 1600–1770*. London: Yale Universty Press.

35. Corbin, A. (1995). *Time, desire and horror. Towards a history of the senses*. Cambridge: Polity Press.

36. Corbin, A. (1999). *Village bells: Sound and meaning in the nineteenth-century French countryside*. London: Macmillan.

37. Cowan, M. (2006). Imagining modernity through the ear. *Arcadia, 41*(1), 124–146.

38. DeNora, T. (2000). *Music in everyday life*. Cambridge: Cambridge University Press.

39. Dittrich, M. (2001). Sound of silence. In TNO-TPD (Ed.), *TPD in 2000—projecten* (p. 18). Delft: TPD.

40. Douglas, S. J. (1987). *Inventing American broadcasting: 1899–1922*. Baltimore: The Johns Hopkins University Press.

41. Douglas, S. J. (1999). *Listening in: Radio and the American imagination, from Amos 'n' Andy and Edward R. Murrow to Wolfman Jack and Howard Stern*. New York: Times Books.

42. Edelmann, H. (1989). *Vom Luxusgut zum Gebrauchsgegenstand. Die Geschichte der Verbreitung von Personenkraftwagen in Deutschland*. Frankfurt am Main: Verband Öffentlicher Verkehrsbetriebe.

43. Felt. (1932). Advertisement: Composite body silence. *Automotive Industries, 66*(1), 3.

44. Fisher Body. (1935). Advertisement: First class. *Good Housekeeping*, January 3, page unknown.

45. Fisher Body. (1953). Advertisement: Quiet. *Life,* October 19, 178.

46. Fisher Body. (1958). Advertisement: Sound barrier. *Life,* March 24, 86.

47. Fisher Body. (1966). Advertisement: Escape. *Life*, May 20, 99.

48. Flink, J. J. (1992). The ultimate status symbol: The custom coachbuilt car in the interwar period. In M. Wachs & M. Crawford (Eds.), *The car and the city* (pp. 154–166). Ann Arbor: University of Michigan Press.

49. Folkerth, W. (2002). *The sound of Shakespeare*. London, New York: Routledge.

50. Foucault, M. (1976). *Mikrophysik der Macht*. Berlin: Merve.

51. Freimann, R. (1993). Das Auto—Klang statt Lärm. In A.-V. Langenmaier (Ed.), *Der Klang der Dinge* (pp. 45–57). München: Verlag Silke Schreiber.

52. Fricke, N. (2009). Warn- und Alarmsounds im Automobil. In G. Spehr (Ed.), *Funktionale Klänge* (pp. 47–64). Bielefeld: Transcript.

53. Gitelman, L. (1999). *Scripts, grooves, and writing machines: Representing technnology in the Edison era*. Stanford, CA: Stanford University Press.

54. Gitelman, L., & Pingree, G. B. (Eds.). (2003). *New media 1740–1915*. Cambridge, MA: MIT Press.

55. Haring, K. (2006). *Ham radio's technical culture*. Cambridge, MA: MIT Press.

56. Heßler, R. (1926). *Der Selbstfahrer. Ein Handbuch zur Führung und Wartung des Kraftwagens*. Leipzig: Hesse & Becker.

57. Hort, W. (1929). Der Strassenlärm. *Der Motorwagen*, *32*(9), 185–190.

58. Howes, D. (2004). *Empire of the senses. The sensual culture reader*. Oxford: Berg Publishers.

59. Husch, J. A. (1984). *Music of the workplace. A study of Muzak culture*. PhD Dissertation, University of Massachusetts, Amherst.

60. Jackson, A. (1968). Sound and ritual. *Man: A Monthly Record of Anthropological Science*, *3*, 293–299.

61. Jackson, D. M. (2003). *Sonic branding. An introduction*. Houndmills: Palgrave Macmillan.

62. Jewett. (1924). Advertisement: Enclosed comfort. *National Geographic*, April, page unknown.

63. Johnson, J. H. (1995). *Listening in Paris*. Berkeley: University of California Press.

64. Katz, M. (2002). Aesthetics out of exigency: Violin vibrato and the phonograph. In H.-J. Braun (Ed.), *Music and technology in the 20th century* (pp. 174–185). Baltimore: The Johns Hopkins University Press.

65. Katz, M. (2004). *Capturing sound: How technology has changed music*. Berkeley: University of California Press.

66. Kirchner, W. (1925). Die Karosserie von Gestern und Morgen. *Allgemeine Automobil-Zeitung*, *26*(32), 19–20.

67. Koshar, R. (2002). Germans on the wheel: Cars and leisure travel in interwar Germany. In R. Koshar (Ed.), *Histories of leisure* (pp. 215–230). Oxford and New York: Berg.

68. Kouwenhoven, E. (2002). Autoband moet fluisteren. *Algemeen Dagblad,* May 4, 43.

69. Kramer, G. (1994). An introduction to auditory display. In G. Kramer (Ed.), *Auditory display* (pp. 1–77). Reading, MA: Addison-Wesley.

70. Krebs, S. (2012). "Sobbing, whining, rumbling"—listening to automobiles as social practice. In T. Pinch & K. Bijsterveld (Eds.), *The Oxford handbook of sound studies* (pp. 79–101). Oxford: Oxford University Press.

71. Krebs, S. (2012). Standardizing car sound—integrating Europe? International traffic noise abatement and the emergence of a European car identity, 1950–1975. *History and Technology,* 28(1), 25–47.

72. Küster, J. (1907). *Chauffeur-Schule. Theoretische Einführung in die Praxis des berufsmäßigen Wagenführens.* Berlin: R. C. Schmidt.

73. Küster, J. (1919). *Das Automobil und seine Behandlung* (7th ed.). Berlin: R. C. Schmidt.

74. Lanza, J. (1994). *Elevator music. A surreal history of Muzak, easy-listening, and other moodsong.* New York: Picador.

75. Marks, L. U. (2002). *Touch. Sensuous theory and multisensory media.* Minneapolis: University of Minnesota Press.

76. Mauss, M. (1936). Les techniques du corps. *Journal de Psychologie Normale et Pathologique,* 32(3–4), 271–293.

77. Mayer-Sidd, E. (1931). Reparaturen nach schriftlicher Anleitung. *Das Kraftfahrzeug Handwerk,* 4(3), 57.

78. McIntyre, S. L. (2000). The failure of Fordism. Reform of the automobile repair industry, 1913–1940. *Technology and Culture, 41*(2), 269–299.

79. Merki, C. M. (2002). *Der holprige Siegeszug des Automobils 1895–1930.* Wien: Böhlau.

80. Mody, C. (2005). The sounds of science: Listening to laboratory practice. *Science, Technology & Human Values, 30*(2), 175–198.

81. Mom, G., Schot, J., & Staal, P. (2008). Civilizing motorized adventure: Automotive technology, user culture, and the Dutch Touring Club as mediator. In A. de la Bruhèze & R. Oldenziel (Eds.), *Manufacturing technology: Manufacturing consumers. The making of Dutch consumer society* (pp. 141–160). Amsterdam: Aksant.

82. Nieuwenhuis, P., & Wells, P. (2007). The all-steel body as a cornerstone to the foundations of the mass production car industry. *Industrial and Corporate Change, 16*(2), 183–211.

83. Ostwald, W. (1924). Das wetterfeste Auto. *Allgemeine Automobil-Zeitung, 25*(11), 21.

84. Otte, R. (1926). Holz oder Stahl im Karosseriebau? *Motor, 14*(10), 129–132.

85. Petit, H. (1929). Carosserie souple ou rigide? *La Vie Automobile, 25,* 43–44.

86. Picker, J. M. (2003). *Victorian soundscapes.* Oxford: Oxford University Press.

87. Plymouth. (1937). Advertisement: Ride in. *Life,* August 30, 1.

88. Polanyi, M. (2009). *The tacit dimension.* Chicago: The University of Chicago Press. (Originally published Garden City, NY: Doubleday, 1966).

89. Pontiac. (1935). Advertisement: Silver streak. *Good Housekeeping,* January 6, 129.

90. Poulussen, P. (1987). *Van burenlast tot milieuhinder. Het stedelijk leefmileu, 1500–1800.* Kapellen: DNB/Uitgeverij Pelckmans.

91. Rath, R. C. (2003). *How early America sounded.* Ithaca, London: Cornell University Press.

92. Rosen, G. (1974). A backward glance at noise pollution. *American Journal of Public Health, 64*(5), 514–517.

93. Sandberg, U. (2001). Abatement of traffic, vehicle, and tire/road noise—the global perspective. *Noise Control Engineering Journal, 49*(4), 170–181.

94. Schafer, R. M. (1994). *The soundscape. Our sonic environment and the tuning of the world.* Rochester, VT: Destiny Books. (Originally published as Schafer, R. M. *The tuning of the world.* New York: Knopf, 1977.)

95. Schick, A. (1994). Zur Geschichte der Bewertung von Innengeräuschen in Personenwagen. *Zeitschrift für Lärmbekämpfung, 41*(3), 61–68.

96. Schulte-Fortkamp, B., Genuit, K., & Fiebig, A. (2007). A new approach for developing vehicle target sounds. *Sound and Vibration, 40*(10), 2–5.

97. Schulze, G. (1992). *Die Erlebnisgesellschaft.* Frankfurt: Campus Verlag.

98. Sheller, M. (2004). Automotive emotions: Feeling the car. *Theory, Culture & Society, 21*(4/5), 221–242.

99. Siefert, M. (1995). Aesthetics, technology, and the capitalization of culture: How the talking machine became a musical instrument. *Science in Context, 8*(2), 417–449.

100. Smith, B. R. (1999). *The acoustic world of early modern England. Attending to the O-factor.* Chicago: The University of Chicago Press.

101. Smith, M. M. (2001). *Listening to nineteenth century America.* Chapel Hill: University of North Carolina Press.

102. Smith, M. M. (2003). Listening to the heard worlds of antebellum America. In M. Bull & L. Back (Eds.), *The auditory culture reader* (pp. 137–163). Oxford: Berg.

103. Smith, M. M. (Ed.). (2004). *Hearing history: A reader*. Athens, GA: The University of Georgia Press.

104. Smithuijsen, C. (2001). *Een verbazende stilte. Klassieke muziek, gedragsregels en sociale controle in de concertzaal*. Amsterdam: Boekmanstudies.

105. Steketee, M. (2006). Hightech autoruit. *Elsevier Thema Auto,* April, 68–69.

106. Sterne, J. (2003). *The audible past. Cultural origins of sound reproduction*. Durham, NC: Duke University Press.

107. Stockfelt, O. (1994). Cars, buildings and soundscapes. In H. Järviluoma (Ed.), *Soundscapes. Essays on vroom and moo* (pp. 19–38). Tampere: Tampere University Press.

108. Thompson, E. (1995). Machines, music, and the quest for fidelity: Marketing the Edison phonograph in America, 1877–1925. *Musical Quarterly, 79*(1), 131–171.

109. Thompson, E. (2002). *The soundscape of modernity. Architectural acoustics 1900–1933*. Cambridge, MA: MIT Press.

110. Thuillier, G. (1977). *Pour une histoire du quotidien au XIXe siècle en Nivernais*. Paris: La Haye.

111. Toyota Motor Corporation (Ed.). (2008). *Aygo: Betriebsanleitung*. Cologne: Toyota Motor Corporation.

112. Truax, B. (Ed.). (1978). *The world soundscape project's handbook for acoustic ecology*. Vancouver, BC: ARC Publications.

113. Truax, B. (1984). *Acoustic communication*. Norwood, NJ: Ablex Publishing.

114. Truax, B. (1996). Soundscape, acoustic communication and environmental sound composition. *Contemporary Music Review, 15*(1), 49–65.

115. van de Weijer, B. (2007). Kalm tuffen of bronstig cruisen. *De Volkskrant,* February 3, 9.

116. Verslag. (1934). *Verslag van het "Anti-lawaai Congres," georganiseerd te Delft, op 8 november 1934 door de Koninklijke Nederlandsche Automobiel Club in samenwerking met de Geluidstichting*. Delft: KNAC/Geluidstichting.

117. Weber, H. (2008). *Das Versprechen mobiler Freiheit*. Bielefeld: Transcript.

118. Wenzel, S. (2004). Vom Klang zum Lärm. *Neue Zeitschrift für Musik, 165*(2), 34–37.

119. Wetterauer, A. (2007). *Lust an der Distanz. Die Kunst der Autoreise in der Frankfurter Zeitung*. Tübingen: Tübinger Vereinigung für Volkskunde.

120. Weymann. (1925). Advertisement: Sensationelle Neuerung. *Allgemeine Automobil-Zeitung, 26*(28), 60.

121. White, R. B. (1991). Body by Fisher. The closed car revolution. *Automobile Quarterly, 29*, 46–63.

122. Wiethaup, H. (1966). Lärmbekämpfung in historischer Sicht. Vorgeschichtliche Zeit— Zeitalter der alten Kulturen usw. *Zentralblatt für Arbeitsmedizin und Arbeitsschutz, 16*(5), 120– 124.

123. Zeitler, A., & Zeller, P. (2006). Psychoacoustic modeling of sound attributes. *Proceedings SAE World Congress* (Detroit, MI, April 3–6).

2 The Experience of Sonic Interaction

Karmen Franinović and Christopher Salter

What are the specific experiences we have when we interact with sound? How is this interaction different from other sensory modalities like vision, touch, or smell? How do artists, designers, and researchers design, structure, and compose such interactions? In this chapter, the authors aim to situate these questions within the burgeoning field of sonic interaction design. While the most common definition of sonic interaction design includes the exploitation of sound as one of the principal channels conveying information, meaning, and aesthetic/emotional qualities in interactive contexts (see Introduction), this chapter shows how experiences engendered through interdisciplinary creative practices with sound are at the center of a broader shift in thinking about interaction as a system of actions triggering reaction and, instead, as an embodied, temporally emergent context in which new forms, aesthetic and emotional meanings and patterns arise.

2.1 Introduction

The title "The Experience of Sonic Interaction" is a deliberate provocation that aims to focus debates away from the informatic characteristics of technologically mediated interactions with sound. Although from a technical point of view interaction is usually defined as input triggering some form of output, more recent theories of so-called embodied interaction or enactive perception suggest a much more directly situated, less abstract, and, hence, highly experiential framing to the word [7, 37]. Yet, the question of experience in regard to emotional or aesthetic qualities that occur in the process of interaction is fraught with complexities. In *Art as Experience,* for example, philosopher John Dewey defined aesthetic experience as that of heightened vitality; an active and alert interaction with the world [6, p. 18]. In Dewey's understanding, experience is a process due to the continual coupling between human organisms and the environment in what he termed "the very process of living." Dewey's theory of interaction is deeply embodied, that is, linked to sensorimotor perception with an environment, a theory that ties his work directly to questions around what the American psychologist Eugene Gendlin described as experienced or "felt meaning." Like Dewey's theory, Gendlin's phenomenologically inflected concept of "felt meaning" is also rooted in the concrete interaction between organism and environment—an interaction, he argues, that can only be bodily situated in specific contexts. Interaction with the environment is itself a direct form of knowing. "Our bodies sense themselves in living in our situations. Our bodies do our living. Our bodies are interaction in the environment; they interact as bodies, not just through what comes with the five senses. Our bodies don't lurk in isolation behind the five peepholes of perception" [11].

With Gendlin, meaning is an act in which the perceiver's interaction with the external world is guided by direct action in local situations—what the late biologist Francisco Varela labeled as enactive cognition. As Varela wrote, "in a nut-shell, the enactive approach underscores the importance of two interrelated points: that perception consists of perceptually guided action; and that cognitive structures emerge from the recurrent sensorimotor patterns that enable action to be perceptually guided" [37, p. 12]. Here, interaction is predicated not only on a body responding to a specific, concrete circumstance but also through spatiotemporal processes in which that body comes to act in/on and, thus, to know the world. In other words, the concept of enactive perception is dependent on how the world helps guide or modulate action that, in turn, continuously results in the body realigning or remaking that world.

Given these direct connections between interaction and corporeal experience, with the body modulating action in the world, our goal here is to describe and analyze a wide range of art, design, and architectural projects and practices that directly involve creation, shaping, and manipulation, that is, *interaction*, with sound. In particular, we are interested in how corporeally experiential responses within such interaction are catalyzed and directly shaped by the specific material, temporal, and spatial qualities inherent to the sonic medium itself. For example, the vibratory, unstable, and malleable character of sound provides specific affordances or handles for interaction that are fundamentally different from visual cues. Understanding the specifics of sound as a medium and how those specifics affect our understanding of interaction will thus allow us to analyze the unique phenomenological, aesthetic, and social aspects of sonic experience.

2.1.1 The Human-Computer Interaction Perspective

Until recently the most common understanding of the word interaction in relation to sound has been derived from the specific context of human-computer interaction and, more specifically, the interdisciplinary subgroup of experts involved in the area of research into musical interaction called NIME (New Instruments for Musical Expression). Formed in 2001 as a breakout group from the SIGCHI[1] constituency to focus more exclusively on issues of musical import, the central focus of the NIME approach through its annual conference lies specifically in the area of designing new sensor-augmented devices that seek to elicit more nuanced and expressive forms of human-computer interaction than traditional interface devices such as keyboards or screens [25].

Central to the NIME tenet of sonic interaction is the augmented controller, which can be seen as a new kind of instrument. Although this instrument model varies depending on the context (recent controllers include everything from cell phones to giant metal stretched strings), sonic interaction in the NIME orthodoxy is based on a highly codified model of musical expression that involves real-time sensor input for the real-time control of musical parameters, techniques for the conditioning, analysis, and feature extraction of sensor signals, and mapping strategies for creating relationships between input and output parameters. In many ways, these three steps suggest that a model for sonic expression by way of interaction with a sensor-augmented device through gestural input consists of the following formula: *input (sensing) > mapping > output (sound synthesis) = musical expression*. Although this is appropriate within the larger contemporary context of shifting attention away from traditional keyboard and mouse input devices to potentially more body-centered controllers, such as the Nintendo Wii or Microsoft Kinect, there are some fundamental problems with the NIME model as a sustainable one for understanding interaction with sound from an aesthetic and experiential point of view. The main issue is the almost formulaic understanding of interaction as a series of input-output processes: a gesture or action triggers an appropriately stored or mapped series of sonic responses that may be adjusted based on the range of expression of the input. This assumes an already fixed set of relations among the user/interactor, the object/instrument/sound-making body, and the environment in which the interaction with sound takes place. According to embodied cognitivists, this definition of interaction is fundamentally *abstract*; it does not take into consideration the possibilities of the environment potentially altering the interaction over a time frame nor the ways in which the experience of interaction, the *production of expression*, may change in relation to that which is being interacted with or the context in which such interaction takes place.

Abstract here signifies what Varela called "the information processing problem of recovering pre-given properties of the world" [36, p. 12]. In contrast, our understanding of interaction suggests that the perceiver's actions, and not an already pregiven, predesigned world, are the starting point for perception and, consequently, that such sensorimotor action also continuously modulates the local situation. Thus, although an electronic sensor does not necessarily understand the context in which it senses, the sensorimotor apparatus of the human body looks specifically for the handles that enable it to explore, discover, and shape, that is, interact with the environment it finds itself within.

2.1.2   Sonic En(inter)action

For purposes of argument in this chapter, we want to promote a model of interaction in the manner in which the Dutch biologist and media theorist Arjen Mulder defines it: "interaction is a mode of bringing something into being—whether a form, structure, a body, an institute or a work of art—and, on the other hand, dealing with it" [19]. Extending Mulder's definition to the phenomenon of sound, we situate interaction in the following ways: (1) a spatiotemporal-material process, (2) an act of *poiesis* or making that is an active process or formation, (3) situated, concrete, and embodied, (4) performative and emergent in that it constitutes itself in time through different agencies operating in tandem with each other, and (5) nonrepresentational (i.e., not reducible to an act of mimesis, imitation, or purely symbolic processing).

Spatiotemporal/Material Process

As a medium, sound has particular spatiotemporal and material characteristics that distinguish it from other sensory modalities. Any sounding object or structure is inevitably entangled in its spatial-environmental context due to acoustic principles such as resonance, reverberation, diffraction, and refraction. The same is true for its temporally emergent nature, which includes such things as the production of patterns and rhythms, behavior over different time scales, and the ways in which sound is modulated by action over and in time. A third and final characteristic includes the inherent materiality of sound as a medium, in particular, how interaction changes, shapes, or transforms sonic material. This shaping of sound's materiality includes questions of malleability and modulation, the inherent ability for sound to be tactilely as well as acoustically perceived, and the ways in which interaction with sound helps mold further embodied action.

Interaction as Poiesis

The Greek root *poieo,* which is the etymological root of both poiesis and poetry, is literally translated as "to make," and *poiesis* signifies making as an active process or formation. By using the word *poiesis*, we want to argue that interacting with the materiality of sound in a spatial-temporal manner is akin to a form of creative making that involves acts of touch, listening, and movement. In other words, our interest lies in describing how processes of sonic interaction extend creative action through participation and play.

Situated and Embodied

As already stated, one of the central elements of sonic interaction design is the role that embodied action and perception plays, or how action can be guided by sound in a concrete, lived manner. The body is continually navigating through space, attending to cross-modal phenomena and seeking out what James J. Gibson called invariant structures[2] in the environment that will help guide sensorimotor action [12]. But the extraction of what one would normally think of as stable, invariant visual features such as brightness, density, volume, shadow, occlusion, and the edges of objects is, in fact, highly subject to changing circumstances in direct relationship to the perceiver's position and orientation. As is well known, the perception of complex frequencies in sound is highly dependent on the perceiver's position in relation to the sound source as well as the physical features of the environment. Even though sound moves around the body, this phenomenon does not mean that we pick up a fixed set of "features" from that sound. We adjust our head based on the direction of and distance from the source in order to compensate for the acoustic shadows produced by the ears' pinnae and the shoulders' and head's position, actions that, as any starting psychoacoustics student knows, act to filter the overall frequency content of a sound.

Performative and Emergent

By emphasizing the *action* in interaction, we want to propose a way of thinking about the term in a fundamentally *performative* way. Our use of the word *performative* here signifies something that has not yet been constituted and must then (ontologically) arise over and through time. If performance is the action of playing with how something comes to be or is conjured forth in/over time, then the word suggests something that has no fundamentally given essence or fixed set of characteristics. Interaction "brings something into being" as Mulder states [19]. At the same time, one brings forth a world that, although materially preexisting, is being continuously altered as the subject acts in that world. What is interesting in relation to sound is the question of who or what performs the interaction. Sound is inevitably an expression of different kinds of agencies, for example, the human hand, the wind, or the technical apparatus of a loudspeaker. To produce sound, one simply has to have any kind of body that can disturb a medium and put it into oscillation: one's vocal cords, feet, or an object that resonates or vibrates. Indeed, performance has a particular material connotation—sound acts on the world, through what sociologist of science Andrew Pickering calls its *material agency,* a force on other material beings [24].

Nonrepresentational

Sound, as anthropologist Georgina Born states, is "alogogenic, unrelated to language, nonartifact, having no physical existence, and nonrepresentational. It is self-referential, aural abstraction" [4, p. 19]. Although we will focus very specifically on the question of sound's tangible existence, we want to shift not only from sound's representation within the form of a score or text but away from another kind of representation: a fixed plan that is unaware or unrelated to its position within an environment. In other words, in relating sound and our interaction with it in material, performative terms, we wish to distance ourselves from representational modes of thought that suggest an abstract, perceiver-independent world of information and symbols.

We have divided the remainder of this chapter into three parts, each of which attempts to combine theoretical framings for sonic interaction design with a diverse range of artistic and design projects that we feel directly embody, play with, or, at the very least, are resonant to such theoretical handles. The projects discussed range from key historical examples to recent work by younger designers, composers, artists, and researchers. Section 2.2 examines sound as material and the ways in which specific technologies and actions alter its materiality. We draw on early artistic work in which new technologies, such as magnetic tape recorders, allowed artists to directly manipulate and interact with sound and more recent projects that attempt to affect human action through sonic feedback. Section 2.3 examines the spatial aspects of sonic interaction, particularly the ways in which architectural and urban sites are used to shape the political-social-aesthetic perception of built environment. Section 2.4 examines the temporal aspects of sonic interaction and responsiveness, specifically how time perception in interacting with sound is modified or altered through such concepts as temporal shape, continuous versus discrete interaction, and the multiple time scales active in sonic interaction. Finally, based on the practices discussed, we conclude by summarizing the performative, embodied, and material approaches to the design of new kinds of interactive sonic experiences.

2.2   Sound as Malleable Material

Historically, technologies such as telecommunication and recording devices have often been accused of disembodying sounds from the mechanisms that caused them.

Indeed, R. Murray Schafer's coining of the term "schizophonia" [31] aimed to describe such temporal or spatial fracturing between a sound and its generating source. Schafer associated this phenomenon with its negative effects such as a decreased quality of everyday soundscapes, but simultaneously, such technologically catalyzed dissociation also fostered novel expressions within electroacoustic music. Acousmatic sound, created through the advent of magnetic tape recording and, later, the manipulation of electronic signals, led to electronic music's rapid integration in concert hall formats in the 1950s in which seated listeners would *let their ears be guided* into fascinating sonic landscapes. Yet, at the same time, an alternative and more performative approach to sound as creative material was emerging within art and music circles.

2.2.1 Molding Sound

Music captured on a material support such as magnetic tape enabled a more *direct manipulation of sound*. Experiments with audiotape in the 1950s and 1960s from such composers as John Cage, Pierre Schaeffer, Pierre Henry, Nam Jun Paik, and Steve Reich, among others, made sound physically malleable in a novel way. These artists explored the potential of cutting and splicing tape and reassembling it anew, creating startling sonic effects and new forms of music through repetition and modulation of parameters such as speed, phase, and similar characteristics. For example, in his 1963 *Random Access Music*, trained composer and video artist Nam Jun Paik decontextualized the technical apparatus of a tape recorder by removing the tape head from the recording device. Visitors could "interactively" run the tape head over the audiotapes arranged in abstract shapes on the wall, producing sound based on the speed of manipulation as well as continuity and intensity of their gestures. By changing the control of the head from an automatic mechanism to the human hand, a functional piece of technology was converted into an expressive instrument. The unpredictability of visitors' gestures thus created sounds that the artist could not precompose or predict, giving rise to a different kind of aesthetic experience.

Paik's unpacking and rearrangement of a technological device served to offer gallery visitors a rich and dense sonic experience through their manipulation of sound, yet electronic technologies were soon to radically advance other forms of sonic exploration. In his lecture "Four Criteria of Electronic Music" (1972), Karlheinz Stockhausen passionately described the possibility of generating completely new "unheard sound" by the means of digital processing tools [33]. With this emphasis on algorithmic techniques, such creative processes neglected the human gesture, reducing it to simple movements of buttons, sliders, and knobs. Perhaps this reduction of human action to the machine-like precision of button pressing and slider expression reached

Figure 2.1
Mark Hauenstein and Tom Jenkins's *Audio Shaker* (2004).

its apex in the so-called laptop music scene at the end of the 1990s where, as composer Bob Ostertag argued in his 2002 salvo "Human Bodies / Computer Music," "the physical aspect of the performance has been further reduced to sitting on stage and moving a cursor by dragging one's finger across a track pad in millimeter increments" [22, p. 12].

Since the late 1960s, however, sensing and actuating techniques have brought the expressiveness of the musical gesture back into focus. New hybrid interfaces demanded gestural virtuosity, and sound was once again a material to play with through the medium and force of bodily gesture in the work of many composers/performers. For example, *The Hands* (1984), a gestural controller-instrument worn on the hands and developed by the late composer and performer Michel Waisvisz, placed sound literally between the musician's fingers: catching, stretching, and compressing it as the same sound transited through the air. Although the musical and expressive richness potential of *The Hands* could be achieved and explored only by an expert user such as Waisvisz himself, in other projects he created easy-to-use interfaces inspired by children's intuition and inclination toward exploring the physical world. Waisvisz's *Crackle Family* project (1976) presented a vision of the future dining experience in which a number of actions with everyday objects such as pouring tea or using cutlery were sonified. The use of everyday rather than expert musical movements allowed Waisvisz to engage novel sonic expressions with no instructions for use.

Given the goal of reaching the widest audience participation possible, the intuitiveness of interaction proves to be an essential ingredient of embodied sonic experience.

An example is Mark Hauenstein's and Tom Jenkin's *Audio Shaker*, an ordinary looking container resembling a cocktail shaker used to mix sounds rather than liquids (figure 2.1). Users can open the object, speak into it to record sound, shake it to mix and then literally, pour out the sound mix. The sounds keep the reference to the original recorded ones but are subsequently transformed according to the intensity and repetition of shaking gestures. When the sounds are poured out of the shaker, the speed is directly mapped to the tilting of the object, giving the user the direct sensation that the sound drips or flows out of the vessel. In this project, the digital processes and software interfaces move to the background in order to allow the user to explore the auditory potential of the object through physical interaction. The device is an expressive instrument that turns the immaterial and transient into the malleable and tangible: the sound shakes and vibrates between the user's hands, and pours out like water. Directly mapped to gestures such as shaking and pouring, the resulting sonic feedback makes the interface easy to use and yet meaningful, expressive, and physically engaging.

Thus, the intuitiveness of the physical interaction can be created by increasing the perceived materiality of sound. One of the clearest examples is David Merrill and Hayes Raffle's *The Sound of Touch* interface in which the palette of material textures can be touched with a number of wand-like interfaces to record, store, and manipulate sound. The sounds produced can be captured with a microphone in the wand by hitting, caressing, or brushing materials such as wool, metal screen, or broom bristle (see chapter 7, this volume, for more details). The recorded samples can then be manipulated in real time by moving the wand in contact with the texture, this time using the palette as a tool for modulating the sample. Each recorded sound embodies a gesture and is brought to sounding by another gesture. One does not simply compose sonic textures but creates them with a palette of action sounds, which are designed through the relationship among gesture, artifact, and sound. Soundings emerge from the materiality of the world around us and that of our own bodies moving through and touching that world. In a kind of metabolic, open-ended enactive process, *The Sound of Touch* explicitly relates the materiality of sounds to objects and to the gestures that produce them.

2.2.2 Molding Action

The world makes itself available to the perceiver through physical movement and interaction. I argue that all perception is touch-like in this way: perceptual experience acquires content thanks to our possession of bodily skills. What we perceive is determined by what we do (or what we know how to do).

—Alva Noe, Action in Perception [20, p. 5]

Sound can be molded by our actions, and this, in turn, can modify our auditory perception, consequently engaging us to act, move, or dance, thus closing the loop of action-perception through sound. When we let our actions be guided by sound, challenging physical tasks can be accomplished. Such use of auditory feedback to support user performance is exemplified in *The Ballancer* project by Mathias Rath and Davide Rocchesso, where a device is used to balance a virtual ball on a physical stick by listening to the ball rolling across an identical virtual stick presented on the screen [26]. Created by the means of physics-based models that simulate the event of a rolling ball, the sounds are continuously coupled to the action of tilting the physical stick. Users can identify the speed and the movement of the ball across differently textured areas of the virtual stick by listening to the self-produced sounds and thus can quickly adjust their movements to keep the ball stable. As it shapes the action continuously, the interface requires the user's attention and bodily adjustments at every moment. The project demonstrates that human performance can be strongly affected by sound and strengthens the idea that a direct, continuous relationship between action and sound is a core element for embodied interaction.

The tight coupling between action and the vibrations it generates can also be used to affect the sensorimotor behavior of the user. Oddly, the vibrations that we actively produce when in contact with physical objects often pass unnoticed. If one taps one's fingers on a table and someone else keeps a hand on the table, that person will feel the vibrations of the surface, whereas the person producing them will mainly perceive the discrete impact at the fingertips. By exploiting such often unnoticed sensations, Yon Visell and colleagues developed the *EcoTile* (2007–2012), a floor-tile platform for audiohaptic simulation of different types of grounds [38]. The name references Gisbon's ecological approach to perception—a theory based on perception as action rather than as a purely cognitive process of a priori, informatic-based representation [13]. The floor is composed of tiles that provide vibrational and sonic feedback in order to generate audiohaptic sensations of walking across different ground surfaces (figure 2.2). The movement of the feet is captured through force-sensitive resistors attached to the tiles, and depending on the pressure of the user's body, the vibrating actuators attached to the tiles respond with different signals. These vibrations activated by the feet lend new material qualities to the floor, perceptually turning the wooden tiles into sand or snow surfaces.

What interests us here is that only when the users' actions are directly and continuously coupled to sonic response do the sensations acquire meaning, chiefly by being associated with previous walking experiences. When the sound is simply presented to

Figure 2.2
Latest version of *EcoTile* floor by Yon Visell, Karmen Franinovic, and Moritz Kemper (2012).

the person standing on the wooden tile, he or she feels as if the floor is shaking or as if someone is hitting on the tile from beneath. Only when the vibration is interactively and synchronously coupled to a user's movements of walking does he or she begin to experience a wooden floor changing its material properties. The vibrations turn the wood into a responsive and dense material, which becomes a different ground type only when being acted on. Interaction here is a process of active discovery in which the perception of material reality can be continuously changed, adapted, and reinvented. Similarly, Georg Essl and Sile O'Modhrain's *PebbleBox* device (2004) (see chapter 6, this volume) transforms perceived properties of a physical object through auditory feedback. In it, the user can move her or his hand in a box full of pebbles and can generate sounds of different materials such as ice cubes or sand [21].

Depending on the sound produced, the user feels the tactile sensation of immersing his or her hands inside a box full of fluid or ice cubes, despite the fact that only the pebbles are being touched.

Whereas the *PebbleBox* explores the ways in which haptic and auditory senses interact, the *EcoTile* employs the potential of vibratory feedback to affect human motion. As we shape the sonic vibrations of the ground, the vibrations in response affect our kinesthetic experience. It is known that walking across differently compliant surfaces results in adjustment of muscle in the legs [8]. This type of motor response can be modulated through the *EcoTile* because sonic vibration can induce sensations of compliance under the force of our feet [39]. In other words, vibrational feedback can induce the perception of virtual events such as sinking into the ground while one is stepping on the tile, and this in response causes the adjustment of the muscles in the legs to compensate for the compliance of the ground material. These muscular microresponses are a kind of "felt proof" that the action-perception loop can be affected through interfaces that exploit the multimodal characteristics of sound.

In the first part of this chapter we discussed how, in an interactive setting, the perceived materiality of sound is shaped by the relationship between sound and action. Combining the physicality of gesture with changing feedback can alter the perceived properties of a material object. The wooden tile can be perceived as soft as snow or as granular as sand. What we see and what we feel may not coincide, and yet our body combines different sensorimotor information to produce sensations that are deeply and physically felt. If one does not move across the *EcoTiles*, one will not feel the morphing of the ground matter under one's feet. If one does not move a hand in the *PebbleBox*, one cannot feel the liquidness or softness of the pebbles. This interactivity of sound emerges into the world only if our bodies actively engage with their physical surroundings. Our perceptions come into existence through our action, and the other way round. Action and perception must be understood as one and designed as one because they are lived as one. Thus, designers have the possibility to enable a new domain of sensory and social experience to emerge when interacting with technology.

2.3 Sociality and Spaciousness

The malleability of sound offers new opportunities for the transformation of social relations within architectural and urban contexts. Sound emanates from an object, expanding it beyond its physical boundaries into the surroundings, and yet remains tethered to an individual, or collective, gesture. Thus, it makes bodily movements spacious and social, as the gesture of one person reaches out through sounding waves to touch another, inviting for exchange and play. The dynamics of such solo and collaborative acts in a real-world context offer the opportunity to challenge the existing sociocultural structures. The particular qualities of interactive sound, such as tactility and directness, may allow us to modulate the supposedly fixed relationships among environment, sound, and individuals in novel ways, thereby enabling novel constellations to emerge and evolve.

2.3.1 Enacting Privacy

The vibrational nature of sound gives rise to our tangible as well as intimate experience of it. We can feel the vibrations, audible or not, of sounding objects whenever we touch them. But they can be intimate and tactile even when we are not in contact with the sounding body—in a dance club, for example, the powerful throbbing of subwoofers connects bodies and spaces, joining them in a singularly dense vibrating matter. The vibrational quality is accentuated in objects that embed loudspeakers directly into their material structure because the sound source is in direct contact with the user's body. These attributes can be used to recreate vibrotactile sensations caused by an existing physical system, such as in Apple's *Mighty Mouse,* where the action-sound relationship is designed to simulate the traditional clicking wheel mechanism. As a result, the users do not perceive the sound of the wheel as added but feel as if they were interacting with an analog scroll-wheel mechanism.

A more inventive use of embedded sonic vibrations can engender new kinds of aesthetic and social experiences. Kelly Dobson's *ScreamBody* (1998–2004), for example, stores and reproduces a very personal vocal action: that of screaming. The user screams into an amorphically shaped object worn on his or her chest during any situation in which screaming would be awkward, such as waiting in a public location (figure 2.3). The scream is silenced and stored in the *ScreamBody* so that the recorded voice can be replayed on another occasion. A strong and sudden squeeze of the worn artifact releases the scream. This gesture reenacts the actual screaming gesture adding to the emotional engagement of the user (see figure 2.3). Such physical and emotional reenactment is made possible by the shape of the wearable device—its placement on the chest of the user affords the compression of the object and, at the same time, the convulsion of the user's body that pushes the scream out. The vibration of the embedded speaker shakes the chest of the user and by doing so further enhances the reality of the scream's becoming from the past into the present. The object therefore has the potential to store and recreate emotions such as anger and desperation, literally shaking the user's bones with the scream. Thus, the sonic interaction with the

Figure 2.3
Kelly Dobson's *ScreamBody* (1998–2004).

Figure 2.4
Social Mobiles (2002) by Crispin Jones and Graham Pullin with IDEO.

ScreamBody excites the user's auditory, tactile, and kinesthetic senses in multiple ways, allowing him or her to record, play, express, and share private emotional states.

By the means of gestural sonic interaction, *Social Mobiles (SoMo)* (2002) developed by Crispin Jones and Graham Pullin with the design consultancy IDEO addressed the annoying uses of mobile phones in public locations. The *SoMo5*, one of the five developed prototypes, for example, can be used to throw an annoying sound toward a person who is talking too loudly on his or her handset (figure 2.4). The user pulls a catapult-like device mounted on the phone, aiming it toward the offending person in order to activate a warning sound. When the catapult is released, an abrupt sonic interruption is emitted from the other person's phone, causing auditory annoyance for the loudly speaking person. The catapulting gesture's spatial directness and sonic consequences create the feeling that something physical has been thrown at the annoying person. The expression and the physical release of anger are thus enabled through such a sonic gesture. If a similar sound were caused by a different gesture, for example, by pushing a button, the experience would be far less satisfactory. In addition, the combination of physical and digital sounds makes the simulation more engaging. The first sound, caused by the impact of the catapult part hitting the main body of the phone, is followed by a second, digitally created sound in the aggressor's phone. The composition of mechanical and digital sounds, both caused by the same human gesture, results in a satisfactory infliction of sonic pain to the annoying speaker.

Thus, the performance of a sonic event, whether simulated or real, can lend strong materiality to an emotional episode. Whereas Dobson's *ScreamBody* reenacts past personal events, the *SoMo5* generates violent acts toward another person without physically harming him or her. These experiences are both embodied in their physical gesture as well as situated in a public context—exposing such actions to the public eye. The physicality of the sounding artifact externalizes one's opinions or emotions, making them more tangible.

Because of the visibility inherent in tangible interaction, others are affected by the performance. For example, by using *SoMo5*, a person publicly displays personal judgments regarding someone else's behavior. This shared quality of sound can be used in a collaborative and participatory manner using public space in the real sense of its meaning.

2.3.2 Relational, Playful, and Participatory

In numerous public projects, interactive sound has played a distinctive role in critiquing our everyday situations and urban experiences. Art researchers in mobile sound technology creatively have engaged citizens in public space production feeding from the steady colonization of our cities by new wireless and sensor-based technologies [10, 32]. Others work within site-specific locations, turning them into playful fields of sonic intensities within the urban tissue that may allow citizens to appropriate public space. Such urban interventions may be described as *citizens' media* or strategies in which "collectivity is enacting its citizenship by actively intervening and transforming the established mediascape" [28]. The goal is to challenge legitimized identities and social codes by confronting participants with their own established behaviors and identities through the creation of their own media, those of their group and of their environment. The medium of sound is charged with potential for such interactions because of its ability to connect, engendering senses of intimacy, ambiance, and shared experience.

However, sound can also act to reinforce the identities of a group or individual, as in Erika Rothenberg's public artwork *Freedom of Expression National Monument* (1998–

2004), which aimed to create space for the expression of one's opinion. It consisted of a monumental red megaphone on an elevated stand that was provided as a platform and visible marker from which any member of the public might make statements. In practice, many of the statements that it evoked were impolite or meaningless—one person who tried to talk about war was largely ignored, leading her to conclude that nobody actually listens any more [14]. Although microphones in public space typically serve the function of facilitating the expression of an opinion or performance by one party, interactive sound can also be used to question those opinions and identities and put them in dynamic relationships.

This challenge was explored in Zero-Th's *Recycled Soundscapes* (2004), a platform provided to the public for the purpose of active listening, recording, and recomposition of surrounding sounds in an open urban setting [9] (figure 2.5). The red, body-size Beludire sculpture allowed for the auditory observation and capturing of distant and transient sounds, which would otherwise be difficult to hear (figure 2.6). As distant ambiances and sound sources were made more proximate, the user of the Beludire felt closer to, and experienced a more intimate relation with, visually

Figure 2.5
Recycled Soundscapes (2004) by Zero-Th.

distant strangers. Reflexively, the device's conspicuously surveilling nature did not leave many passers-by indifferent to its use. This relation was sometimes understood by the observed individuals as an invitation to communicate and to play and at other times as an intrusion into their private spheres. In each case, it seemed as though the presence of the user of the Beludire was projected out and distributed among the surroundings. Once captured, the amplified sounds were echoed following a soft gong sound from speakers integrated within the other two sculptures, called the Sonic Bowls. This replayed sounds communicated to the player that his or her recordings have been stored in these objects. They could then be used to compose the new soundscape out of recorded sounds by spinning the reflective black dishes, the moving parts of the Sonic Bowls. Thus, participants generated the soundscape out of a multiplicity of recorded voices molded through their physical actions—they might hear a bird, a statement very different from their own, or their own words mixed with other voices. Such a strategy of remixing challenged participants' expectations and established identities and has irritated those who were eager to find their own statements within the sonic memory of the location accessible through the Sonic Bowls.

Recycled Soundscapes was capable of becoming for some a closed and insulating system and of forging connections to those that were removed from their spatial vicinity for others. The Beludire opened and extended the sonic as well as the social space of the installation into the existing surroundings. Such a pulsating, expanding, and contracting space was enabled by the choice of the interactive medium—sound, which as dynamic, transient, and pervasive phenomena involved all the passers-by in the location, even those who did not have any interest or time to interact. Using the Beludire, players would reach out into the sonic domains of passers-by, including them in performative ecology of the place. Playing the Sonic Bowls, participants molded voices and noises left imprisoned in public sculptures until the moment of participants' interaction. Both players and observers hesitated and mused around emerging questions: What makes a conversation in an urban area private? What can one do with the transient information floating in the public? How close can one come to a stranger?

Although embodied sonic interaction creates an opportunity to play with public behaviors, it also makes it more difficult for passers-by to get involved because they may feel observed and judged as their actions, emotions, and thoughts become tangible. *Recycled Soundscapes* uses play as a strategy for engaging action and loosening the prejudices about the inappropriateness of certain behavior in public space. Playing

Figure 2.6
Beludire sculpture in *Recycled Soundscapes* (2004) by Zero-Th.

with what is already there—sounds transiting beyond the notions of public and private—participants critically engage with the existing relationships and pose questions related to the use of public space in a more direct manner. Such transparent relationships between the environment and its inhabitants increase the sense of responsibility for our sonic actions on the world.

2.3.3 Sound's Architectural Interactions

In contrast to the acousmatic tradition, the Dream House is formed at the moment an individual enters the sonic field immersed as in a fluid, sounds oscillate across a range of frequencies though the movements of the body. [16, p. 73]

Just as interaction with sound carves out new sociospatial practices, it also enables acoustic-spatial affordances that directly shape perceptual action. Historically, artists have used this spatial encounter between sound and architecture to explore how sound activates and resonates existing structures, creating what Barry Blesser and Linda-Ruth Salter have called "aural architecture." "Aural architecture, with its own beauty, aesthetics, and symbolism, parallels visual architecture. Visual and aural meanings often align and reinforce each other. For example, the visual vastness of a cathedral communicates through the eyes, while its enveloping reverberation communicates through the ears" [3, p. 3]. This emphasis on amplifying specific architectural features through sound has been made experientially explicit in many artistic projects that use techniques of acoustic excitation, resonance, and spatialization to demonstrate how sound opens up hidden elements of built space, increases auditory awareness, and, perhaps most interestingly, serves to generate new sensorimotor-based experiences through listening.

A key historical example of the ways in which the perception of aural architecture influences bodily perception of space is undoubtedly the colossal *Philips Pavilion* designed by Le Corbusier, Iannis Xenakis, and Edgard Varèse for the 1958 Brussels World's Fair. Rather than simply placing a series of systems of speakers into an architectural shell, Xenakis and Le Corbusier developed a hyperbolic paraboloid geometry for the building that directly influenced the behavior of sound and vice versa. In particular, Xenakis sought to create a space where architectural surfaces would help define the manner in which sound moved through the space and, consequently, the techniques by which the human body would physically interact with and navigate within such a space. "Shapes are doubly primordial due to reverberation and the quality of the sound they reflect, and also by the way space itself conditions the human body. We live in such spaces, we listen with our ears (we hear space with our ear) and we see it with our eyes, therefore both senses as well as the body's movement within a given space are involved" [40, p. 144].

Combining Varèse's interest in "projecting music into space" and Xenakis's desire for a kind of parallel sonic architecture running alongside the built architecture, the *Philips Pavilion* was one of the earliest examples to combine electronically controlled and generated image, light, sound, and a locomotive public through the space within a total architectural context. It was also one of the first to use sophisticated ideas of spatialization, particularly stereophony and multichannel projection in order to achieve Varèse's compositional strategy for an expanding and contracting sonic space of architectural masses and planes.

Xenakis's design of the sound routes ("routes de son") included a spatialization system with more than three hundred individual loudspeaker units that aimed to maximize the ways in which different frequencies in Varèse's 8-minute-long *Poème électronique* would react to the space. As Xenakis argued, "the acoustics of a space is linked to the way that space is formed, to the shape of its covering." While low frequencies were handled by the speakers near the base of the structure, a large array of smaller speakers was mounted on the interior surfaces of the building, following the parabolic curves of the structure so that high frequencies would interact better on the warped forms as they zipped up and moved across the geometric paths of the building. As sound traveled across the surfaces, listeners would continually reorient their bodies and shift listening perspective because there was no single aural vantage point or "sweet spot" from which to completely hear the complex timbral and spatial ranges present in Varèse's work.

Xenakis also continued to explore the concept of an interactive mobile listener in his later *Polytopes*, a series of "pan musical" works combining light, sound, and architecture begun in 1967 with the *Polytope de Montréal* and continuing until the immense temporary *Diatope* building that marked the inauguration of the Centre Pompidou in Paris in 1978. With the first *Polytope de Montréal*, realized for the 1967 World's Expo, Xenakis erected a similar hyperbolic paraboloid-shaped architecture like the *Philips Pavilion*, installed in the central atrium of the French pavilion but this time constructed of steel wires and augmented with 1,200 lighting fixtures. Seeking again to use music as an instrument for the "conquest of geometric space" [40, p. 134], the composer spatialized his prerecorded, acoustically generated score played back on tape by means of a series of four groups of speakers hung within the atrium. Like many experimental attempts in the late 1960s (e.g., Stockhausen's *Gruppen*), Xenakis similarly spatially distributed the score for *Polytope de Montréal* via loudspeaker arrays across the multilevel atrium of the pavilion, thus turning seated and passive concert goers into moving, active perceivers based on their listening position in the space.

In a different fashion, starting in the mid-1960s La Monte Young's and Marian Zazeela's *Dream House* project used just-intoned sine waves and room acoustics in order to vibrate different areas of room space. "The sound took on physical mass or better, the actual physical movement of sound waves became apparent in a way that was exhilarating for some, painful for others, but in any case inescapable" [30, p. 32]. Like Xenakis's *Polytopes*, such tangible experience was enabled through human motion in the room—moving throughout the space or slightly turning one's head resulted in a continuous change of sound. The visitors' auditory perception of sound was modulated as they moved through the space, lending physicality to the invisible and generating the sensation of being immersed in a dense, vibratory matter. Through the visitors' explorative and active behavior, sound acquired a body while audition became tactile.

If a sound's path and behavior with architecture yields radically different relations between space and body, recent projects from Mark Bain, Jacob Kierkegaard, and the Berlin-based tamtam harness its resonant characteristics to activate existing architectural structures, either by way of acoustic transduction (turning structures into speakers by setting objects into resonance) or by using sound to bring forth a parallel invisible aural architecture. For example, the American artist Mark Bain's infamous architecture-interface experiments such as *The Live Room* (1998) reexamined Nicola Tesla's mythic mechanical oscillator or "earthquake machine" by deploying unbalanced motors attached to columns in a building to provoke intensely powerful structural resonances in the material.

In a subtler manner, for his 2007 *Broadway* installation at the Swiss Institute in New York, Danish sound artist Jacob Kierkegaard utilized high-end accelerometers to capture the threshold vibrations produced in six columns running through the eight-story building, including the actual gallery space. Attaching 60 acoustic transducers to the structures, the artist then played the frequencies back through the material of the columns themselves, turning them and the larger gallery environment into sounding objects.

This acoustic unveiling of hidden features in architecture perhaps finds its most perceptually and aesthetically engaging example in the artist duo tamtam's (Sam Auinger and Hannes Strobl) sound installation *Raumfarben 02*, created in the context of the 2009 Klangstaetten–Stadtklaenge festival in Braunschweig, Germany. Researching the acoustics of the fifteenth-century gothic St. Aegidien church by activating specific resonance features of the architecture with tuned prepared sounds diffused through a minimal number of loudspeakers within the space, Auinger and Strobl sought to "make the architecture speak" by bringing forth timbral qualities (*farben* in German) directly shaped by sound's behavior with the architectural configuration of the building.

In *Raumfarben 02*, sound is used to subtly excite the structure and bring forth proportions and volumes: architectural shapes based on the actual geometric proportions of the church. But tamtam's precise design of the interior acoustic environment also brings the inhabitants of this sacred space into a different listening state, highlighting the strong contrast between the sounding space brought on by the church's precise architectural proportions and the exterior sounds of trams, automobiles, and pedestrians. Indeed, in discussing the work, tamtam described that after a certain threshold period of time listening inside the church, visitors to the installation reported the sound of the outside world as "gray and harsh." Here, exciting the space of the church not only leads to the space enhancing the music. Rather than sound using the space as a kind of neutral "sounding container," tamtam's work reminds us that the perceiving body's ability to hear is not intrinsically given but subject to the spatial-social contexts of complex, dynamic, and continually transforming acoustic environments.

2.4   Responsiveness: The Temporal Shape of Sonic Interaction

Over a period of time the computer's displays establish a context within which the interaction occurs. It is within this context that the participant chooses his next action and anticipates the environment's response. If the response is unexpected, the environment has changed the context, and the participant must reexamine his expectations. The experience is controlled by a composition [that] anticipates the participant's actions and flirts with his expectations [15, p. 423].

As we have seen, built space directly influences the behavior and trajectory of a sound's movement through time, from the earliest first-order reflections as it ricochets through space and bounces off surfaces through the slow loss of brightness as spectral energy in its frequency partials is gradually absorbed by the environment itself. In other words, in relationship to sound, the built environment is also a medium that catalyzes and generates evolving action. A sound's interaction with built space denotes evolution and transformation of its physical characteristics over time, whereas, from an architectural viewpoint, interaction denotes how structural phenomena in the environment amplify and attenuate acoustic experience.

There is no interaction with sound without time. The disturbance of a medium through the unleashing of pressure waves already defines sound's temporal essence. In its ideal form imagined by Fourier, the phenomenality of sound is marked by its temporal attributes: period, phase, frequency, and amplitude. Artificial techniques of sound synthesis use time as the building block for manipulating timbral, rhythmic, and dynamic behaviors. Likewise, the direct organization and constitution of sound in time by the human hand or machines acting on sonic bodies and spaces articulate the craft and alchemy inherent in the temporal act that we call composition. But in order to grapple with sound's interaction with, in, and through time, to understand the force of its elasticity and bounce, behavior and ballistics, presence and then subsequent disappearance without trace, we have to grapple with how time shapes sound and how sound shapes time.

2.4.1   Responsiveness Is the Medium

From a seemingly straightforward acoustic perspective, the term responsive signifies the way sound "answers" or responds to the acoustic features of an environment. Yet, response also denotes the way sound acts on and in an environment: the way it enunciates its material qualities and how the environment changes it at the same time. Responsiveness is a temporal phenomenon as characterized through its properties of elasticity, dynamics, and liveness—those forces that, as Andrew Pickering argues, operate on other material beings [24].

From the technical perspective of machine-augmented environment-inhabitant interaction, however, the term *responsive sonic environment* signifies something else entirely. Although there is no agreed-upon definition, responsive sound environments generally refer to a system that "regenerates a soundscape dynamically by mapping known gestures to influence diffusion and spatialization of sound objects created from evolving data" [17].

Outside of a musical context, the original use of the term responsive in the specific context of machine-based interaction derives from the work of computer scientist Myron Krueger in the 1970s. Krueger sought to define a computationally augmented space or what he labeled a "responsive environment" as a "physical space in which a computer perceives the actions of those who enter and responds intelligently through complex visual and auditory displays" [15].

Glowflow (1970), Krueger's first responsive audiovisual environment, already set out to experientially explore interaction or what Krueger would later call "the direct performance of the experience" through the combination of participant action and system reaction. A darkened environment of suspended, transparent tubes contained phosphor particles that, when charged suddenly by light triggered by pressure-sensor based foot-pads, would begin to glow. Sonically, a similar modality of interaction occurred in which participants would influence parameters as well as the spatial position of the sounds produced by a Moog synthesizer within the environment.

The difficulty of identifying the exact response of the system for *Glowflow's* participants led Krueger to articulate a specific set of conditions for his next experiments. Above all, his concept of the computer (and here we include sensors, analog/digital and digital/analog converters, processing units, etc.) was firmly anchored in the machine intelligently responding to the presence of visitors, processing (understanding) that presence, and outputting the response through an auditory display (usually loudspeakers). In Krueger's understanding, response made explicit the time span between human action and system reaction. He sought to clarify this relationship by arguing "the only aesthetic concern is the quality of the interaction" (i.e., the perception of the temporal gap between action and response) [15, p. 424].

Although pioneering in its use of computer science research techniques, *Glowflow* was not the first example of such work. Swept up in the Op Art and Kinetic Art movements in the 1960s, artists and collectives such as Gruppo T (T for Tempo), Gruppo N, GRAV, and others associated with the Nouvelle Tendence scene in 1961, built numerous walk-in audiovisual installations that explored environmental triggering of sounds by way of crude electronic sensing devices such as photocells and microphones. Unlike the projects from Krueger and others, which used human gesture or movement to trigger sound, light, and image, Robert Rauschenberg's 1968 work *Soundings*, (developed in collaboration with the E.A.T.—Experiments in Art and Technology organization) utilized audio itself as an input to reveal hidden silk-screened images of chairs behind a mirrored and Plexiglas facade. As visitors shouted, shrieked, or moved through the space, frequency analysis techniques, which divided the particular sound up into four distinct bands, were used to turn on and off different lights embedded in the facade, revealing and concealing hidden layers over time.

In contrast to the almost semiotic intent of *Soundings,* where sound was used as a trigger to reveal visual signs through a direct coupling of action and reaction, call and response, was Rauschenberg's *Mud-Muse*, an intriguing collaboration with the Teledyne corporation as part of the Los Angeles County Museum of Art's Art and Technology Program from 1968 to 1970, which utilized sound from microphones as well as a *musique concrete*–like collage of everyday sound to control pneumatic pumps in a large tank of mud. Inspired by the fantastic bubbling mud paint pots at Yellowstone National Park in Wyoming, it was Rauschenberg's interest in creating a "dynamic work that would stimulate more than the visual senses and interact with the viewer" [34, p. 286] that led him away from what he labeled the "one-to-one response" of *Soundings* to a more dynamics shaping of time in *Mud-Muse*. "*Mud-Muse* starts from sound: an impulse is turned into an electrical signal and then spreads out . . . depending on its dynamics" [34, p. 287].

The discrete correspondence between a human action and the resulting sonic response versus the far richer but potentially less "legible" diffusion of action into a complex set of responses still haunts to this day the practice of constructing environmentally based interactive acoustic spaces. In fact, despite the increased computational ability to deal with real-time response, many recent works in the arena of sonic interaction design within the environmental or architectural context demonstrate that this tension is still very much alive. For example, Australian composer and researcher Garth Paine's 2002 responsive audio environment *Gestation* featured two separate spaces: the first where movement of visitors influenced "a surround sound field generated in real time using video sensing equipment" and the second in which the position and growth patterns of composited ultrasound videos of fetuses were influenced by the activity in the first environment. Thus, the sonic experience for the visitors is one generated by a "change in relation to the direction, speed of movement and number of people within the space," creating an "intimate, textured sound, a viscous, fluid environment for the making of life" [23].

Gestation once again invokes Krueger's definition of a responsive environment as a temporal zone through the tight coupling of the inhabitants' movement and the resulting sonic space by way of standard apparatuses of sensing devices, computers, and the resulting media output. Even Paine's title and his description of the work directly employ language associated around time-based phenomena: growth, evolution, formation, and gestation to describe the behavior of sound. Yet, his design/composition strategy also emphasizes the typical audio-influenced language of using human interaction in the environment to mix, layer, and texture sound or to turn a space into a real-time instrument.

This phenomenon of the visitor-generated soundscape is also evident in many architecturally based sonic responsive environments that use sensor-triggered audio to recreate in real time (that is, in the present moment with little latency) compositional strategies that any composer, designer, or sound engineer would normally execute sitting behind a mixing desk or in front of an audio editing workstation. Architect Christian Moeller's 1997 interactive light and sound environment *Audio Grove* consisted of a 12-meter-diameter platform with fifty-six 5-meter-tall steel columns augmented by capacitance-based touch sensors in which each pole acted as an interface to trigger a variety of sounds developed by composer Ludger Brummer using physical modeling techniques. "Visitors touching the posts can evoke a soundscape which always results in a harmonic whole whatever the conceivable combination of interactions" [18]. In this installation, moving through the environment and touching the poles sets off individual sounds, but it is the combination of many actions triggering sound simultaneously that generates the more complex sonic mix.

Similarly, the Netherlands-based architectural projects *Fresh Water Pavilion*, *Salt Water Pavilion*, and *Son-O-House*, all collaborations between architects (Lars Spuybroek and Kaas Oosterhuis, respectively) and composer/sound artist Edwin van der Heide, also invoke the language of instrument, real-time mixing, and composition to describe their principles of sonic response. The two parallel Fresh and Salt Water pavilions in the seaside area of Neeltje Jans articulate the sonic responsiveness as dependent on the real-time actions of visitors. "The music in the two different spaces is not a fixed composition but has a generative approach to it and is therefore composed on the moment itself. The rules for how sounds can be combined are predefined; the actual decision of what sounds is made in real time" [35].

Likewise, in the 2004 *Son-O-House*, an architectural-art-sound installation with van der Heide, Spuybroek also invokes the language of real-time compositional strategies to describe the structure's *interactive sounding architecture*. "The structure is both an architectural and a sound installation that allows people to not just hear sound in a musical structure, but also to participate in the composition of the sound. It is an instrument, score, and studio at the same time" [1]. In all of these works, responsiveness and interaction signify the sonic activation of space through visitor movement and motion-sensing systems, computers, and large, distributed arrays of speakers embedded in the architectural structure that serve to mix and layer partially structured and partially improvisational fragments of sonic material. But van de Heide also uses the language of enaction to describe the goal of *Son-O-House*: to use sensors and the resulting sonic behavior as an affordance to "influence and interfere with the perception and the movements of the visitors" [35]. Here, as in the earlier architectural examples of Xenakis, movement is seen as something more than just a trigger for real-time mixing and instead is a catalyst for the evolution and, indeed, perception of sound.

2.4.2   The Time Constants of Auditory Interaction

Son-O-House also reveals the way that different time scales based on the response times of the human body, sensors, and sound all operate in tandem with one another. What appeared at first to be a simple cause-and-effect interval between the onset of an action and the response of the environment in the work of Krueger, Rauschenberg, or others is now revealed to be far more complex due to the way in which different time scales among body and environment simultaneously comingle and interact with each other.

In *Microsound*, composer and researcher Curtis Roads provides a cogent framework for understanding the sliding temporal scales intrinsic to the act of sonic composition. As Roads describes, "a central task of composition has always been the management of the interaction amongst structures on different time scales" [27, p. 3]. Composition and the perception of sound through time traverse a scale from the infinite (the ideal mathematical duration of a sine wave envisioned in Fourier synthesis) through the macro (the scale of musical architecture or form, usually measured in minutes or hours) and further down, to the micro (sound grains lying on the threshold of auditory perception). Yet, how could Roads's descriptions be applied to situations where the experiential repercussions of sonic interaction depend on a delicate counterpoint among an entire crisscrossing and overlapping network of emerging and dissolving temporal scales: the few seconds that it takes the human body to make a gesture, the millisecond time of a sensor to register action, the time windows for processing, filtering, or extracting many gestures over both short time and long time, the movement of sound through space, and the felt experience of the body in response to a compositional choice?

Figure 2.7
Tgarden (2002) by FoAM and sponge.

Several contemporary installation and performance projects have directly attempted to explore how such overlapping and intertwined sonic time scales produced in the interaction between human bodies and computational machines in an environment become felt and perceivable to participants. For example, the 1999–2003 responsive environment *TGarden*, jointly developed by the then San Francisco-based art research group Sponge and the Brussels-based research organization FoAM, created a cross between a theatrical installation and improvisational playspace in order to examine how people individually and collectively interact with and make sense of dynamic and fluid space of responsive media nonverbally (figure 2.7). In its public presentations, *TGarden* consisted of a series of private, enclosed dressing rooms leading to a physically curtained-off performance space. The event could accommodate two to five visitors over a defined but variable time cycle, ranging from 20 to 40 minutes. In order to track the participants' gestures in real time, each costume was embedded with a series of wireless accelerometer sensors that measured the degree of acceleration from small arm gestures to larger, full-body movements. The sensors were either loosely attached or, in some cases, woven into the garments. Each participant was also outfitted with a small commercial portable computer housed in a belt pack. As the participants moved in the hybrid clothing/costumes, the environment produced a series of dense audio/visual responses by way of a networked series of computers, speakers, and video projectors. Technically, the media response was the result of continuous data streams generated by the accelerometers picking up the force and acceleration (the rate of change of the participants' velocity) of the participants' individual motions. These data were then transmitted from each of the individual participants by way of the sensors and microcontrollers, decoded, and preprocessed on the players' portable computers and then wirelessly transmitted to a central computer.

Although *TGarden* clearly departs from the trigger/mix model of sonic interaction, what is particularly interesting is the explicit manner in which interaction is understood to operate over two different scales: body and room/environment time. Interaction that takes place at the body scale (participant gestures and movements) provides the system with continuous movement information and direct and immediate media response, whereas interaction at the environment scale is manifested over longer time frames through sounds not directly coupled to individual movements and gestures.

In what *TGarden* collaborator Joel Ryan calls the "time constants" of the system or the durations characterizing distinct behaviors the system exhibits, different physical actions are associated with different sonic behaviors [29]. For example, jumping, bouncing, gait and rapid changes in acceleration each create a different response in the sonic material associated with the interaction between player and sound environment, specifically based on the feature extraction of physical motion picked up by the accelerometers. Thus, in contrast to visitor motion triggering playback models of sound, *TGarden* directly couples expended physical energy to the overall sonic behavior (bouncing, jumping, scratching, friction) of the environment.

This experience between the time of the body and the time of an environment is made perhaps even more explicit in the 2007 media theater performance *Schwelle* [2]. *Schwelle* is an evening-length theatrical event exploring the varying threshold states of consciousness that confront human beings in everyday life, such as the onset of sleep or the moments before physical death (figure 2.8). Part II of the project consists of a theatrical performance between a solo dancer/actor and a "sensate room." The exerted force of performer Michael Schumacher's movement and changing ambient data such as light and sound are captured by wireless sensors located on both Schumacher's chest and arms as well as within the theater space. The continuously generated data from both dancer and environment are then used to influence an adaptive audio scenography: a dynamically evolving sound design that creates the dramatic impression of a living, breathing room for the spectator.

The data from the sensors are statistically analyzed so the system can react to changes in the environment rather than to absolute values and then scaled dynamically before being fed into a mathematical dynamics model inspired by physiologist J. F. Herbart's theory on the strength of ideas over time. The dynamic scaling ensures

Figure 2.8
Schwelle (2007) by Chris Salter, Michael Schumacher, and Marije Baalman.

that when there is little change in the sensor data, the system is more sensitive to it. The output of the Herbart system is mapped to the density as well as to the amplitude of various sounds that comprise a room compositional structure of sixteen different layers of sound material. This mapping determines the degree to which each sensor influences which individual or group sounds over time.

The interesting aspect of the use of the dynamics model in *Schwelle* lies in its behavior, which influences the temporal evolution of sound and light. Although it is possible that there is a direct reaction based on a new impulse into the system, the model's reaction depends on the current state of the system. If, for example, there is lingering energy from previous impulses, a new impulse will not impact the system as much as when there is little energy left in the system. Thus, the experience of watching and listening to *Schwelle* is one of suspension between the physical time present in the performer's body and the varying time behaviors created by the sensor data extracted and weighed by the dynamical model that the room exhibits.

During certain periods, the activity of the room is directly coupled to the dancer's actions, appearing and sounding jittery and nervous as demonstrated by flickering light and fragmented, particulate-like sound coming from multiple speakers in the actual theatrical space. At other times, the room's sonic atmosphere, like Schumacher himself, settles into a slow, steady state with little movement, inducing a state of boredom combined with an undefined, trance-like feeling in the audience. Over the duration of the performance, these quasi-tangible-intangible room qualities settle into patterns directly following the performer's gestures and actions and establishing a clear cause-and-effect relationship, whereas at other times the environment seems totally unrelated to his presence. Thus, the supposed given causal link between action and reaction in the behavior of the system is broken, giving way to a more complex multiplicity of temporal scales: Schumacher's own bodily time coupling and decoupling with the internal machine processes all made aurally perceptible through the fluctuating patterns of human agency and machine agency (the continuously changing sensor data emerging from the environment).

2.4.3 Performance, Responsiveness, and Mixed Agency

In *TGarden* and particularly *Schwelle*, interlocking and continuously interacting time scales are made perceivable and felt by the fact that there is a preestablished macro scale of the event, the specified real-time unfolding of the performance within a given time period, by which a spectator/listener orients his or her own bodily experience of time. This convention of the experienced passage of real time in a live performance plays a central role in helping to construct the expectation that something happens from one moment to the next. Henri Bergson described the act of perception as the filtering out of select "images" (any kind of perceptual stimuli) from the continual material flux of the world and the way such "images" transmit temporality—that is, movement and affect—to the body. The fact that the frame of live performance makes a spectator acutely aware of time's behavior partially establishes a perceptual frame for how different interacting orders of time (body, machine, material, and so on) that are not necessarily sequentially linked might, in fact, be felt pulled out by perception and directly experienced at the bodily level.

In his *Folded String* performance, musician, sound artist, and scientist Joel Ryan provides a glimpse of the way such varying time scales can become perceivable at a level of direct bodily affect. Ryan's performance consists in the performer amplifying a long helical coil of steel piano wire with a piezo contact microphone and then playing the wire using a wooden saragni bow (figure 2.9).

Bowing, scraping, and plucking it with his fingers and, in general, gesturally exciting the material into modes of vibration of wildly varying rhythm and intensity, Ryan simultaneously transforms these physical results of interaction into a progression of sonic forms, all in real time. The performance begins with the composer slowly stretching the coiled wire from one side of the stage to the other. Already, in this early phase, the loose curves of the wire in part and as a whole begin to manifest their own particular shifting time constants, in the form of a waving, fluid-like behavior. Ryan's

Figure 2.9
Joel Ryan performing with the *Folded String*.

gentle touch prods the material into a low, tamboura-like sound produced by bowing the metal's natural vibration together with tuned digital processing. The performance proceeds with his piloting the wire with ever increasing and extended gestures—lifting, tossing, catching, beating, and scraping it with the wood, running with the bow along its length, bowing on momentary bridges formed by the fingers of the left hand and gathered via the techniques and gestures that the artist learned from fencing. What is particularly revealing is the way that Ryan, a trained physicist as well as musician, exploits what Andrew Pickering calls the *material agency* of the wire so that it can exhibit multiple modes of vibration. At times, the speed of the wire's motion is too fast for the eye to see. At other times, the wire's rhythmic behavior falls into synchrony with Ryan's bowing, giving the spectator/listener the experience of several living agencies performing together.

Ryan's performance is so interesting to us because it creates (in the sense of *poiesis*) many experiences of time: the wire's modalities of vibration, the bodily time of gestures, the brio of musical string playing, the atemporal digital processing, the physical behavior of the jumping object in relationship to stage lighting, and the time produced by the instrument's direct materiality. This concrete sensorimotor nature of interaction gives both performer and spectator access to many different types of real time. Moreover, both the wire's physical modes of interaction in a shifting environment and Ryan's bodily interaction save the performance from a kind computational input/output abstraction of interaction, one that we have seen influences many artistic discussions of sonic responsiveness. Indeed, our lived experience of time becomes all too apparent when Ryan silences everything in an instant by pinching the wire between his fingers, halting its liveness and suspending the present moment.

Varela pointed out that bringing forth a world involves a process of temporal constitution brought on by complex perturbations between system and environment. Through its materialization of time, Ryan's wire performance makes explicit that a far more complex and interacting world of agencies lies beneath the simple visual appearance of an object—a world in which the interaction of human and nonhuman forces is necessary in order to constitute our felt experience of sound.

2.5   Responsiveness and Responsibility

This chapter has focused on the aesthetic, phenomenological, and social experience of sonic interaction. By shifting the theoretical discourse away from listening as consumption of auditory material and toward active engagement with sonic material, we have attempted to describe a wide range of artistic, design, and architectural practices. Specifically, we focused on projects that exemplify recent discoveries from cognitive sciences and philosophy of the body, supporting the view of interactivity as enaction. Inspired by those projects and theories, we have begun to shape a theoretical framework for discussing aesthetic experience with interactive sound based on relating the material, temporal, and spatial qualities inherent in the sonic medium, such as its vibratory, unstable, and malleable character, to the interactive experiences these qualities may afford in both personal and collective situations. Discussing those sonic properties in terms of their potential for interaction has allowed us to shed a new light on sonic experience in general.

The materiality of sound gains a truly embodied quality through physical interaction—away from equating the medium to the physical support on which it is stored and toward the medium of action-sound relations shaping tangible and corporeal experience. The critic and composer Michel Chion argued that the core of sound design for cinema is the relationship between the visual and sonic material rather than a sum of them [5]. Similarly, in sonic interaction, it is the relationship between the gesture and sound that shapes our experience and, therefore, should be seen as one of the central subjects of study of sonic interaction design, in general. Through the projects presented in this chapter, we have seen how metaphors for interaction can be generated through action-sound couplings and the affordances they embody. These examples compellingly demonstrate that sonic feedback coupled to action can change human performance as well as social and aesthetic experience.

Sonic interaction design challenges the traditional gaps between the tangible and immaterial, everyday sound and music, functional action and expressive gesture, public and private space, and human and nonhuman agency. The potential for agency intrinsic to interactivity allows for a form of creative making that involves acts of touch, listening, and movement. Bringing together both sound experts and tyros, participatory projects hold the promise that, in the future, sonic experiences will emerge in various creative forms in everyday life. In this setting, the responsibility for our action and the responsiveness of sound constitute the coupling that may have the power to deeply transform habitual activities. Although the basis for sonic interaction as a discipline is still being formed, the challenges of bringing this medium to the level where it can be present in our daily life at the point where television, concerts, or cinema now exist are yet to be confronted.

Notes

1. SIGCHI is the leading organization in human-computer interaction and stands for the Special Interest Group on Computer-Human Interaction of the Association for Computing Machinery.

2. For more details on the invariants see chapter 4.

References

1. Architecture and Art. http://www.nox-art-architecture.com/.

2. Baalman, M. A. J., Moody-Grigsby, D., & Salter, C. L. (2007). *Schwelle:* Sensor augmented, adaptive sound design for live theatrical performance. In *NIME '07: Proceedings of the 7th international conference on new interfaces for musical expression* (pp. 178–184). New York: ACM.

3. Blesser, B., and Salter, L.-R. (2006). *Spaces speak, are you listening? Experiencing aural architecture.* Cambridge, MA: MIT Press.

4. Born, G. (1995). *Rationalizing culture: IRCAM, Boulez, and the institutionalization of the musical avant-garde.* Berkeley: University of California Press.

5. Chion, M. (1994). *Audiovision: Sound on screen.* Claudia Gorbman (Trans.). New York: Columbia University Press.

6. Dewey, J. (1934). *Art as experience.* New York: Minton.

7. Dourish, P. (2001). *Where the action is: The foundations of embodied interaction.* Cambridge, MA: MIT Press.

8. Ferris, D. P., Louie, M., & Farley, C. T. (1998). Running in the real world: Adjusting leg stiffness for different surfaces. *Proceedings of the Royal Society of London: Biological Sciences, 265*(1400), 989–993.

9. Franinović, K., & Visell, Y. (2004). Recycled soundscapes. In *Proceedings of the international symposium les journées du design sonore.* Paris, France.

10. Gaye, L., Holmquist, L. E., Behrendt, F., & Tanaka, A. (2006). Mobile music technology: Report on an emerging community. In *NIME '06: Proceedings of the 2006 conference on new interfaces for musical expression* (pp. 22–25). Paris, France, IRCAM—Centre Pompidou.

11. Gendlin, E. (1992). The primacy of the body, not the primacy of perception. *Man and World, 25*(3–4), 341–353.

12. Gibson, J. (1968). *The senses considered as perceptual systems.* London: Allen and Unwin.

13. Gibson, J. (1979). *The ecological approach to visual perception.* Boston: Houghton-Mifflin.

14. Kadet, A. (2004). Mouths find freedom in big megaphone. *Tribeca Tribune,* October. Available at http://204.42.140.58/archives/newsoct04/freedom-megaphone.htm

15. Krueger, M. W. (1977). Responsive environments. In *Proceedings of the national computer conference* (pp. 423–433). Dallas, TX.

16. Labelle, B. (2006). *Background noise: Perspectives on sound art.* New York: Continuum International Publishing Group.

17. Livingstone, D., & Miranda, E. (2004). Composition for ubiquitous responsive environments. In *Proceedings of the international computer music conference* (pp. 321–325). Miami, FL.

18. Moeller, C. (2000). http://www.christian-moeller.com.

19. Mulder, A. (2007). The exercise of interactive art. In J. Brouwer & A. Mulder (Eds.), *Interact or die! There is drama in the networks*. (pp. 52–69) Rotterdam: NAI Publishers.

20. Noe, A. (2004). *Action in perception*. Cambridge, MA: MIT Press.

21. O'Modhrain, S., & Essl, G. (2004). Pebblebox and Crumblebag: Tactile interfaces for granular synthesis. In *Proceedings of the conference for new interfaces for musical expression* (pp. 74–79). Hamamatsu, Japan.

22. Ostertag, B. (2002). Human bodies, computer music. *Leonardo Music Journal*, 12, 11–14.

23. Paine, G. (n.d.). http://www.activatedspace.com.

24. Pickering, A. (1995). *The mangle of practice: Time, agency, and science*. Chicago: University of Chicago Press.

25. Poupyrev, I., Lyons, M. J., Fels, S., & Blaine, T. (2001). New interfaces for musical expression. In *CHI '01 extended abstracts on human factors in computing systems* (pp. 491–492). New York: ACM.

26. Rath, M., & Rocchesso, D. (2005). Continuous sonic feedback from a rolling ball. *IEEE MultiMedia*, 12(2), 60–69.

27. Roads, C. (2004). *Microsound*. Cambridge, MA: MIT Press.

28. Rodríguez, C. (2001). *Fissures in the media landscape: An international study of citizens' media*. Creskill, NJ: Hampton Press.

29. Ryan, J., & Salter, C. (2003). *Tgarden:* Wearable instruments and augmented physicality. In *NIME '03: Proceedings of the 2003 conference on new interfaces for musical expression* (pp. 87–90), Singapore: National University of Singapore.

30. Schaefer, J. (1996). Who is La Monte Young? In W. Duckworth & R. Fleming (Eds.), *Sound and light: La Monte Young and Marian Zazeela* (pp. 25–44), Lewisburg, PA: Bucknell University Press.

31. Schafer, R. M. (1994). *The soundscape: Our sonic environment and the tuning of the world*. Rochester, VT: Destiny Books. (Originally published 1977)

32. Shepard, M. (2005). Tactical sound garden [tsg] toolkit. *Regarding Public Space*, 1(1), 64–71.

33. Stockhausen, K. (1972). Four criteria of electronic music. Stockhausen-Verlag, Kuerten, Germany.

34. Tuchman, M. (Ed.). (1971). *Art and technology: A report on the Art and Technology Program of the Los Angeles County Museum of Art, 1967–1971*. New York: Viking Press.

35. van der Heide, E. (n.d.). http://www.evdh.net/.

36. Varela, F. (1999). *Ethical know-how: Action, wisdom, and cognition*. Palo Alto, CA: Stanford University Press.

37. Varela, F., Thompson, E., & Rosch, E. (2002). *The embodied mind: Cognitive science and human experience*. Cambridge, MA: MIT Press.

38. Visell, Y., Cooperstock, J., & Franinović, K. (2011). The ecotile: An architectural platform for audio-haptic simulation in walking. In *Proceedings of the 4th international conference on enactive interfaces*, Grenoble.

39. Visell, Y., Giordano, B., Millet, G., & Cooperstock, J. (2011). Vibration influences haptic perception of surface compliance during walking. *PLoS One, 6*(3), e17697.

40. Xenakis, I. (2008). *Music and architecture*. Sharon Kranach (Trans.). New York: Pendragon.

3 Continuous Auditory and Tactile Interaction Design

Yon Visell, Roderick Murray-Smith, Stephen A. Brewster, and John Williamson

This chapter addresses the design of continuous sonic and tactile interactions for applications such as computer interfaces or other interactive products. We adopt the term *continuous* to refer to systems that both employ input methods depending continuously on control actions of their users and provide concurrent auditory or tactile information about the resulting state or response of the system. Although such action-to-sound couplings are instrumental to many everyday activities that do not necessarily involve computing (e.g., steaming milk for a cappuccino, or driving a car), limited attention has been given to the design and usability of computationally enabled interfaces that integrate them. As noted in the Introduction to this book, designers today are presented with new challenges and opportunities for shaping sonic and tactile interactions with future artifacts. In part because of the historical development of HCI and related design disciplines, there is a lack of knowledge, methods, and patterns to aid this process. This chapter highlights some perspectives and knowledge that may improve our encounters with contemporary and future artifacts that employ sound and tactile feedback in rich and useful ways.

We begin with a review of issues concerning auditory and tactile displays in human-computer interfaces (HCIs), including roles that they have assumed, known advantages and problems affecting them, and notable applications in desktop, mobile, and ambient computing. Significant factors affecting the usability of auditory displays in HCIs are also noted; for a detailed review of perceptual evaluation methodologies, the reader is referred to chapter 5 of this book.

The second part of the chapter reviews issues affecting sound presentation and synthesis for tasks involving continuous gestural interaction with artifacts or systems. Techniques for audio synthesis, control, and display are known in the HCI and auditory display communities, but the rendering of audio or tactile feedback to aid continuous interaction has been less developed. Computer music research has contributed to knowledge in interactive sound synthesis, albeit primarily with the aim of facilitating musical performance. When users are given expressive control over interaction with digital information in more functional settings (e.g., using a device to navigate an unfamiliar city center), a number of new issues are brought to the fore, which we highlight here. One goal is to create a framework enabling designers to make the user aware of subtle variations in the nature of the content they are engaging with and manipulating, enabling *negotiated interaction*, where there is an ongoing bidirectional flow of information during interaction with a device.

The third part of the chapter considers the design of continuous control affordances by considering human capacities for perceiving everyday sounds and vibrations and for using this sensory information to execute everyday tasks (e.g., pouring a drink). A number of real-world and computational examples that have been studied in prior literature are reviewed. These highlight perceptual tasks associated with interacting with everyday artifacts and may provide design patterns for new interactions. The notion of everyday metaphors has been a persistent element of HCI design, even if the idea of design by metaphor presents potential drawbacks [55]. From a complementary viewpoint, a Gibsonian model of task analysis suggests that the utility of an auditory or tactile information source may be gauged by the extent to which it enables a user to interact more effectively, a perspective that motivates this discussion.

The chapter concludes with a brief integrative synthesis of the aforementioned issues in relation to current and future application scenarios for audiotactile interaction design. Considerations related to evaluation and the development of guidelines are raised, and some current trends and open questions for future research and development are suggested.

3.1 Auditory and Tactile Displays in HCI

Auditory and tactile cues have been used successfully to improve interaction in diverse user interfaces. This section reviews such cues and gives examples of their use in HCIs. Generally speaking, both auditory and tactile displays can be said to offer a number of advantages and several disadvantages for interaction.

Advantages

Interdependence of vision, hearing, and touch   Our sensory systems work well together. Our eyes provide high-resolution information around a small area of focus (with peripheral vision extending further). Sounds, on the other hand, can be heard from all directions: above, below, in front, or behind, but with a much lower spatial resolution. Therefore, "our ears tell our eyes where to look": if there is an interesting sound from outside our view, we turn to look at it to get more detailed information. Touch provides high-resolution information across the body surface, especially at the hands, and for nearby objects.

Temporal resolution   As Kramer notes, "Acute temporal resolution is one of the greatest strengths of the auditory system" [54]. In certain cases reactions to auditory stimuli have been shown to be faster than reactions to visual stimuli [8]. Tactile feedback has a higher temporal resolution than vision but lower then audition. Although humans can hear sounds in the range from 20 to 20,000 hertz, the frequency range of sensitivity of the skin to mechanical vibrations is smaller, extending to about 1,000 hertz, with maximum sensitivity [98] and finest spatial discrimination [22] near 250 hertz.

Visually undemanding   Auditory and tactile feedback are particularly important when visual attention is unavailable, such as in mobile computing contexts, when users may be walking or otherwise navigating [73]. In such cases, visual information may be missed while the user is not looking at the device, whereas if it were presented via sound or vibration, it would be delivered "eyes free."

Underutilization of auditory and tactile modalities   The auditory and cutaneous senses are underutilized in our everyday interactions with technology. The auditory system is very powerful; we can discern highly complex musical structures by listening and can discriminate a range of subtle textures through our skin. By taking advantage of these senses, we can enrich our interactions with our technology.

Attentional salience   Users can choose not to look at something, but it is harder to avoid hearing sounds. This makes sound very useful for delivering important information. Similarly, a tap on the shoulder is a powerful cue for grabbing attention.

Familiarity   Some objects or actions within an interface may have a more natural representation in sound or touch. Bly suggests that the "perception of sound is different to visual perception, sound can offer a different intuitive view of the information it presents . . ." [8]. Feeling an object with one's hands allows a different experience than looking at a picture of the same object. Touching, holding, manipulating, or shaking it can provide many cues about the structure, in some cases about aspects that are difficult to discern in other ways [42]. The use of different modalities can allow us to understand information in complementary ways.

Accessibility   Auditory and tactile feedback can make computers more usable by visually disabled people. With the development of graphic displays, user interfaces became much harder for visually impaired people to operate. A screen reader (software for translating textual elements on screen to speech) cannot readily present this kind of graphic information. Providing information in an auditory and tactile form may allow visually disabled persons to better use the facilities available on modern computers.

Disadvantages

Limited resolution   Auditory or tactile senses are not always suitable for high-resolution display of quantitative information. Using sound volume, for example, allows only a very few values to be unambiguously presented [14]. Vision and touch have much higher resolutions. Under optimal conditions, differences of about 1 degree can be heard in front of a listener [7]. In vision, differences of an angle of 2 seconds can be detected in the area of greatest acuity in the central visual field. The most sensitive part of the human body is the fingertip. The two-point contact discrimination threshold is 0.9 mm when the stimuli are placed against the subject's finger in the absence of any movement lateral to the skin's surface, and far finer features can be distinguished when relative movement is involved.

Lack of absolute sensory precision   Presenting absolute data via auditory or tactile feedback is difficult. Many interfaces that use nonspeech sound to present data do it in a relative way. Users hear the difference between two sounds to tell if a value is going up or down. It is difficult to present absolute data unless the listener has perfect pitch (which is rare). Conversely, in vision a user only has to look at a number or a meter to get an absolute value or estimate.

Lack of orthogonality   Changing one attribute of a sound may affect the others. For example, changing the pitch of a note may affect its perceived loudness and vice versa [54]. Tactile perception can be similarly affected [37], as can vision. In the latter case, however, there is ample practical knowledge to guide visual information design, which is less the case for touch and sound.

Transience of information   In some cases, sound disappears when it has been presented; users must remember the information that the sound contained, or some method of replaying must be provided. A visual display, such as a static image, can ensure that the information can be seen again at a glance.

Potential for nuisance   Sounds can become annoying to others nearby. One advantage of tactile displays is that they are only received by the person in contact with them. This can make them very effective for discreet communication.

3.1.1 Past Uses of Auditory Displays

Through our sense of hearing we can extract a wealth of information from the pressure waves entering our ears as sound. Sound provides a continuous, holistic contact with our environment and what is going on around us; we hear interactions with objects close to us, familiar sounds of unseen friends or family nearby, noises of things to avoid such as traffic, or to attend to such as a ringing telephone. Music, environmental sounds, or sound effects provide different types of information than are communicated through speech and can be less specific and more ambient, whereas speech is often precise and requires more focus. Blattner and Dannenberg discuss some of the advantages of multimodal interfaces, writing: "In our interaction with the world around us, we use many senses. Through each sense, we interpret the external world using representations and organizations to accommodate that use. The senses enhance each other in various ways, adding synergies or further informational dimensions" [5].

The classical uses of nonspeech sound can be found in the human factors literature, where it is used mainly for alarms and warnings or monitoring and status information. Buxton [13] extended these ideas and suggested that encoded messages could be used to present more complex information in sound; it is this type of auditory feedback that is considered here. The other main use of nonspeech sound is in music and sound effects for games and other multimedia applications. Such sounds can indicate to the user something about what is going on and try to create a mood for the setting (much as music does in film and radio). As Blattner and Dannenberg note, "Music and other sound in film or drama can be used to communicate aspects of the plot or situation that are not verbalized by the actors" [5]. Research on auditory output for interaction extends this idea and uses sound to represent things that a user might not otherwise see or important events that he or she might not notice. To date, the main approaches to the presentation of nonspeech audio messages are *auditory icons* and *earcons*. Substantial research has gone into developing both, with the main results reviewed below.

3.1.2 Auditory Icons

Gaver first developed the idea of *auditory icons* [33, 34]—natural, everyday sounds that can be used to represent actions and objects within an interface. He defines them as "everyday sounds mapped to computer events by analogy with everyday sound-producing events. Auditory icons are like sound effects for computers." Auditory icons rely on an analogy between the everyday world and the model world of the computer [34]. Gaver uses sounds of events that are recorded from the natural environment, for example, tapping or smashing sounds. He uses an "ecological listening" approach [70], suggesting that people do not listen to the pitch and timbre of sounds but to the sources that created them. When pouring liquid a listener hears the fullness of the receptacle, not the increases in pitch. Another important property of everyday sounds is that they can convey multidimensional data. When a door slams, a listener may hear the size and material of the door, the force that was used, and the size of room on which it was slammed. This could be used within an interface so that selection of an object makes a tapping sound, the type of material could represent the type of object, and the size of the tapped object could represent the size of the object within the interface. Gaver integrated these ideas in the SonicFinder [33], which ran on the Apple Macintosh and provided auditory representations of some objects and actions within the interface. Files were given a wooden sound, applications a metal sound and folders a paper sound. The larger the object the deeper the sound it made. Thus, selecting an application meant tapping it. This produced a metal sound, which confirmed that it was an application. The deepness of the sound indicated its size. Copying was associated with pouring liquid into a receptacle. The rising pitch indicated that the receptacle was getting fuller and the copy progressing. Another early attempt to make richer use of sound in human–computer interaction was the ENO system [3], which was an audio server designed to make it easy for applications in the Unix environment to incorporate nonspeech audio cues. Rather than handling sound at the level of the recorded waveform, ENO allowed sounds to be represented and controlled in terms of higher-level descriptions of sources, interactions, attributes, and sound space.

One of the biggest advantages of auditory icons is their ability to communicate meanings that listeners can easily learn and remember, whereas other systems (such as earcons, described in the next section) use abstract sounds, whose meanings can be harder to learn. Problems can occur with representational systems such as auditory icons, however, because some abstract interface actions and objects have no obvious representation in everyday sound. Gaver used a pouring sound to indicate copying because there was no natural equivalent; this is more like a "sound effect." He suggests the use of movie-like sound effects to create sounds for things with no easy representation. This may cause problems if the sounds are not chosen correctly, as they will become more abstract than representational, and the advantages of auditory icons will be lost.

3.1.3 Earcons

Earcons were first developed by Blattner, Sumikawa, and Greenberg [6]. They consist of abstract, synthetic tones combined in structured ways to create auditory messages. They define earcons as nonspeech auditory messages that are used in computer interfaces to provide information about a digital object, operation, or interaction. Unlike auditory icons, there is no intuitive link between the earcon and what it represents;

the connection must be learned. Earcons are constructed from simple building blocks called *motifs* [6]. These are short, rhythmic sequences of notes that can be combined in different ways. Blattner et al. suggest their most important features are these:

• *Rhythm* Changing the rhythm of a motif can make it sound very different. This was described as the most prominent characteristic of a motif [6].

• *Pitch* There are 96 different pitches in the Western musical system. These can be combined to produce a large number of different motifs.

• *Timbre* Motifs can be made to sound different by the use of different timbres, for example, playing one motif with the sound of a violin and the other with the sound of a piano.

• *Register* This is the position of the motif in the musical scale. A high register means a high-pitched note and vice versa. The same motif in a different register might convey a different meaning.

• *Dynamics* This is the volume of the motif. It can be made to progressively increase (crescendo) or decrease (decrescendo).

There are two basic ways in which earcons can be constructed. The first, and simplest, consist of simple motifs concatenated to create more complex messages. For example, a set of one-element motifs might represent system elements such as "create," "destroy," "file," and "string." These could then be concatenated to build up more complex earcons. Hierarchical earcons can be more complex, but can be used to represent more complex structures in sound. Each one is a node in a tree and inherits properties from the earcons above it. Brewster, Wright, and Edwards [11] carried out an evaluation of compound and *hierarchical earcons* based on the design proposed by Blattner consisting of simple system beeps and richer examples based on more complex musical timbres. In these experiments, participants were presented with earcons representing families of GUI icons, menus, and combinations of both. Results showed that the more complex musical earcons were significantly more effective than either the simple beeps and Blattner's proposed design, with over 80 percent recalled correctly. Brewster et al. found that timbre was much more important than previously suggested, whereas pitch on its own was difficult to differentiate.

3.1.4 Tactile Displays

The area of haptic (touch-based) human computer interaction has grown rapidly over the last few years, opening a range of new applications that use touch as an interaction technique. The human sense of touch is composed of two subsystems: kinesthetic and cutaneous. The kinesthetic perceptual system relates to the information arising from forces and positions of the muscles and joints. Force-feedback haptic devices are used to present information to the kinesthetic sense. Cutaneous perception is mediated by receptors in the skin and includes the sensations of mechanical vibration (over a range of frequencies from about 10 hertz to 1,000 hertz), temperature, pain, and indentation. Tactile devices are used to present mechanical feedback to the cutaneous sense, and it is this latter category that we consider here. Most vibrotactile actuators use electromagnetic actuation to drive a mass in either a linear or rotational oscillatory fashion to stimulate the skin. Current-generation mobile phones, video game controllers, and other devices use low-cost eccentric-mass motor actuators consisting of a mass mounted on a DC motor. Such actuators can be driven to higher amplitudes only by increasing their vibration frequency, preventing amplitude and frequency from being independently specified, which is a limitation [43]. Two other commercially available vibrotactile actuators, the Tactaid VBW32 (www.tactaid.com) and the EAI C2 Tactor (www.eaiinfo.com), are shown in panels (a) and (b) of figure 3.1. Both of these devices are resonant at 250 Hz. The third device shown in the

(a)   (b)

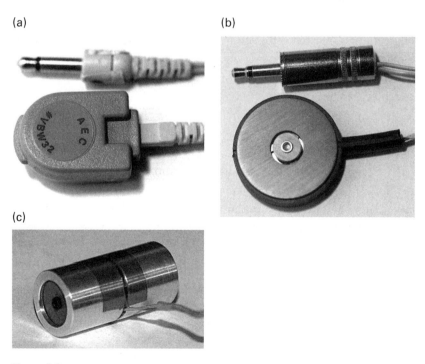

(c)

Figure 3.1
(a) A tactaid VBW32 transducer. (b) An EAI C2 Tactor. (c) Tactile Labs' haptuator.

figure, Tactile Labs' Haptuator (http://www.tactilelabs.com/main), possesses a wide-bandwidth frequency response, from about 50 Hz to 1 kHz, making it possible to present a much larger range of mechanical vibrations than is possible with the afore-mentioned devices because the vibration waveform may be arbitrarily specified. Vibro-tactile displays have lower spatial resolution than distributed tactile displays, such as those based on actuated pin arrays, can provide, but they have greater dynamic range, greater frequency bandwidth, cost less, and can exert larger forces (so can be felt through clothing); they can also be distributed over the body to give spatial cues (for example, mounted in a vest on the user's back or in a belt around the waist). More detailed reviews of vibrotactile actuator technologies and related considerations are available [43, 49, 98, 107].

Because of their compact size and power requirements, tactile actuators offer a discreet, affordable means of providing access to data via the sense of touch, suitable for embedding in computer mice, keyboard game controllers, or mobile telephones. Most examples of the last use simple vibrations to provide non–audio-based indica-tions of incoming calls or messages. The video game industry has utilized them as an inexpensive way to provide feedback via handheld game controllers. Tactile sensations are crucial to success in tasks such as object manipulation, edge detection, palpation, and texture perception [51]. They are also implicated in more expressive and qualita-tive contexts such as nonvisual communication (e.g., a firm handshake or a caress on the hand) and perceptions of product quality. Several researchers have investigated finger- [81] or stylus-based [57] interactions for touch screens or handheld computers, and there is considerable commercial interest in such displays (e.g., Immersion's Vibe-Tonz line of products—www.immersion.com). Vibrotactile displays have been incor-porated into canes used by visually impaired people. The UltraCane (www.ultracane.com) uses ultrasound to detect objects in a user's environment and displays the loca-tion and distance to targets by vibrating pads on the handle of the cane. Several other tactile aids for people with visual disabilities are reviewed in a survey by Lévesque [61], and tactile displays for sensory substitution are reviewed by Visell [107].

3.1.5   Tactons and Haptic Icons

A number of authors have investigated symbolic encoding of information via tactile stimuli [10, 62, 99, 104], often referred to as tactons or haptic icons, by analogy to earcons or auditory icons. They are structured abstract messages that can be used to communicate nonvisually. Such messages can be constructed using techniques similar to those used to construct auditory icons or earcons and can be presented through diverse devices, ranging from devices worn on the body to exercise machines [30] or floor surfaces [109]. Tactons can be created by manipulating the parameters of cutaneous stimulation to encode information. For example, Brown, Brewster, and Purchase [12] encoded two pieces of information into a Tacton to create messages for mobile telephones. The type of call or message (voice call, text message, or multimedia message) was encoded in the rhythm while the priority (low, medium or high) of a call or message was encoded in the roughness (via amplitude modulation). An initial study [12] on these nine Tactons showed that participants could identify them with over 70 percent accuracy, with rhythms being identified correctly 93 percent of the time. These results, and others in this area [62, 76], show that tactons can be a powerful, nonvisual way of communicating information.

3.2   Designing Continuous Audiotactile Feedback for Interaction

As illustrated in the review in section 3.1, auditory display in HCI has historically been used primarily to supply discrete cues in response to user interactions or changes of state in a software program. Humans have, however, evolved to act and control their environment in a continuous fashion, and the auditory and tactile responses in everyday interactions tend to involve nuanced feedback depending subtly on human actions. The process of a human controlling a computer or machine can be represented as an interaction loop (figure 3.2). The human has goals or reference states he or she would like to bring the system to and can act in various ways to achieve this. He selects actions based on feedback received from the system, and, consequently, his level of control may depend on the quality of the feedback that is received [82].

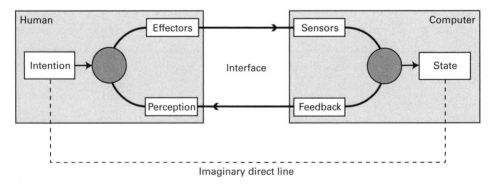

Figure 3.2
Human-machine interaction loop.

Classical human-operator control tasks, such as automotive or flight control or teleoperation, have traditionally involved continuous control [46, 94]. These fields developed at a time in which analog computers were predominant in practical applications and prevailing models of interaction were continuously variable in nature. In such settings, HCI can also be profitably regarded as a control task, making it possible to apply techniques and results from control theory.[1] Doherty and Massink provide a review of elementary manual control theory from an HCI perspective and a basic discussion of its use for the analysis and design of interfaces [25].

The introduction of graphic user interfaces based on the use of a mouse and desktop paradigm introduced elements of continuous interaction—in particular, the idea of direct manipulation—but auditory feedback tended to be linked to discrete events such as button clicks, incoming mail alerts, or error messages. Although this was acceptable for the applications of the time, which were aimed at representing clear consequences of symbolic operations on unambiguous, low-dimensional data states, such an interaction model has inherent limitations. Modern software applications can involve higher levels of complexity, including dense, heterogeneous streams of data, either generated within the user's own device or streamed from the Internet, at a wide range of time scales. These data can be considered to constitute an inherently uncertain, complex context consisting of the combined state of the user and the system. Hence, the amount and type of information to be conveyed may demand more complex displays than are needed for simpler desktop computing tasks.

3.2.1 Sensing Movement and Generating Feedback

From the time of our birth we become accustomed to the auditory feedback that results from our actions and to the association between different sounds and different locations in the world. Outside of some entertainment applications, interacting with computers has typically provided little auditory feedback to continuous activities. However, in recent years, inertial and location sensing have become commonplace, appearing in millions of devices such as mobile phones, music players, and in computer games such as the Nintendo Wii. This has, almost overnight, created a wide range of opportunities for the design of systems that provide auditory feedback as a function of users' movements coupled with data on the device. Designers and architects are also increasingly integrating complex sensing and multimodal (visual, auditory, tactile) information display systems in features of the built environment, and a wide range of interactive walls, tables, and floor surfaces have been described in recent literature [1, 40, 50, 77, 86, 109, 115]. Such interfaces can pose similar challenges to those encountered in mobile computing, particularly when they are meant to be accessible to individuals or passers-by who may have little familiarity with the systems involved. Another recent development is the release of Microsoft's Kinect motion sensor (combining a color videocamera, depth sensor, and microphone array to capture multiperson bodily movements), which is likely to encourage further experimentation by designers and artists because of the simplicity and flexibility of the package.

The first commercially available mobile phone able to recognize simple motion gestures was the Samsung SCH-S310, released in 2005. SonyEricsson's W580i and W910i mobile phones and, more recently, the Apple iPhone allow the user to shake to shuffle the music, or change tracks. Designing such interactions is relatively new, and the techniques used so far have been based on classifying certain conditions or motions as "virtual buttons" and providing feedback on recognition of these discrete, symbolic movements. Such interactions can be difficult for users to learn and can break down when perturbed by disturbances that are common in mobile environments. A current challenge is to create devices that are able to sense users' activities, infer contextual states or locations, and respond with information that may be of interest. However, there can be much uncertainty concerning these inferred intentions, and the feedback provided may be more useful if it reflects such ambiguities. In addition, in mobile settings, it cannot be exclusively visual because of users' need to focus on their environment while walking outside, around obstacles, or through traffic. Audio and vibrotactile feedback can be used to represent time-sensitive information in a more transparent fashion than is currently common while allowing users to remain in control. The challenge, however, is to achieve this in ways that feel natural, that are not irritating to people nearby, that demand users' attention only when required, and that otherwise improve the utility of the device.

3.2.2 Sound Design and Synthesis for Continuous Interaction

A wide array of techniques have been developed for real-time sound synthesis, from the playback of segments of prerecorded audio to the calculation of the oscillations of complex physical systems in real time. The goal of this chapter is not to survey these methods, as there are many good reviews [21, 88, 95]. Virtually any synthesis method other than trivial sampled waveform playback may be suitable for continuous interaction. Here, we focus our discussion on systems that support continuous interaction with auditory displays that are rich enough to represent the evolution of complex system states and the uncertainty in those states.

Applications to Manual Control Systems

Auditory interfaces have been used in a range of manual, continuous control tasks in earlier literature. Some examples are reviewed by Poulton [80]; these are generally based around pitch modulation [68] or interruption rate (for example, for error display [27]). An early, completely integrated audio feedback system for aircraft control is described by Forbes [31]. It was sufficient to allow inexperienced pilots to follow straight paths in a mechanical flight simulator. In the early 1970s Vinje developed an audio feedback system for vertical or short-take off and landing (V/STOL) vehicle hovering to augment conventional instruments [106] (cited by Thomas [94, p. 263]). The heading, height, and position of the aircraft were all sonified. This display was reported to have reduced the pilot's subjective workload. Technological limitations at the time of these studies constrained the sophistication of the resulting systems.

Physical Models of Everyday Sounds

With the introduction of affordable digital signal processing, more complex sound synthesis became practical. More refined methods based on physical modeling were developed, and this has made it possible to render rich sonic feedback in real-time interactions. Such an approach can be of benefit to users learning a new interaction or using the system in adverse or variable conditions.

In more recent work, physically modeled synthesis has come to the fore, as described in detail in section 3.4. For example, the Foley automatic system [102] was developed for the synthesis of realistic contact sounds in computer animation. A number of open-source physically based sound synthesis libraries, including JASS [103], STK [21], and SDT [89], exist and provide models suitable for continuous auditory feedback. All of these techniques and models can be applied to display the state of continuously varying dynamical systems if the variables of interest are associated with parameters of the models or with events or energy flows to the models.

Physical modeling is also increasingly used in game design[2] for the virtual environment the player interacts with, providing increased levels of realism and reducing development time, as the designer does not need to prespecify every possible interaction. This also means that users can sometimes complete tasks in ways designers had not anticipated, a feature that might prove to be valuable in broader human-computer interaction scenarios.

Physical metaphors and models for interactive sound synthesis can be used to simulate the evolution of physical processes, such as water pouring into a basin. Taken further, one can associate a physical metaphor to both the sound and the interaction itself, relating a user's actions to dynamic states of a virtual physical system linked to the computer application. In section 3.3.2 we discuss a number of physical metaphors, which can be linked to variables of interest in creative ways.

Shoogle

Shoogle [114] is an example of physical-modeling-based interaction. It displays information by simulating a model consisting of "balls" moving inside a "box." Motion sensing—using accelerometers—captures the movement of a mobile device (for example, a phone) and transfers this motion as virtual forces to the ball-in-box model. When the virtual balls collide with the boundaries of the box, the collision is rendered in both vibration and sound. This gives the illusion of a container with a varying number of moving parts inside that can be stimulated to determine the quantity and composition of these parts (figure 3.3).

The sonification model is granular synthesis—the production a large numbers of sonic events governed by an automatic model. In this case, the events are generated nonstochastically but as the result of a physical simulation. Prerecorded impact sounds are played on impact between balls and box boundaries, with an intensity proportional to the kinetic energy of the impact. This simple but physically realistic model results in the qualitatively convincing sense of manipulating objects in a box. By swapping the sample pool used for playback, the effect of different materials can be presented: wood versus glass or metal versus water. One of the key advantages of the modeling approach is that it responds to input; it does not assail the user with sensory information but reacts in a plausible and communicative manner. The level of feedback is directly tied to the stimulation applied. Information is coded into the model,

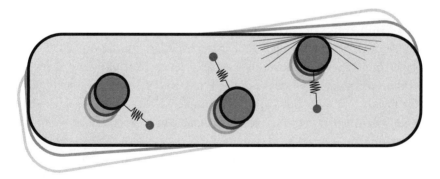

Figure 3.3
Ball-in-box simulation in Shoogle.

not into the sounds or vibrations, which happen as a consequence of simulated motion. This makes it simple to consistently link display modalities (for example, the vibration and the audio feedback) in a single model without convoluted metaphors or symbolic codings.

Parallels to Musical Interaction

A device that generates sounds in a continuous fashion as we interact with it is like a musical instrument. Its interactive response arguably affords a measure of creative expression to its user. A number of concepts relevant to the design of such interactions have been developed in research on gestural control of musical sound synthesis [16, 69, 90]. This includes the sense of effort needed to produce a response, the role of a gesture in parametrically modulating or energetically exciting a sound source, or the way in which sensor data are mapped to sound synthesis parameters. A key difference characterizing the functional contexts described here is that, notwithstanding the importance of aesthetics in product design, the primary role of an auditory feedback is often to aid users in communicating nonexpressive intentions, acquiring information, or controlling the state of a system. One challenge for designers of such feedback systems, is that nonintentional gestures may also be sensed (e.g., changes in pose or disturbances related to walking or traveling in a vehicle), and these may communicate little or nothing about the intention of the user but might end up generating feedback. Furthermore, although people are prepared to invest years of practice to learn to play an instrument, few are prepared to spend this much time learning a new interface! Functional devices need to have low entry costs in terms of difficulty. Ideally, they should also be able to accommodate increasing skill levels, exposing additional functionality as their interfaces are mastered by their users.

3.2.3   Continuity in Negotiation and Resolution of Uncertainty

By taking advantage of users' natural familiarity with the dynamics of physical systems, one can couple continuous auditory or tactile feedback with algorithms that collect and infer the relevance of information sources via an interactive device, computer, or mobile phone, allowing designers to make the user aware of subtle variations in the nature of any content they may be engaging with or manipulating. This can be an important building block for enabling *negotiated interaction*, where there is a bidirectional flow of information during interaction.

An HCI interprets user actions and carries out the users' intentions. In practice, a system cannot interpret intention with absolute certainty; all systems have some level of ambiguity. The interface, in the most general sense, communicates the state of hidden variables that are negotiated between the user and the system across the interaction modalities available. From this point of view, the system interface can be considered to be a continuous control system, with sensory inputs, a display, and an inference mechanism. This last continuously interprets the intentions of a user from his or her control actions and feeds the result back to him, enabling the user to negotiate a satisfactory interpretation of his or her intention.

Such a negotiated approach contrasts with prevailing techniques that take the form of discrete action accept/reject cycles. In such interfaces, exchanges of the form select, confirm, and perform are common. The confirmation step indicates that the original decision was taken without sufficient evidence (otherwise it would be redundant), but because the original selection was forced into a single discrete action, further evidence cannot be accumulated without additional interaction. A key idea of negotiated interaction is that information can be allowed to flow between the participants in an interaction without rigidly structuring the temporal sequence. Decisions can be made whenever appropriate evidence is observed, and the accumulation of evidence can be spread over time in a flexible manner.

Representing Uncertainty over Time

Explicitly representing ambiguity and displaying it for the user is intended to increase the interaction quality. As Mankoff, Hudson, and Abowd have described [63], ambiguity is an important issue when a system's interpretations are complex and hidden from users. By contrast, users can intuitively manipulate controls such as physical buttons because they have no hidden complexity.

Explicit uncertainty in feedback provided by an interface is a step toward this goal by making the true state of inner processes of the system visible to the user. There is evidence from research on human sensorimotor control that displays representing uncertainty lead to appropriately compensatory behavior [53]. For example, Koerding and Wolpert demonstrated that in a target tracking task, where a disturbance was artificially added to the user's control input, uncertain display (Gaussian point clouds, where the position is represented by points sampled from a Gaussian distrubtion) leads to behavior that was consistent with optimal integration of the uncertain sensory inputs and prior beliefs about the system. Operator models that incorporate Kalman filters, which optimally track the expected state of a system, have previously been used to model uncertainty in user control behavior [46].

In the Shoogle device, feedback of this kind was provided through granular synthesis, a probabilistic sound generation method, based on drawing short (10- to 500-millisecond) packets of sound, called grains or granules, from source waveforms according to a probability distribution [19, 88, 101, 116]. This technique provides an elegant way for displaying time-varying probabilistic information to users [112].

3.2.4   Benefits and Risks of Continuity for Interaction Design

Continuity brings added value in making interaction expressive and opening the space of variations and deviations, even in repetitive tasks. For example, even in user inter-face target acquisition tasks (typically modeled via Fitts' law), there can much expression in what happens between the selecting actions, and this information is usually neglected in conventional, discrete interactions.

Continuity is a key to embodiment [20, 26, 71], as we are accustomed to perceiving audible consequences of our everyday actions. Noë presents a case that perceiving is "not something that happens to us, or in us. It is something we do" [71], and we "enact our perceptual experience." Closed-loop relationships between our actions and audible or tactile feedback, if designed well, have the potential to increase a user's sense of engagement. If the coupling among sensing, information, and audio synthesis is rich enough, it may engender a sense of growing virtuosity, as users deftly engage with information streams, probing, guiding, and controlling the system. McCullough, however, points out how many modern interactive systems disrupt the flow of interaction, and in his investigation of the possibility of craft in the digital realm [65], he argues that our computational environment should be a rich medium that can be "worked" by a digital craftsperson, as a carpenter works with the grain of his wood.

Continuous interaction also poses risks to design if the implementation is clumsy. The *Midas touch* problem is that if every action creates audible feedback, people can rapidly become irritated and are likely to reject the system. Appropriate natural methods for "declutching" the audio feedback are likely to be critical to acceptance.

3.3   Perceptual Guidelines

As noted, traditional HCI methods are of definite, but limited, utility for the design of continuous auditory and tactile interactions. This section looks beyond the HCI domain and considers whether design guidelines may be drawn from prior research on human abilities to acquire perceptual information during the observation or manipulation of physical systems or artifacts that provide auditory or tactile feedback. We review a range of examples selected from prior literature that collectively illustrate several types of perceptual information that can be communicated through auditory and vibrotactile sensory channels during interaction with relatively simple mechanical systems over time. Moreover, because these largely relate to intrinsic perceptual abilities, they suggest that users may be able to perform a wide range of interactive tasks intuitively, without significant training, provided designers respect their innate perceptual-motor skills in shaping the affordances and feedback modalities of the interactive system or artifact.

One type of perceptual information that is highly salient to auditory and tactile modalities arises through the temporal patterns that structure many sound events. Sensory information encoded in such patterns has been found to be salient to human identification of sound sources and their intrinsic dynamical properties, such as configuration, mass, or rotational inertia. A second type of perceptual cue results from interdependencies between users' movements and the dynamics of mechanical systems that they interact with. Ample evidence from everyday experience and empirical studies indicates that the acoustic signatures that result can provide information to aid users in learning and executing skilled manipulation or control.

In the literature on ecological perception, perceptual cues have been described as originating with two main types of invariance principle [67]. Instances of the first type of invariance are called structural constraints. They determine how the acoustic pattern of a sound (say, a bouncing ball) is shaped by the physical and geometrical laws governing the event that produced them (such as elastic rebound from a rigid surface). The second type consists of transformational invariants. These are said to arise from perceptual-motor contingencies between a gesture and the sensations that accompany it. For example, the sound of scratching the ground with a stick depends on the forces applied and the velocity of movement. People are able to discover and learn to utilize patterns of acoustic or tactile information to perform a wide range of perceptual and motor tasks. Several examples described below provide indications as to how sonic interactions may be effectively designed by analogy to physical, mechanical systems, and they may point toward models for the conception of novel ways of enabling access to digital information via movement and sound.

Interaction design strategies that rely on physical analogies can yield unified, consistent models for producing feedback from multiple modalities (e.g., sound, tactile, or visual feedback). They provide advantages over alternatives that require ad hoc mappings across modalities. For example, in some real-world systems, auditory and vibrotactile information reflect high-temporal-accuracy, low-spatial-resolution aspects of the evolution of a physical system, whereas more slowly varying aspects may be represented visually. The availability of a single common model can make designing such perceptually coherent multimodal displays simpler. Several design examples are described in the last part of this section.

3.3.1 Temporal Patterns in Everyday Sound Events

During passive listening to everyday sounds, a variety of acoustic cues are available to aid in the interpretation of the events that are heard. As is clear from everyday experience, untrained listeners can readily detect and identify common events and activities in their surroundings by ear—be they knocks at the door or the steps of someone ascending or descending a staircase [35]. Although much of the literature on auditory perception has aimed to explain listeners' performance in terms of sensory features of acoustic signals, such as loudness, attack time, or average frequency content, results from several studies on the perception of everyday sound events have underlined the importance of complex temporal patterns in guiding listeners' perception, especially those patterns related to mechanical processes that produced the sounds. In experiments on free identification and sorting of everyday sounds [35, 105], those with easily recognizable temporal patterns are usually identified at the highest rates—greater than 95 percent accuracy in the experiments of Van Derveer [105]. Moreover, confusion errors in identification and clustering in sound-sorting tasks tend to evidence perceptual grouping by similarity of temporal pattern. Hammering, clapping, and footsteps are well recognized and grouped together, whereas the sound of filing fingernails is also readily identified but tends to be grouped separately. Such studies have also tended to reveal that most untrained listeners resort to using simpler sensory features, such as loudness or center frequency, only when the sound source cannot be identified. Taken together, these results suggest that listeners are most adept at attending to properties of complex sound sources that are determined by the macroscopic dynamics of the source and that, in many cases, their predisposition is to attend to temporal patterns that best evidence these aspects.

Tumbling, Breaking, and Bouncing Objects

The simplest auditory events, such as single impacts, often lack the necessary structure to fully illustrate the utility of these temporal cues. However, such idealized sounds are comparatively unusual. Real sound events more often involve multiple collisions that are patterned in time by the structure of the objects and physical processes involved—think of the clatter of a tea cup falling onto the table. Despite their added complexity, they can be more revealing of the identity and parameters of their processes of origin. Warren and Verbrugge studied listeners' ability to distinguish the sounds of bottles of different sizes and shapes falling to the floor, and either bouncing or breaking [110]. In a first experiment, listeners were found to be able to distinguish the two types of event with a success rate of 99 percent. Subsequently, the authors determined that a distinctive instantaneous cue in the sounds—a noise burst that is

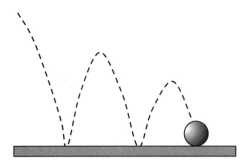

Figure 3.4
Auditory perception reveals the characteristic pattern produced by a bouncing object.

typical of the onset of breaking—had little effect on performance. After this part of the recordings was removed, listeners' success rate at discriminating bouncing from breaking remained high, at 96 percent. Finally, by editing sequences of impact sounds so as to most closely resemble the temporal pattern of breaking or bouncing, the authors were able to cause the resulting recomposed sounds to be identified with breaking or bouncing with high probability (respectively, 87 percent and 92 percent).

For a homogeneous, spherical object falling and bouncing repeatedly off of a flat surface (figure 3.4), simple physical considerations readily yield invariant expressions describing the characteristic rhythm of bouncing. At the kth bounce, the object strikes the ground with kinetic energy $E_k = mv_k^2/2$. Immediately after the bounce, the energy is given by $E_{k+1} = RE_k$, and the velocity (having changed sign) has magnitude $|v_{k+1}| = \sqrt{R}|v_k|$, where $R < 1$ parametrizes the elasticity of the object-ground collision. If energy loss due to air resistance is neglected, v_{k+1} is also the velocity immediately prior to the next bounce. The time interval between collisions is easily seen to be $\tau_{k+1} = \frac{2}{g}v_{k+1}$. The basic pattern is thus characterized as a temporal process whose interimpact time intervals decay according to the invariant relation $\tau_{k+1}/\tau_k = \sqrt{R}$ and impact energies according to the invariant $E_k/E_{k+1} = R$.

Real-world objects are not perfectly uniform, and their pattern of collisions with the ground depend on shape in more complicated ways. Carello, Anderson, and Kunkler-Peck investigated listeners' abilities to use auditory cues from an extended object to discern object shape. Specifically, they asked listeners to judge the length of cylindrical rods dropped onto the floor from a fixed height via the resulting clatter [17]. The task was performed without practice or feedback. The authors found subjective length to vary monotonically with actual length. They found that two invariant physical quantities, related to the inertia tensor of the rods, could, to a large extent, explain participants' length estimates (regression with $r^2 = 0.97$). In studies that have investigated listeners' abilities to discern the length [35] of a bar that is simply tapped, listeners' estimates are found to be less discriminating, possibly due to the relatively uninformative nature of the impacts [18].

Brain Activity Connects Hearing and Performance

The close link between action recognition through listening on one hand and action performance on the other is reflected in the connections and patterns of activity in the brain, as are most functions related to perception and movement. The coordination of the two is a large topical area involving physiology, psychology, and neuroscience [4]. Although a discussion would exceed the scope of the present chapter, it is interesting to note some of the more pertinent developments of the past decade. One type of neural connection between action observation and action performance has been found to be mediated by so-called mirror areas, linking sensory observations with brain activity in areas of the motor cortex that would be involved if the observer were the one performing [74]. Kohler et al. found and recorded activity in neurons in the monkey premotor cortex that discharge both when the animal performs an action, such as ripping paper or dropping a stick, and when it simply listens to the same act [52]. Thus, audition need not be considered as a process that is dependent solely on recognizing patterns of actions but can be regarded as capable of recruiting activity in the parts of the brain that would be needed to effect the same actions. There is additional evidence, obtained from experiments that use transcranial magnetic stimulation to estimate the excitability of corticospinal pathways from motor areas in the brain to the muscles needed to perform actions, that auditory observation is capable of priming an individual's motor apparatus to prepare to execute an observed action [2] and hence that hearing actions can be regarded as preparatory to their motor performance. Additional research has uncovered connections between motor corticospinal excitability and speech perception and visual observation, as reviewed by Fadiga, Fogassi, Pavesi, and Rizzolatti [29].

Listening to a Process Complete

Perceptual researchers have examined many other situations in which perceivers are asked to infer the time to completion (TTC) of an unfolding process from the available sensory information without recourse to explicit mental calculation [15, 41, 100]. Many of these settings have involved passive observation; for example, listeners may be asked to attend to the sound of an approaching acoustic source in order to predict its time of closest passage. As discussed in section 3.1, related research in HCI has

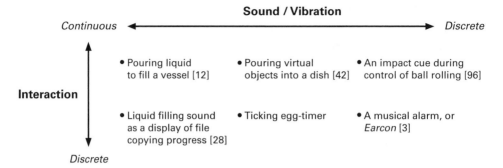

Figure 3.5
Several examples of task progress indicators arranged along axes indicating their level of continuity in interaction and sound presentation.

studied auditory display techniques for enabling users to unobtrusively monitor the progress of a background computational process, such as the transfer of a large file, via an auditory display [23, 36, 75]. Lee and co-workers hypothesized that human perceptual capacities in such settings can be explained in terms of derived temporal cues, called τ variables [56, 57], that can be defined as functions of the available sensory features. Although many of the examples cited above involve passive observation, such estimation tasks are also involved in contexts, such as those discussed in section 3.2, in which continuous auditory cues are used for the manual regulation of a continuous system that evolves toward a target state. The water-filling task described above is another such example. The concept of τ invariants, which refer to perceptual variables that indicate the time before an impending event, has also been applied to model predictive behavioral control in settings that do not necessarily involve audition. Motor control tasks that have been studied in this way involve the control of automobile braking [56], execution of table tennis forehand shots [9], or the catching of falling objects [66, 93].

Figure 3.5 arranges several examples of auditory task progress indicators involving different levels of continuity in both interaction and sound presentation.

3.3.2 Audiotactile Cues in Response to Movement

As emphasized in ecological accounts of sound perception, any description of human audition that is solely devoted to explaining the mechanisms underlying auditory awareness and recognition of sound sources and their properties is incomplete, as it ignores the primary purpose of perception, which is, as argued in such theories, to guide activity [18]. The chief value in gathering information about objects and events in our surroundings is said to be contained in the behavioral opportunities that they offer, which Gibson termed their affordances [38, 39]. The latter concept is, of course, widely applied within human computer interaction. There are important differences between situations in which users are passive observers and settings in which they are actively engaged in the generation of sound events that are heard. The perceptual problems posed by inferring the properties of objects in an active setting can be regarded as more complicated: the sensory stimuli may be highly variable, and the sonic effects of a user's motor actions must be predicted so that he or she can distinguish the attributes intrinsic to the system from those imposed through the gesture. Conducting experiments that allow for interaction is also more challenging because when participants are permitted control over a stimulus, it is difficult to ensure that they all experience it under the same conditions. See chapter 5 of this book for further discussion.

However, just as a less-constrained physical event can provide more cues to an observer about the properties of the object (think of a clattering impact as opposed to a clean strike), one that allows for interaction can make additional information available to an observer. For example, it is easier to discern the hardness of a plate if one is allowed to strike it. On one hand, by interactively probing such an artifact, a user may consider variations in its response as his or her control gesture is varied. Equally importantly, a new type of cue is often available to observers in the active setting, arising from the relations among a user's actions, the sensations that are experienced during the course of the interaction, and the physical perserverance of the system being interacted with. We discuss a few examples illustrating such cues.

Tilt-to-Pour: Acoustically Guided Vessel Filling

Filling a container with liquid is an everyday activity associated with a familiar acoustic pattern created by the flow of water and the resulting excitation of the column of air occupying the rest of the vessel's volume (figure 3.6). The resonant pitch of the air cavity rises with the water level, providing an indicator of the progress of the filling operation toward the desired level. Cabe and Pittenger found that listeners are able to perceive the changing water level from this auditory information and use it to control the filling of unknown vessels [15]. The researchers identified a temporal cue derived from the acoustic pattern that is available to observers as an aid in estimating the remaining time required to fill the vessel. They experimentally verified that participants were able to perform a filling time estimation task that required them to use such a cue. This time-to-completion cue is described in terms of the vessel's fundamental resonant frequency f, which increases with the liquid height $\ell = h - L$, where

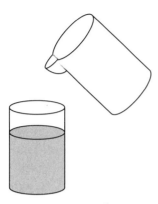

Figure 3.6
The familiar task of filling a vessel with liquid.

L is the height of the air column above the liquid. For a cylindrical vessel of radius R, the frequency is given by

$$f = \frac{c}{4L + 2.44R},$$ (3.1)

where c is the speed of sound in air. f increases monotonically as the vessel is filled and approaches a value of $f_0 = c/(2.44R)$ when the vessel is full. By analogy with prior work on visual estimation of time to completion, Cabe and Pittenger define a completion time variable τ through the ratio

$$\tau = f/(df/dt) = \frac{L + 0.62R}{dL/dt} = \frac{L}{dL/dt} + T_C$$ (3.2)

where the time derivative df/dt is the rate of change in the fundamental resonant frequency. For a constant fill rate, dL/dt, τ is the time that the water level would take to rise to the constant (overfill) level $0.62R$ above the fill line. A pourer would have to stop the filling process an amount of time $T_C = 0.62R/(dL/dt)$ seconds before τ in order to exactly fill the vessel. Participants were found to be able to estimate the necessary filling time when permitted to listen to an initial segment of a constant fill rate recording. Estimated time to fill tracked actual fill time closely, with errors that depended on both initial fill height and fill rate. Pittenger and colleagues also found that providing participants with training about the accuracy of some of their estimates greatly reduced errors, both for the vessels with which training had been provided and with new vessels for which it had not [78, 79]. As noted in section 3.1, similar cues were used by Gaver to accompany file copying processes in the Sonic-Finder [33].

Flops are instrumented drinking glasses containing virtual objects [32] and providing luminous and acoustic feedback (figure 3.7). In a study on the influence of sound and task on emotions elicited by these artifacts, Lemaitre et al. [58, 59] investigated a familiar activity in which users poured specified numbers of virtual objects from them.

When the glass is tilted, virtual objects inside of it move toward the lip of the glass, following a sliding dynamics described by $\ddot{x}(t) = \frac{2}{3} g \sin \alpha(t) - b\dot{x}(t)$, where x is position along the glass, α is the tilt angle, b is viscous damping, and g is the acceleration due to gravity. After objects are poured beyond the lip of the glass, they each produce impact sounds as they collide with a virtual surface beneath it. Lemaitre et al. found that task difficulty dominated acoustic features (such as sharpness) in determining users' emotional response when feedback was generated by active manipulation of the device [58, 59].

Tilt-to-Roll

Control of an artifact or system can be facilitated when a mental model of its physical behavior is available. In some circumstances, knowledge about the physics of a system can be inferred from auditory or tactile feedback that is received in response to control actions. Yao and Hayward investigated a novel perceptual task in which participants felt vibrations generated by a virtual ball rolling in a real tube as they manipulated the latter (figure 3.8) [117]. The haptic feedback consisted of plausibly realistic periodic rolling vibrations accompanying the rotation of the ball and vibrational impact transients coinciding with the collision of the ball with the end of the tube. This feedback enabled participants to estimate the length of the tube by interacting with it. Participants had no prior training or instruction in the interpretation of these vibrations. The authors described several simple cues available to allow estimation of the motion of the virtual ball from the vibrations, including the period of the ball's rotation, the time to impact with the rod end, and the energy of the impact. For each of these, awareness of the inclination angle $\alpha(t)$ of the rod over time is required. The estimation task is feasible because of the invariant relation

$$\ddot{x}(t) = (g/1.4)\sin(\alpha(t)) \tag{3.3}$$

between the acceleration $\ddot{x}(t)$ of a solid, rolling ball along the length of the incline and the angle $\alpha(t)$ of the manipulated rod. The relation depends on g, the acceleration due to gravity, but is invariant with respect to unknown parameters such as the ball's

Figure 3.7
Flops [32, 58] instrumented drinking glasses containing virtual objects and providing luminous and acoustic feedback.

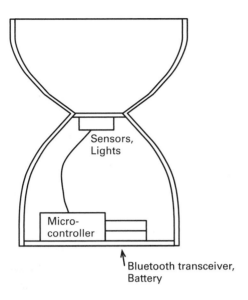

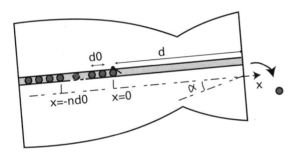

Figure 3.7
(continued)

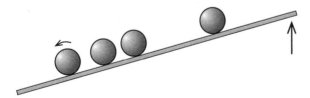

Figure 3.8
Control of rolling as a basic tactile-auditory task.

mass or radius. The difference with respect to the sliding dynamics of objects in the Flops device is due to the rotational inertia of the ball.

Rath and Rocchesso investigated another tangible electronic interface for controlling a virtual rolling ball, called the *Ballancer*. It is comprised of a handheld wooden bar that is manually tilted in order to control the movement of the ball, which was accompanied by auditory feedback in the form of rolling sounds and, optionally, by a visual display on a monitor [83, 84]. The latter were generated in real time and were intended to be more realistic than the abstracted rolling vibrations used by Yao and Hayward. The sounds were synthesized using a model of the contact physics between the ball's surface and asperities on the inclined plane. The authors designed a task in which users were required to maneuver the ball into a randomly selected target region. When compared with a purely visual display, significant advantages were seen with the addition of auditory feedback. However, only marginally significant differences were observed between trials employing a realistic rolling sound and others that used a simple feedback consisting of a sawtooth-wave oscillator whose frequency was proportional to the ball velocity [85]. One interpretation that can be derived from these results is that the simpler auditory display was just as effective because it provided participants with equivalent information, in the sense of communicating the invariant dynamic sensorimotor law of equation 3.3. The interaction method studied by Rath et al [84]. was later adapted to a functional setting in which users tilt a mobile phone in order to scroll within a menu (similar to prior work on tilt-based menu navigation by Rekimoto [87] and others).

Control of Wheel Spinning

The Spinotron interface of Visell, Franinović, and Meyer [108] was designed to investigate tasks involving manual control of the rotation of a virtual dynamical system via a pump-like interface (figure 3.9). The latter could be used to control the rotation of a virtual ratcheted wheel via the dynamics resembling that of a child's spinning top. The ratcheted wheel was represented solely through auditory feedback in the form of a temporal train of impact events, whose rate was determined by the motion of the ratchet. Lemaitre et al. found that participants were able to use this feedback in order to maintain the wheel's velocity within a specified range [59]. They also found that a more complex virtual system, consisting of a ball rolling in an eccentrically rotating bowl driven by the same input device, with auditory feedback that simulated rolling noise, was too complex for users to be able to reliably control.

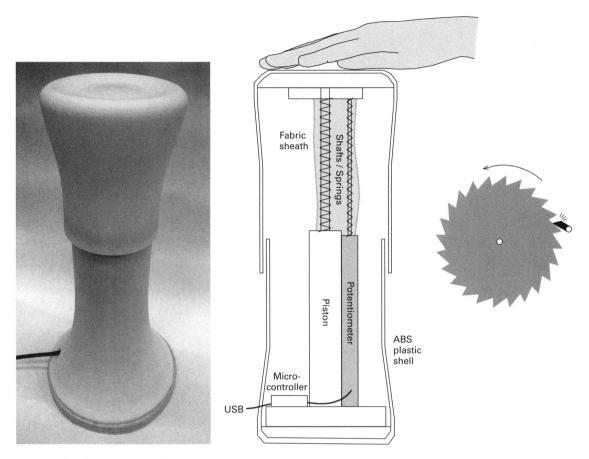

Figure 3.9
The Spinotron [108], a basic interface for the control of rotation via sound.

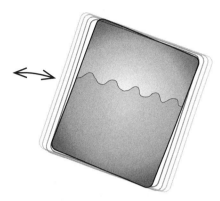

Figure 3.10
Shaking an opaque vessel to perceive its contents.

Shake, Slosh, and Stir: Numerosity, Viscosity, Weight, and Volume Estimation

A basic task in interaction with tangible closed or opaque containers concerns the estimation of the amount of material they contain. There are many circumstances in which one might wish to enable users to acquire similar information when interacting with invisible digital data, inspired by familiar tasks like this one: You lift a jar of coffee beans and shake it to see if there is enough remaining until tomorrow (figure 3.10). The encounter offers several sensory clues that could assist this judgment: the weight of the jar in your hand, the shift of the mass inside as the jar is shaken, the muted sound and rattle of the coffee impacting the walls of the jar, to name a few. One might benefit from drawing on several redundant cues, as humans are notoriously prone to being mislead in haptic judgments of weight [48, 111]. In addition, one might want to find out more about the contents of the jar—for example, the size and coarseness of the ground beans.

Studying a related problem in perception, Pittenger et al. found that people can estimate the number of steel balls held in opaque containers from acoustic or haptic information generated in shaking them [78] (a setting that is highly evocative of the Shoogle interactive device mentioned earlier). Additional studies by some of the same authors determined that people are able to judge the relative size of small granules (less than 1 mm diameter) or larger beads (a range of sizes between 2.4 and 10 mm) by shaking or stirring with a rod [79, 78]. Haptics was found to be most effective during stirring, whereas audition performed better during shaking. The larger-sized objects were more accurately ordered by size than the small granules were. Sekiguchi, Hirota, and Hirose designed a hand-held artifact affording shaking interactions, with collisions displayed via taps from a solenoid actuator against the walls of the artifact [91, 92]. They found that within the limitations of their device, users were able to reliably estimate the number N of objects in the virtual box, up to a maximum number of about $N = 5$, from these vibrations. Interactions of this type were applied to the Shoogle mobile information device described in the previous section and section 3.3.4 below.

Fluid-containing vessels can provide further cues to aid an observer in performing such a task because of the dynamics of the fluid, which can be induced to oscillate or slosh when agitated. The resonant frequency of small sloshing oscillations depends on the vessel shape and the fill level [24]. Jansson, Juslin, and Poom investigated whether people can judge the amount of liquid left in 1-liter cartons (ranging from nearly empty to nearly full) by lifting or shaking them, using either auditory or haptic information [47]. The authors found judgments of liquid volume to depend on the exploratory procedure participants used, with lifting tending to produce strong over-estimates; shaking did not lead to overestimation but did result in larger variances in judgments. As early as 1964, Stevens and Guirao published research on the perceptual scaling of viscosity of silicone fluids held in jars [96]. Participants judged the fluid viscosity by stirring or shaking and turning the jar it was held in, using haptic or haptic and visual feedback. They found that estimated velocity b scaled as $b \propto B^v$, where B is actual viscosity, and $0.42 < v < 0.46$. Magnitude estimation was found to be relatively unchanged under the different sensory conditions and exploratory procedures that were permitted (i.e., stirring or shaking and turning).

Perceptual Laws and Sensory Substitution

Sensorimotor perceptual laws have been most often applied in vision research. Individuals can be said to exploit invariant relations linking the signal captured through the retina, the movements of the body, and the structure of the environment [39, 72], through effects such as motion parallax and occlusion. Arguably, however, such laws need not be modally specific (and some have suggested that they are essentially amodal [18]), provided the same invariant law can be equally expressed through a nonvisual sensory channel that substitutes for the visual one (see Visell [107] for a recent review). One possible explanation is that information of this type is contained in a sensorimotor contingency rather than being coded in the specific properties of the stimulus.

Lenay, Canu, and Villon studied the role of sensorimotor contingencies in the association of patterns of proximal sensations to objects in the environment [60]. They

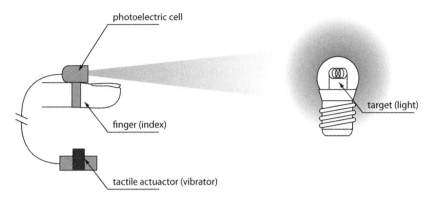

Figure 3.11
The minimal sensory substitution experiment of Lenay and his colleagues (figure courtesy of the author).

created a device worn on the finger consisting of a single photoreceptor, with an angle of sensitivity of approximately 30 degrees, coupled to a single vibrotactile actuator with a simple on/off capability (figure 3.11). The environment used in their experiments consisted of a single point light source in an otherwise dark space. In the absence of motor control over the device, its user is unable to "see," because all that can be inferred is whether the light source concerned stands in front of the receptor or not. If the user is granted control over the orientation of the sensor (via the direction the finger points to) he or she can infer the precise direction to the light source, and if enabled to translate, the sensor can infer the position of the light using parallax. One way to explain this is that, through interaction, the user infers the invariant relation

$$L = b(\sin(\alpha) - \cos(\alpha)\tan(\alpha + \beta)) \tag{3.4}$$

Here, L is the distance to the light source, b is the length of the arm, α and β are the arm and wrist angles, respectively. The stimulus point of origin can be readily distinguished via this formula. Another point that can be taken away from this example is that when a suitable sensorimotor contingency exists, even a very impoverished sensory cue (derived, for example, from a very simple auditory stimulus) can be sufficient to enable a perceiver to perform a perceptual task with a high level of efficiency. This can arise in cases in which the most useful information is contained in the dynamic sensorimotor law and is not in the movement-independent aspects of the stimulus.

Table 3.1
Summary of several examples involving estimation or control tasks accompanied by auditory and vibrotactile feedback

Configuration	Action	Estimate required	Feedback	Invariant
Ball on incline [83, 117]	Tilt	Speed v, dist. x	Rolling noise	Accel. $\ddot{x}(t)$ vs. angle $\alpha(t)$
Liquid in vessel [15]	Pour	Fill height L, time τ	Pouring, resonance	Time $\tau(t)$ vs. freq. $f(t)$
Falling object [18]	Drop	Size, shape	Clattering hits	Inertia I, elasticity K vs. time t_i, energy $E(t_i)$
Objects in container [113]	Shake	Number, mass, volume	Rattle, splash	—
Ratcheted wheel [59]	Pump	Angular speed $\dot{\theta}$	Clicks	—

3.3.3 Perceptual Invariants: Implications for Design

The examples discussed in this section illustrate at least two issues relevant to the areas of sonic interaction design that are addressed in this chapter. One is that humans are adept at perceiving through manipulation and are able to accomplish a wide range of perceptual estimation tasks related to structural or sensorimotor invariants. During the course of observation or manipulation, there are typically multiple channels of concurrent sensory information available to a perceiver. Audio and tactile cues often arise from and convey information about contact interactions between objects or relative motion between them, as in the production of friction noise by an object sliding on an incline plane or the rattle produced by dried beans shaken in a can. If a new artifact presents cues that preserve invariants respecting basic physical laws, such as constancy and energetic or temporal causality, users may be more likely to be able to perform similar feats of estimation and control, perhaps (as noted in the introduction to this section) even without significant training. Several examples are summarized from this standpoint in table 3.1, which identifies five salient characteristics: the physical *configuration* of the system involved; the *action or movement* involved in the perceptual task; the *physical quantity to be estimated* in the perceptual task; the *feedback modality*; and the *perceptual or perceptual-motor invariant* that facilitates task completion. Such characteristics could be mixed and matched to create a number of additional possibilities to inspire design.

A second implication is that the design of new artifacts that provide continuous sound or tactile feedback may benefit from the application of metaphors inspired by real world manipulation activities, just as the design of interactions with data objects in graphical user interfaces has been informed by symbolic representations of familiar objects (e.g., the file folder). A few artifacts illustrating design along such lines were noted above. The Ballancer has been used for an interaction involving a mental model in which selection of items in a list was controlled by a rolling ball. The Shoogle device (described above) adopts a metaphor in which messages received on a mobile device are embodied as virtual balls within a box to be shaken.

3.3.4 Applications to Interactive Retrieval of Digital Information

There is a diverse range of settings in which people are able to make use of auditory feedback generated during the manipulation of everyday objects or computational artifacts inspired by them, and some experiments have shed light on the ability of humans to perceive certain kinds of information intrinsic to these systems. It is natural to ask whether interaction designers can also benefit from this knowledge and use it to develop interfaces that allow users to better control digital devices such as phones and computers.

Here, we describe a few examples of human-computer interactive scenarios involving devices or tangible interfaces that sense user gestures and provide access to digital information through synthesized sound. Each is based on a physical metaphor (associated with a simple mechanical system) that mediates the correspondences among the control gesture, synthesized sound, and digital information.

Coupling Message Content to a Shaking Metaphor

Shoogle [114], as described in the previous section, provides one such example, as it allows users to shake a vessel containing virtual objects in order to hear and feel the number of messages or the type of contents they hold. It avoids the intrusion of passive notification alerts because, like an opaque food container, the device must be haptically queried for its contents. The system can support interaction via shaking directions, for example, used to stimulate different types of objects that provide distinct auditory feedback. One version automatically classifies text messages using a language model and then maps the classes onto separate activation planes. Stimulating the device in different directions probes messages of different styles.

Interactive Data Sonification

The interactive auditory display of digital data sets was investigated by Hermann and Ritter, who describe their approach as "model-based sonification" [45]. It associates elements of the digital dataset to virtual physical objects, such as points of a given mass connected by springs with certain values of the stiffness, that can be excited via a user interface. Hermann, Krause, and Ritter proposed a tangible computing interface —a handheld ball with integrated sensors—for this purpose [44]. The system was excited by shaking movements performed with the interface, and dynamic variables of the virtual system representing the dataset (e.g., the positions and velocities of the masses) were used to control a sound synthesis algorithm in order to provide the sonification.

Mobile Spatial Interaction

Strachan and Murray-Smith explored the use of granular sound synthesis to provide audio and vibrotactile feedback for interactions based on navigational bearing (figure 3.12) [97]. A user of a mobile phone with an electronic compass was able to point the device at locations in the surrounding space in order to access digital content, such as text messages, images, or services, as evidenced through auditory feedback. The sensing of both location and compass bearing were prone to significant and temporally changing errors, and, in accord with the discussion of section 3.2, the feedback was designed to reflect this. The current location and bearing were estimated using Monte Carlo techniques, which rely on a large collection of hypothetical locations or bearings, each of which is assigned a probability of being correct given the observations. The hypotheses were coupled to a granular sound synthesis display that varied from focused to blurred, depending on the level of uncertainty. Accelerometers were used to monitor device orientation, which was used to control the horizon distance. To explore distant areas a user would tilt the device forward, and to explore nearby locations he or she would tilt the device back again, toward his or her chest. This way of interacting makes it possible to obtain nearby information without the explicit need to move to a particular spot, as is the case with most location-aware systems.

The authors considered potential applications of the system for interacting, probing, and querying information at locations in the surrounding environment, which can be defined relative to egocentric locations on the body (e.g., the hip on which the mobile phone is usually worn) or exocentric sites in the surroundings (e.g., to retrieve digital information about a building ahead of him) [64]. Such interactions can form the basis for multimodal interaction with geocached information (e.g., a Wikipedia article about a building up ahead) and might pave the way for new kinds of highly interactive mobile Internet.

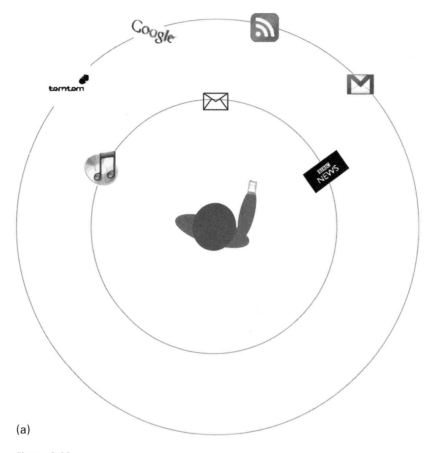

(a)

Figure 3.12

Illustration of the Mobile Spatial Interactive Device of Strachan and Murray-Smith [97]. (a) A user has a number of objects arranged around himself that he can point at to activate. The objects are arranged in a layered fashion, creating more space. (b) The local area has a number of objects left there by other users with which it is possible to interact.

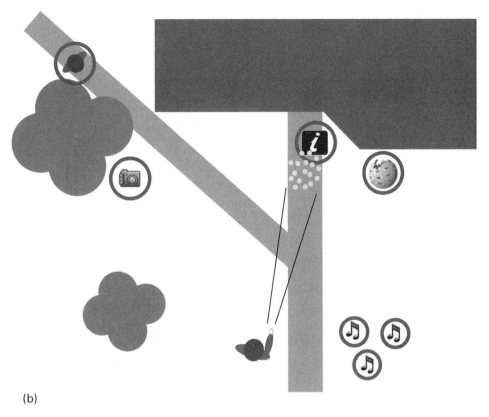

(b)

Figure 3.12
(continued)

Content-Sensitive Audio Feedback during Document Scrolling

Eslambolchilar and Murray-Smith investigated how exploring a large or complex document might profit from rich auditory feedback synthesized as a function of both the physical manipulation of the device, as captured by its sensors, and the content of information within the document [28]. In it, synthesized impact-based sounds were used to provide feedback linked to the speed of browsing through a document, the zoom level, and the local language content of the document. Statistical language models were used to analyze the lines of text in the document, and the inferred language was used to modulate impact sounds that were generated as the user scrolled across those lines.

3.4   Conclusion and Outlook

This chapter explored the concept of continuity as a quality characterizing the control affordances of an interactive system, the evolution of an auditory display or other sound event, and the perceptual-motor processes that arise when humans are present in the interaction loop. Vibrotactile displays and events were also included in the discussion because many of the same considerations can be seen to apply to them. The discussion was organized around fundamental issues affecting the utility of such a system to a human operator. Several aspects of human perception are relevant to the acquisition of information from continuously evolving auditory feedback. Temporal patterns in real-world sound events are, for example, relevant to the identifiability of everyday sounds and to basic estimation tasks that can arise from them. The acoustic pattern of many everyday sounds can be linked to the evolution of a physical process toward a goal (as in water filling a vessel or a ball rolling down an incline). Another example concerns properties of collections of objects brought into interaction with each other (as when a ball is dropped, or beans in a container are shaken).

When continuous, manual control is involved, additional questions are brought to the fore. The design of usable interaction techniques for such systems can be challenging because of the large space of possible control gesture trajectories. However, the coupling of gesture to sound can also be seen to introduce perceptual-motor constraints that may facilitate control of a system and understanding of its structure or contents. We discussed several examples illustrating gesture-sound couplings with this property. Most were modeled on manual interaction with real or virtual dynamical systems.

Because of limitations of technology (such as sensor noise affecting GPS data) and of human operator performance, continuous control tasks also involve uncertainties. An auditory or tactile interface can benefit from making these explicit. It may be advantageous functionally and aesthetically to display such uncertainties in ways that respect metaphors or other choices that have informed the overall design. In the interactive mobile device described in section 3.3.4, this was achieved by modulating the amount of random variation in a cloud of sound granules.

3.4.1   Open Questions

It is clear that the field of sonic interaction design is still at an early stage of development despite the continued advancement of electronic technologies. Commensurate methodological advances will be needed in order for designers and users to best profit from such technologies.

Clearly, there is a need for guidelines to aid in the design of usable interfaces of the type considered here. However, foundational tools such as manual control theory and sound synthesis are widely used in HCI or product design. Further work is needed in order to make them accessible to HCI researchers, perhaps through the development of new software toolboxes and methodologies adapted for common sound design tasks.

Although some common methods for the evaluation of input techniques, such as performance models based on Fitt's law, are applicable to multimodal interfaces like those described here, there is a need for new metrics for evaluating complex interactions that involve nontrivial, yet highly structured, control dynamics.

Prior literature on auditory display evaluation has primarily been devoted to cases involving passive listening or interaction of a discrete nature, whereas auditory feedback accompanying manual interaction has, notwithstanding the examples cited here, scarcely been studied. More investigation is needed.

Several of the results and case studies presented here point toward the importance of context of use, tasks and cognitive load, and related factors in contributing to the quality of experience provided by an artifact, but many questions remain concerning how continuous audiotactile interactions may best be designed to respect and profit from their surroundings.

Further work is needed to clarify appropriate and promising application roles for such interfaces. Mobile human-computer interaction is one domain in which continuous sonic interaction may be of benefit, as visual attention is often required to guide movement or other concurrent motor tasks.

Research on human sensory and motor capabilities has proved beneficial to the past several decades of development in human factors and HCI. As emphasized in section 3.3, knowledge about fundamental human capacities for interacting and acquiring information through sound is being developed in several different disciplines. Although scientific research in many areas (such as the neural correlates between audition and motor action) is ongoing, it may be some time before the implications of such results for the design of new interfaces are fully apparent.

It is hoped that the material presented in this chapter may convince the reader of the potential for such interfaces and that researchers in related areas of design, science, and engineering are encouraged to contribute further to areas such as those outlined above and to unforeseen questions that will inevitably arise as new paradigms, technologies, and applications take hold.

Acknowledgments

This work was supported by EPSRC grants EP/E042740/1, EP/F023405/01, a fellowship from the Scottish Informatics and Computing Science Alliance (SICSA), European FP6 NEST project CLOSED (no. 029085), and the ESF COST Action: Sonic Interaction Design (no. IC0601).

Notes

1. *Control theory* is an interdisciplinary branch of engineering and mathematics that deals with the behavior of dynamical systems. The desired output of a system is called the reference. When it is desired that one or more output variables of a system follow a certain reference over time, a controller manipulates the inputs of a system to obtain the desired effect.

2. www.havok.com.

References

1. Augsten, T., Kaefer, K., Meusel, R., Fetzer, C., Kanitz, D., Stoff, T., et al. (2010). Multitoe: High-precision interaction with back-projected floors based on high-resolution multi-touch input. In *Proceedings of the 23nd annual ACM symposium on user interface software and technology* (pp. 209–218). ACM.

2. Aziz-Zadeh, L., Iacoboni, M., Zaidel, E., Wilson, S., & Mazziotta, J. (2004). Short communication. Left hemisphere motor facilitation in response to manual action sounds. *European Journal of Neuroscience, 19*(9), 2609.

3. Beaudouin-Lafon, M., & Gaver, W. W. (1994). ENO: Synthesizing structured sound spaces. In *Proceedings of the 7th annual ACM symposium on user interface software and technology*.

4. Berthoz, A. (2002). *The brain's sense of movement*. Cambridge, MA: Harvard University Press.

5. Blattner, M., & Dannenberg, R. B. (Eds.). (1992). *Multimedia interface design. Frontier series.* New York: ACM Press.

6. Blattner, M., Sumikawa, D., & Greenberg, R. (1989). Earcons and icons: Their structure and common design principles. *Human-Computer Interaction, 4*(1), 11–44.

7. Blauert, J. (1997). *Spatial hearing*. Cambridge, MA: MIT Press.

8. Bly, S. (1982). *Sound and computer information presentation*. Unpublished PhD Thesis UCRL53282, Lawrence Livermore National Laboratory.

9. Bootsma, R. J., & van Wieringen, P. C. W. (1990). Timing an attacking forehand drive in table tennis. *Journal of Experimental Psychology. Human Perception and Performance, 16*(1), 21–29.

10. Brewster, S. A., & Brown, L. M. (2004). Tactons: Structured tactile messages for nonvisual information display. In A. Cockburn (Ed.), *Proceedings of Australasian user interface conference* (pp. 15–23). Dunedin, NZ: Australian Computer Society.

11. Brewster, S. A., Wright, P. C., & Edwards, A. D. N. (1994). A detailed investigation into the effectiveness of earcons. In G. Kramer (Ed.), *Auditory display* (pp. 471–498). Reading, MA: Addison-Wesley.

12. Brown, L. M., Brewster, S. A., & Purchase, H. C. (2005). A first investigation into the effectiveness of tactons. In M. Bergamasco & A. Bicchi (Eds.), *Proceedings of worldhaptics 2005* (pp. 167–176). Pisa, Italy: IEEE Press.

13. Buxton, W. (1989). Introduction to this special issue on nonspeech audio. *Human-Computer Interaction, 4*(1), 1–9.

14. Buxton, W., Gaver, W., & Bly, S. (1991). Tutorial number 8: The use of non-speech audio at the interface. In S. Robertson, G. Olson, & J. Olson (Eds.), *Proceedings of ACM CHI'91*. New Orleans: ACM Press, Addison-Wesley.

15. Cabe, P. A., & Pittenger, J. B. (2000). Human sensitivity to acoustic information from vessel filling. *Journal of Experimental Psychology and Human Performance, 26*(1), 313–324.

16. Cadoz, C. (1988). Instrumental gesture and musical composition. In *Proceedings of the international computer music conference* (pp. 1–12).

17. Carello, C., Anderson, K. L., & Kunkler-Peck, A. J. (1998). Perception of object length by sound. *Psychological Science, 9,* 211–214.

18. Carello, C., Wagman, J. B., & Turvey, M. T. (2005). Acoustic specification of object properties. In J. D. Anderson & B. F. Anderson (Eds.), *Moving image theory: Ecological considerations* (p. 79). Carbondale, IL: Southern Illinois University Press.

19. Childs, E. (2002). A sonification of probability distributions. In *ICAD 2002*. Achorripsis.

20. Clark, A. (1997). *Being there: Putting brain, body, and world together again.* Cambridge, MA: MIT Press.

21. Cook, P. R. (2002). *Real sound synthesis for interactive applications.* Natick, MA: A K Peters.

22. Craig, J. C., & Sherrick, C. E. (1982). Dynamic tactile displays. In W. Schiff & E. Foulke (Eds.), *Tactual perception: A sourcebook* (pp. 209–233). Cambridge: Cambridge University Press.

23. Crease, M., & Brewster, S. A. (1998). Making progress with sounds—The design and evaluation of an audio progress bar. In *Proceedings of ICAD* (vol. 98). Citeseer.

24. Dodge, F. T. (2000). *The new "Dynamic behavior of liquids in moving containers."* San Antonio, TX: Southwest Research Institute.

25. Doherty, G., & Massink, M. (1999). Continuous interaction and human control. In *European conference on human decision making and manual control.*

26. Dourish, P. (2004). *Where the action is: The foundations of embodied interaction.* Cambridge, MA: MIT Press.

27. Ellis, W. H. B., Burrows, A., & Jackson, K. F. (1953). *Presentation of air speed while deck-landing: Comparsion of visual and auditory methods.* Technical Report 841, UK RAF Flying Personnel Research Committee.

28. Eslambolchilar, P., & Murray-Smith, R. (2006). Model-based, multimodal interaction in document browsing. In S. Renals, S. Bengio, & J. G. Fiscus (Eds.), *Machine learning for multimodal interaction, 4299,* 1–12.

29. Fadiga, L., Fogassi, L., Pavesi, G., & Rizzolatti, G. 1995. Motor facilitation during action observation: A magnetic stimulation study. *Journal of Neurophysiology, 73*(6), 2608.

30. Ferber, A. R., Peshkin, M., & Colgate, J. E. (2007). Using haptic communications with the leg to maintain exercise intensity. In *Proceedings of robot and human interactive communication.*

31. Forbes, T. W. (1946). Auditory signals for instrument flying. *Journal of the Aeronautical Sciences, 13,* 255–258.

32. Franinović, K., & Visell, Y. (2008). Sonic and luminescent drinking glasses. In *International biennale of design, St. Etienne.* Flops.

33. Gaver, W. (1989). The sonicfinder: An interface that uses auditory icons. *Human-Computer Interaction, 4*(1), 67–94.

34. Gaver, W. (1997). Auditory interfaces. In M. Helander, T. Landauer, & P. Prabhu (Eds.), *Handbook of human-computer interaction,* 2nd ed. (pp. 1003–1042). Amsterdam: Elsevier.

35. Gaver, W. W. (1988). *Everyday listening and auditory icons.* PhD thesis, University of California, San Diego.

36. Gaver, W. W., Smith, R. B., & O'Shea, T. (1991). Effective sounds in complex systems: The ARKola simulation. In *Proceedings of the SIGCHI conference on human factors in computing systems: Reaching through technology* (pp. 85–90). New York: ACM.

37. Geldard, F. A. (1957). Adventures in tactile literacy. *American Psychologist, 12,* 115–124.

38. Gibson, J. J. (1979). *The ecological approach to visual perception.* Boston: Houghton Mifflin.

39. Gibson, J. J., & Carmichael, L. (1966). *The senses considered as perceptual systems.* Boston: Houghton Mifflin.

40. Guimbretière, F., Stone, M., & Winograd, T. (2001). Fluid interaction with high-resolution wall-size displays. In *Proceedings of the 14th annual ACM symposium on user interface software and technology* (pp. 21–30). ACM.

41. Hancock, P. A., & Manster, M. P. (1997). Time-to-contact: More than tau alone. *Ecological Psychology, 9*(4), 265–297.

42. Hayward, V. (2008). Haptic shape cues, invariants, priors, and interface design. In M. Grünwald (Ed.), *Human haptic perception—Basics and applications* (pp. 381–392). Basel: Birkhauser Verlag.

43. Hayward, V., & MacLean, K. E. (2007). Do it yourself haptics: Part I. *IEEE Robotics & Automation Magazine, 14,* 89.

44. Hermann, T., Krause, J., & Ritter, H. (2002). Real-time control of sonification models with a haptic interface. In *Proceedings of the international conference on auditory display (ICAD)* (pp. 82–86).

45. Hermann, T., & Ritter, H. (1999). Listen to your data: Model-based sonification for data analysis. *Advances in intelligent computing and multimedia systems* (pp. 189–194). Baden-Baden.

46. Jagacinski, R. J., & Flach, J. M. (2003). *Control theory for humans: Quantitative approaches to modeling performance.* Mahwah, NJ: Lawrence Erlbaum Associates.

47. Jansson, G., Juslin, P., & Poom, L. (2006). Liquid-specific stimulus properties can be used for haptic perception of the amount of liquid in a vessel put in motion. *Perception-London, 35*(10), 1421.

48. Jones, L. A. (1986). Perception of force and weight: Theory and research. *Psychological Bulletin, 100*(1), 29–42.

49. Jones, L. A., & Sarter, N. B. (2008). Tactile displays: Guidance for their design and application. *Human Factors, 50*(1), 90.

50. Kaltenbrunner, M., & Bencina, R. (2007). ReacTIVision: A computer-vision framework for table-based tangible interaction. In *Proceedings of the 1st international conference on tangible and embedded interaction* (pp. 69–74). ACM.

51. Klatzky, R., & Lederman, S. (2003). Touch. In A. Healy & R. Proctor (Eds.), *Handbook of psychology, Vol. 4: Experimental psychology* (pp. 147–176). New York: John Wiley & Sons.

52. Kohler, E., Keysers, C., Umilta, M. A., Fogassi, L., Gallese, V., & Rizzolatti, G. (2002). Hearing sounds, understanding actions: Action representation in mirror neurons. *Science, 297*(5582), 846.

53. Kording, K. P., & Wolpert, D. (2004). Bayesian integration in sensorimotor learning. *Nature, 427,* 244–247.

54. Kramer, G. (1994). An introduction to auditory display. In G. Kramer (Ed.), *Auditory display* (pp. 1–77). Reading, MA: Addison-Wesley.

55. Laurel, B., & Mountford, S. J. (1990). *The art of human-computer interface design.* Boston: Addison-Wesley Longman.

56. Lee, D. N. (1976). A theory of visual control of braking based on information about time-to-collision. *Perception, 5*(4), 437–459.

57. Lee, J. C., Dietz, P., Leigh, D., Yerazunis, W., & Hudson, S. E. (2004). Haptic pen: A tactile feedback stylus for touch screens. In *Proceedings of UIST 2004* (pp. 291–294). Santa Fe, NM: ACM Press Addison-Wesley.

58. Lemaitre, G., Houix, O., Susini, P., & Visell, Y. & Franinović, K. (2012). Feelings elicited by auditory feedback from a computationally augmented artifact. IEEE Transactions on Affective Computing, issue PP (preprint), vol. 99, Jan. 2012

59. Lemaitre, G., Houix, O., Visell, Y., Franinović, K., Misdariis, N., & Susini, P. (2009). Toward the design and evaluation of continuous sound in tangible interfaces: The Spinotron. *International Journal of Human-Computer Studies, 67*(11), 976–993.

60. Lenay, C., Canu, S., & Villon, P. (1997). Technology and perception: The contribution of sensory substitution systems. In *Second international conference on cognitive technology, Aizu, Japan [aJKO]*.

61. Lévesque, V. (2005). *Blindness, technology, and haptics*. Technical Report TR-CIM-05.08, McGill University Centre for Intelligent Machines.

62. MacLean, K., & Enriquez, M. (2003). Perceptual design of haptic icons. In *Proceedings of Eurohaptics* (pp. 351–363).

63. Mankoff, J., Hudson, S. E., & Abowd, G. D. (2000). *Interaction techniques for ambiguity resolution in recognition-based interfaces* (pp. 11–20). UIST.

64. Marentakis, G., & Brewster, S. A. (2005). A comparison of feedback cues for enhancing pointing efficiency in interaction with spatial audio displays. In *MobileHCI '05: Proceedings of the 7th international conference on human computer interaction with mobile devices & services* (pp. 55–62). New York: ACM Press.

65. McCullough, M. (1998). *Abstracting craft: The practiced digital hand*. Cambridge, MA: MIT Press.

66. McIntyre, J., Zago, M., Berthoz, A., & Lacquaniti, F. (2001). Does the brain model Newton's laws? *Nature Neuroscience, 4*(7), 693–694.

67. Michaels, C. F., & Carello, C. (1981). *Direct perception*. Englewood Cliffs, NJ: Prentice-Hall.

68. Milnes-Walker, N. (1971). A study of pursuit and compensatory tracking of auditory pitch. *Ergonomics, 14*, 479–486.

69. Miranda, E. R., Kirk, R., & Wanderley, M. M. (2006). *New digital musical instruments: Control and interaction beyond the keyboard*. Middleton, WI: A-R Editions.

70. Neuhoff, J. G. (Ed.). (2004). *Ecological psychoacoustics*. San Diego: Elsevier Academic Press.

71. Noë, A. (2005). *Action in perception*. Cambridge, MA: MIT Press.

72. O'Regan, J. K., & Noe, A. (2002). A sensorimotor account of vision and visual consciousness. *Behavioral and Brain Sciences 24*(5), 939–973.

73. Oulasvirta, A., Tamminen, S., Roto, V., & Kuorelahti, J. (2006). Interaction in 4-second bursts: The fragmented nature of attentional resources in mobile HCI. In *Proceedings of ACM CHI 2006* (pp. 919–928). Portland, OR: ACM Press Addison-Wesley.

74. Oztop, E., Kawato, M., & Arbib, M. (2006). Mirror neurons and imitation: A computationally guided review. *Neural Networks, 19*(3), 254–271.

75. Pacey, M., & MacGregor, C. (2001). Auditory cues for monitoring a background process: A comparative evaluation. In *INTERACT'01, Tokyo, Japan* (p. 174).

76. Pasquero, J., Luk, J., Little, S., & MacLean, K. (2006). Perceptual analysis of haptic icons: An investigation into the validity of cluster sorted MDS. In *14th Symposium on haptic interfaces for virtual environment and teleoperator systems* (pp. 437–444). IEEE.

77. Peltonen, P., Kurvinen, E., Salovaara, A., Jacucci, G., Ilmonen, T., Evans, J., et al. (2008). It's mine, don't touch! Interactions at a large multi-touch display in a city centre. In *Proceedings of the twenty-sixth annual SIGCHI conference on human factors in computing systems* (pp. 1285–1294). ACM.

78. Pittenger, J., Belden, A., Goodspeed, P., & Brown, F. (1997). Auditory and haptic information support perception of size. In *Studies in perception and action IV* (p. 103).

79. Pittenger, J. B., & Mincy, M. D. (1999). Haptic and auditory information support perception of size: Fine granules. In *Studies in perception and action V: Tenth international conference on perception and action, August 8–13, Edinburgh, Scotland* (p. 68). Mahwah, NJ: Lawrence Erlbaum Associates.

80. Poulton, E. C. (1974). *Tracking skill and manual control.* New York: Academic Press.

81. Poupyrev, I., & Maruyama, S. (2003). Tactile interfaces for small touch screens. In *Proceedings of UIST 2003* (pp. 217–220). Vancouver, Canada: ACM Press.

82. Powers, W. T. (1973). *Behavior: The control of perception.* Hawthorne, NY: Aldine.

83. Rath, M., & Rocchesso, D. (2005). Continuous sonic feedback from a rolling ball. *IEEE MultiMedia, 12*(2), 60–69.

84. Rath, M., & Rocchesso, D. (2005). Continuous sonic feedback from a rolling ball. *IEEE Multimedia Special on Interactive Sonification, 12*(2), 60–69.

85. Rath, M., & Schleicher, R. (2008). On the relevance of auditory feedback for quality of control in a balancing task. *Acta Acustica United with Acustica-Stuttgart, 94*(1), 12–20.

86. Rekimoto, J. (2002). SmartSkin: An infrastructure for freehand manipulation on interactive surfaces. In *Proceedings of the SIGCHI conference on human factors in computing systems: Changing our world, changing ourselves* (pp. 113–120). ACM.

87. Rekimoto, J. (1996). Tilting operations for small screen interfaces. In *ACM Symposium on User Interface Software and Technology* (pp. 167–168).

88. Roads, C. (2002). *Microsound.* Cambridge, MA: MIT Press.

89. Rocchesso, D., & Fontana, F. (2003). *The sounding object.* Mondo Estremo.

90. Ryan, J. (1991). Some remarks on musical instrument design at STEIM. *Contemporary Music Review, 6*(1), 3–17.

91. Sekiguchi, Y., Hirota, K., & Hirose, M. (2003). Haptic interface using estimation of box contents metaphor. In *Proceedings of ICAT2003* (pp. 197–202).

92. Sekiguchi, Y., Hirota, K., & Hirose, M. (2005). The design and implementation of ubiquitous haptic device. In *World haptics conference* (pp. 527–528).

93. Senot, P., Zago, M., Lacquaniti, F., & McIntyre, J. (2005). Anticipating the effects of gravity when intercepting moving objects: Differentiating up and down based on nonvisual cues. *Journal of Neurophysiology, 94*(6), 4471.

94. Sheridan, T. B., & Ferrell, W. R. (1974). *Man-machine systems: Information, control, and decision models of human performance.* Cambridge, MA: MIT Press.

95. Smith, J. O. (2010). *Physical audio signal processing for virtual musical instruments and audio effects.* W3K Publishing.

96. Stevens, S. S., & Guirao, M. (1964). Scaling of apparent viscosity. *Science, 144*(3622), 1157.

97. Strachan, S., & Murray-Smith, R. (2009). Bearing-based selection in mobile spatial interaction. *Personal and Ubiquitous Computing, 13*(4), 265–280.

98. Summers, I. R., (Ed.). (1992). *Practical aspects of audiology: Tactile aids for the hearing impaired.* London: Whurr Publishers.

99. Tan, H. Z., Durlach, N. I., Rabinowitz, W. M., & Reed, C. M. (1999). Information transmission with a multi-finger tactual display. *Perception & Psychophysics, 61*(6), 993–1008.

100. Tresilian, J. R. (1999). Visually timed action: time-out for "tau"? *Trends in Cognitive Sciences, 3*(8), 301–310.

101. Truax, B. (1988). Real-time granular synthesis with a digital signal processor. *Computer Music Journal, 12*(2), 14–26.

102. Van Den Doel, K., Kry, P. G., & Pai, D. K. (2001). FoleyAutomatic: Physically-based sound effects for interactive simulation and animation. In *Proceedings of the 28th annual conference on computer graphics and interactive techniques* (pp. 537–544). New York: ACM.

103. van den Doel, K., & Pai, D. K. (2001). JASS: A Java audio synthesis system for programmers. In *Proceedings of the international conference on auditory display.* Citeseer.

104. Van Erp, J. B. F., & Van Veen, H. (2001). Vibro-tactile information presentation in automobiles. In *Proceedings of eurohaptics* (pp. 99–104). Citeseer.

105. VanDerveer, N. J. (1979). *Ecological acoustics: Human perception of environmental sounds*. PhD thesis, Cornell University, New York.

106. Vinje, E. W. (1971). *Human operator dynamics for auditory tracking*. PhD thesis, Department of Aerospace Engineering, University of Connecticut.

107. Visell, Y. (2009). Tactile sensory substitution: Models for enaction in HCI. *Interacting with Computers, 21*(1–2), 38–53.

108 Visell, Y., Franinović, K., & Meyer, F. (2009). Demonstration cases for sonic interaction design. *CLOSED Project deliverable 3.3*, EC FP6-NEST-Path no. 29085.

109. Visell, Y., Law, A., & Cooperstock, J. R. (2009). Touch is everywhere: Floor surfaces as ambient haptic interfaces. *IEEE Transactions on Haptics, 2*(3), 148–159.

110. Warren, W. H., & Verbrugge, R. R. (1984). Auditory perception of breaking and bouncing events: A case study in ecological acoustics. *Journal of Experimental Psychology, 10*(5), 704–712.

111. Weber, E. H. (1834). *De pulsu, resorptione, auditu et tactu. Annotationes anatomicae et physiologicae*. Leipzig: CF Koehler.

112. Williamson, J., & Murray-Smith, R. (2005). Sonification of probabilistic feedback through granular synthesis. *IEEE MultiMedia, 12*(2), 45–52.

113. Williamson, J., & Murray-Smith, R. (2009). Multimodal excitatory interfaces with automatic content classification. In E. Dubois, P. Gray, & L. Nigay (Eds.), *The engineering of mixed reality systems* (p. 233). London: Springer-Verlag.

114. Williamson, J., Murray-Smith, R., & Hughes, S. (2007). Shoogle: Multimodal excitatory interaction on mobile devices. In *CHI '07: Proceedings of the SIGCHI conference on human factors in computing systems* (pp. 121–124). New York: ACM Press.

115. Wu, M., & Balakrishnan, R. (2003). Multi-finger and whole hand gestural interaction techniques for multi-user tabletop displays. In *Proceedings of the 16th annual ACM symposium on user interface software and technology* (pp. 193–202). ACM.

116. Xenakis, I. (1971). *Formalized music: Thought and mathematics in composition*. Bloomington: Indiana University Press.

117. Yao, H. Y., & Hayward, V. (2006). An experiment on length perception with a virtual rolling stone. In *Proceedings of Eurohaptics* (pp. 325–330).

4 Pedagogical Approaches and Methods

Davide Rocchesso, Stefania Serafin, and Michal Rinott

Continuous interaction and multisensory feedback are key ingredients for successful interactive artifacts of the future. However, the complexity of the systems of sensors, actuators, and control logic that are necessary for exploiting such ingredients poses tremendous challenges for designers who are mostly used to visual thinking and discrete interactions. Specifically, designers not acquainted with sound lack a number of meaningful skills required to deal with sonic interaction projects:

- *Means* to present them to others
- *Language* to discuss them with others
- *Skill set* to prototype them
- *Processes* to iterate them

In this chapter we present a number of methods adopted and adapted to enable thinking about sonic interactions, generating ideas and prototyping them at different levels of fidelity and specificity. These methods are focused on the special challenges and possibilities of interactive sound.

4.1 Basic Design Methods

The birth of design as a discipline is usually attributed to the Bauhaus school, founded in Weimar in 1919 and later moved to Dessau and Berlin, where it was closed in 1933. Although the life span of the Bauhaus was relatively short, its impact on design practices and theories was huge. Since the foundation of the school under the direction of Walter Gropius, it was clear that a discipline had to be grown out of education, and the importance of introductory courses (grundkurs) was immediately evident. What was less clear, at the beginning, was what to teach and how to teach. It took many years and the effort of several educators to develop a basic design method that would produce mature designers. The early classes of Johannes Itten were a sort of sensory training for students. Then, László Moholy-Nagy introduced some technological elements to widen the range of possible phenomena and configurations that could be experienced by students. It was only with the classes of Josef Albers that a drive for objectivity entered design education. The sensory awareness of the designer was cultivated through exercises, trial and error, and confrontation with peers. It was Albers himself who pushed this method further over 27 years of teaching in the United States. Specifically, Albers's decade at Yale University (1950–1960) refined the basic design method of education as research, and its peak was reached in studying the interactions of colors [1]. The exercises assigned by Albers and the solutions given by his students demonstrate a synthesis of many decades of efforts, in both artistic and scientific contexts, to understand color perception. This synthesis of art, science, and technology was even more explicit in the New Bauhaus, founded in 1937 by Moholy-Nagy in Chicago, and it inspired the creation of the Hochschule für Gestaltung in Ulm (1953). Especially under the direction of Tomás Maldonado, the basic design classes developed a method based on problem solving, where objectives and constraints are clearly expressed in the exercises. At the same time, the introductory classes were specialized according to specific curricula (visual communication, product design, and others), and several variants of basic design started to emerge.

In the early twenty-first century some theorists and educators have been reinterpreting basic design [2–4] and proposing it as a key pedagogical approach even in design contexts that are much larger than those faced by the design schools of the twentieth century. In contemporary contexts the designer has to face interaction as an important, if not pivotal, element of configuration. The sensory, cognitive, and social phenomena that a designer should consider are complex and multifaceted. The complexity of the interaction design space can be tackled by thinking in terms of basic interaction phenomena constructively. The fundamental gestalts are the basic, immediate, and inherently meaningful actions of a person, such as pushing, pulling, and shaking [5], that are exploited in interaction. Such gestalts may result from abstraction of actual interactions [6] or be derived from the movement primitives considered in motor sciences [7]. A difficulty is that these gestalts are not properties of objects but are rather emerging properties of user-object interaction unfolding in time. A method of inquiry may proceed by analyzing actions, extracting interaction gestalts, and designing exercises around a specific interaction gestalt [8].

In the context of musical instrument design, Essl and O'Modhrain [9] proposed the grouping of actions according to some shared physical behavior that can be abstracted from the specific physical object. Their PebbleBox described in chapter 6 is a prototypical design of such approach.

Workbenches such as the PebbleBox served the purpose of developing basic design practices in contexts where interaction is primarily mediated by sound. This helped to define basic sonic interaction design as the practice of research through education that is being developed in those schools and laboratories that have prominent interest in the sonic manifestations of objects.

A central problem in basic interaction design is the choice of appropriate raw materials. This is no longer as simple as it was for Albers to experiment with configurations of colored surfaces. And it is not only a problem of choosing an effective toolkit of sensors, actuators, and microcontroller boards. There is often the choice between designing the materials themselves or using readymades and augmenting them with technologies. Although the first choice allows a finer degree of experimental accuracy and a sort of semantic neutrality, the second is often faster, cheaper, and highly expressive. An oscillatory balance between function and expression is found to be important in interaction design practices because it allows an understanding of the expressive features of objects in use while at the same time it elicits new uses, or misuses, of objects [6]. In basic sonic interaction design, sound synthesis models and algorithms are to be considered among the raw materials to work with. They play the same role that colored paper sheets played in Albers's exercises.

Another crucial issue is how to evaluate basic designs and how reliable such evaluations are. The methods of psychophysics and experimental psychology, although valuable and applicable when reliability and repeatability of results are mandatory, are not usually included in basic design practices. An experiment, while being a difficult and time-consuming endeavor, can only help nail down a precise scientific question. That a question of this kind arises as a crucial element in a design process is the exception rather than the rule. Conversely, direct experimentation, shared appreciation, and discussion are invariably present in design practices. This is what makes basic design very close to experimental phenomenology, where the process of knowledge acquisition is distilled in a few selected self-speaking demonstrations [10]. Indeed, introspection and intersubjectivity are the key tools of experimental phenomenology or descriptive experimental psychology dating back to the method of understanding by demonstration advocated by Franz Brentano in the nineteenth century [11]. Basic design and experimental phenomenology, in this respect, both use the practice of shared observation as the only possible way of assessing the properties of objects. The fact that this sharing may include naive subjects may increase the robustness of results. Bozzi [12] proposed the interobservational method, where an experiment is performed by jointly exposing a small group of subjects to the stimuli. Because the members of the group have to agree on a report, problems of outliers and degree of expertise are largely reduced. At the same time, joint observation and discussion contribute to make the description of facts more stable and rich. In the design practice, it is clearly more convenient to let the team of designers play the role of subjects and perform such interobservation. Even though experimental phenomenologists recommend the direct participation of the experimenter without a privileged position with respect to the subjects, a potential bias is recognized in reducing the group of subjects to the students of a class or to just the team of designers. Such bias was clearly present, for example, in the color-shape tests performed by Kandinsky with his students in the Bauhaus [13]. However, a justification for this convenient choice may obviously be found in the difference in objectives between experimental psychology and design.

In interaction design, especially where sound and haptics are important, the dissemination of interactive experiences is problematic. Video examples can sometimes replace first-hand experiences, but discussions around prototypes or interactive sketches are invariantly present in basic interaction design practices. Sometimes, videos can become prototypes themselves, especially to overcome the difficulty of augmenting a prototype object with interactive sound (see section 4.3.2).

4.2   Sensitizing to Sonic Interactions

Whether teaching interaction design within other design disciplines (graphic, industrial, multimedia) or teaching in an interaction design program, these design students do not typically have background knowledge and competence in sound design. A challenge encountered in teaching sonic interaction design to visually oriented students has been to motivate them and enhance their interest in exploring the possibilities offered by sonic feedback.

We first describe different exercises that we propose to the students in order to understand the importance of sound in real and mediated environments. Such exercises range from sound walks to description of sounds in physical objects to sound-only stories. We then describe exercises that are targeted to the development of sonic feedback for artifacts.

4.2.1   Performing Sound Walks

One of the first exercises we propose to students who are not used to working with sound or thinking about sound is to perform a sound walk around a specific location [14]. Sound walks were originally proposed by Murray Schafer as an empirical methodology to identify and describe a soundscape of a specific location [15].

When performing a sound walk, people are asked to navigate in a delimited area with open ears, remembering all the sounds heard. We ask students to perform such exercises in pairs, where one person is blindfolded and the other one acts as the guide.

This exercise has proven to be an ear opener and a good starting point to enhance students' motivation in performing more elaborate assignments. The exercise is always followed by a discussion in the classroom to share the different experiences. These introductory experiences with sounds resemble the practices of sensory training developed by Johannes Itten in the Bauhaus.

4.2.2 Listening and Describing Audio Dramas

Another exercise aimed at enhancing the appreciation of sound is the exposure to audio dramas. An audio drama is a collection of timed non–speech-based sound effects combined in a soundtrack. While listening to the soundtrack, students are asked to associate meaning and create a story. The outcome of this exercise is that students realize that audio-only content, even when not containing speech, can be used to evoke a narrative structure.

One particularly interesting audio drama is *The Revenge*, a radio play without words written and performed by Andrew Sachs in 1978. *The Revenge* was commissioned by the BBC with the precise goal of investigating whether nonverbal sounds can render a meaningful entertainment [16].

4.2.3 Writing a Short Audio Drama

After having listened to existing auditory content, coming either from the real world or from a recorded soundtrack, students are taught to create their own content. This is achieved by asking them to design a short audio drama, involving the collection of content, either from existing sound libraries or generated by the students themselves.

First, students are introduced to the concept of Foley (the process of live recording of sound effects) and Foley artists and are encouraged to creatively record different sonic material. They are then introduced to basic sound-editing tools and allowed to creatively explore how to combine, merge, and transform the available material in order to create a story of about 3 minutes. Once the assignment has been completed, it is followed by a class discussion. Here, some of the students present their productions, and the audience is asked to describe what they hear. The different interpretations of the perceived story are discussed.

This approach can be stretched further by asking actors to perform by following the proposed sound track [17]. The analysis of the performance makes students aware of how sounds affect gestures and how, conversely, gestures may affect the mental representations elicited by sound.

4.2.4 Exploring Audiotactile Interaction

The exercises described up to this point do not include any interactivity. Their main goal is to motivate students to start working with sounds and to get them familiar with manipulating and editing sonic content.

In the following series of exercises, we focus on audiotactile interaction: the tight connection between sound and touch. The first exercise is inspired by an experiment conducted by Lederman and Klatzky [18]. The goal of the experiment was to investigate the ability of subjects to recognize different objects while blindfolded, only using their sense of touch. While performing this experiment, they noted the stereotypical nature with which objects were explored when people seek information about particular object properties. For example, when subjects are asked to recognize the texture of an object, they move their hands laterally; when they seek to know which of two objects is rougher, they typically rub their fingers along the objects' surfaces. Lederman and Klatzky called such an action an exploratory procedure, by which they meant a stereotyped pattern of action associated with an object property. The authors suggest that this way of interacting with real objects should also be adopted when one is designing interfaces based on touch [18].

Our exercise starts by dividing students into pairs. One student is asked to close her eyes while the other student is asked to pick a surrounding object and give it to the blindfolded student. The blindfolded student is asked to recognize both the given object and some of its properties such as weight, material, texture, shape, and size. The person who provided the object is then asked to note which kinds of gestures the other person is performing while interacting with the object. In the second part of the exercise, while the student is still blindfolded, she is asked to identify the different sounds associated with the object. First, the sound-producing gestures are reported, that is, the sounds the student produced while interacting with the object to identify its different properties. Then, all other possible sounds obtained when interacting with the object are identified. As the last part of the exercise, the students are asked to brainstorm on how the given object can be enhanced with other sound-producing gestures, for example, by shaking it or hitting the object. Students are asked to reproduce the sonic interactions between gestures and sounds by using either physical objects or their own voices.

After being "sensitized" to sound through exercises such as those described in the previous sections and through the presentation and discussion of inspirational examples from the field, students can often envision interesting and evocative concepts for sonic interactions. The next section deals with ways to sketch and prototype these ideas easily.

4.3   Sketching and Prototyping Sonic Interactions

Creating interactive prototypes is, in general, a complex task. In the relatively young field of interaction design, a number of methodologies have been developed that attempt to circumvent the complexity of fully functional prototypes yet still answer the need of testing out ideas during the design process. These ideas of "just enough" prototyping, "smoke and mirrors" techniques, and "experience" prototypes are central to the maturation of interaction design as a design discipline (as opposed to an engineering one). They enable students' focus to move from the technology to the experience it entails in the stages of the process where this focus is needed. Houde and Hill [19] have proposed that prototypes for interactions can address three dimensions: role, look and feel, and implementation, where *role* refers to questions about the function that an artifact serves for the user, *look and feel* denote questions about the concrete sensory experience of using an artifact, and *implementation* refers to questions about the techniques and components through which an artifact performs its function. In these terms, these methods forgo the implementation dimension to focus mainly on the look and feel dimension and to different extents also on the role dimension. Such methods include presenting users with screens made of paper (e.g., Post-its)—a method aptly named paper prototyping—and creating fake prototypes that work by having someone behind the scenes pull the levers and flip the switches, called the "Wizard of Oz" technique.

This section describes two main methods aimed at providing students and practitioners with a means to sketch and prototype sonic interactions and thus to present and discuss sonic interaction concepts before the actual implementation of a working prototype. First, the sketches and prototypes enable the creators to get a feeling of the experience they entail. Second, they provide a means for others to experience them. This enables the creators to perform meaningful observation and receive feedback at early stages of the design process.

The methods described are relevant and useful both as a step in a design process leading to a working prototype (e.g., in a hands-on–type course) and as a final product of a design assignment (e.g., in a more conceptual design course). As an obvious extension, they are relevant for practitioners of design dealing with interactive objects and environments.

The two methods presented enable the description of sonic interactions regardless of their complexity (from simple sonic events to tightly coupled, continuous sonic interactions). This is done by separating the design from the implementation, thus enabling designers to think about sonic behaviors and communicate them to others before (and regardless of) implementation. The two methods are related but to some extent complementary. They can be used in sequence within a project, or only one can be chosen according to the fit of its attributes to the project nature. Both methods are valuable before physical prototyping with interactive sounds.

4.3.1 Vocal Sketching

Vocal sketching involves the use of the voice, along with the body, to demonstrate the relationship between action and sonic feedback. Vocal sketching, in essence, is as simple and straightforward as it sounds: the designer uses his or her voice to produce the sound that would be generated in the sonic interaction. The vocal performance is usually accompanied by some physical action. This "performance" activity is so simple and natural that many vocal sketches are created within conversations without the vocalization being regarded as a sketch. In using it within an educational context, we propose to make this activity more conscious and defined and thus make it more valuable within the design process.

The following attributes of vocal sketching make it a useful tool for the early stages of designing sonic interactions:

Intuitive A testament to the intuitive nature of this method can be found by watching children play with toys. The engine sounds of toy cars, made by the vocal tract, change these—for those involved in the activity—from inanimate plastic objects to powerful vehicles (all the more so for toy guns transformed into deadly weapons). Another rich behavioral reference is the preverbal play between parents and infants: vocal sounds are often used in accompaniment to different forms of movement and action (e.g., a beep when touching the nose).

Available Vocal sketching requires nothing but the willingness to make sounds. Although people differ in their control over their vocal apparatus, everyone can create expressive sounds with his or her voice [20]. Issues of social comfort arise and can be lessened by facilitation methods described later in this section.

Communicative Vocal sketching can be performed alone but is more likely to be used in a group of two or more people. It is a way to describe the sounds that the designer may hear "in her head." However, forcing the sounds out of the head and into a real vocalization obliges the designer to make a specific description and enables a discussion around it.

Group-friendly   Vocal sketching is a method that really shines when used by a group as a shared tool to plan and describe a sonic interaction. The group members can use their multiple voices to overcome the limitations of the voice and create multitrack performances. In a workshop setting focused on vocal sketching [21], participants used their multiple voices to describe a temporal interaction in which the sonic feedback changed from disharmony to harmony over time. It is probable that the shared production of a vocal sketch by a whole group increases the commitment of the designers to this solution.

Enactive   Vocal sketching is related to body storming, a method of "physically situated brainstorming" [22] in which the designer acts out the design ideas with his or her body or tries to gain insight from a bodily experience that is related to the end user's experience [23]. Especially when vocal sketches are made for tightly coupled interactions, vocal sketching happens in parallel to the body actions that create these sounds.

Vocal sketching poses some challenges; these and some possible remedies are described below:

Social comfort   Not everyone is comfortable with making nonverbal sounds to demonstrate ideas. The willingness to do this depends on personality and contextual factors; extroverts will probably be more likely to enjoy this, and people seem to prefer to vocally sketch in smaller groups and with people they feel comfortable with. The main method of alleviating this discomfort has been some form of warm-up activity prior to vocal sketching. This activity should require people to make sounds within a framework that they are not responsible for, such as a silly game with very defined rules. The person hosting this activity needs to give a personal example to set the stage for others. In a workshop setting [21], most participants acknowledged that they felt some discomfort in making sounds. All stated that this discomfort decreased as the workshop progressed. It should be noted that such discomfort is also found when sketching by drawing is considered. Some people consider themselves poor sketchers and refrain from making freehand sketches. However, designers are usually trained at drawing and, in most cases, enjoy showing off their drawing abilities.

Ephemerality   Vocal sketches are easily created and easily disperse. However, if a vocal sketch is to become a guiding element in the design process, it needs to be captured, by video for example. Using a technology as simple as a video camera can to some extent take away the simplicity and spontaneity of vocal sketching. Capturing a vocal sketch, however, can be used as a starting point for computational sound models; ways to extract data from vocal sketches are currently being investigated in the community of sonic interaction design [24].

Voice limitations   The vocal tract is limited; we cannot make any sound we want. A few obvious limitations are the temporal limitation caused by our breath span, the "single-track" nature of our voice, and the subset of sounds we can produce. Vocalizing in a group can alleviate most of these limitations. Learning to use the voice more professionally can help expand the range of sounds that we can make, as human beatbox practitioners demonstrate extremely well.

No "reality check"   Vocal sketching enables high degrees of creativity in thinking about sonic behaviors. The obvious challenges here are the unknown feasibility of the design solutions and the fact that the difficulty of implementation is not a factor in the design process.

4.3.2   Sonic Overlay of Video

We use the term *sonic overlay* to refer to a form of video prototyping in which an interaction is filmed and the sonic elements are added over the footage at a later stage, creating a video of a fake sonic interaction.

Video prototyping has a rich history in the field of interaction design. In 1980, Robert Spence used cardboard models and filmed interaction to illustrate bifocal display as a novel information visualization technique [25]. In 1994, Bruce Tognazzini and his team at Sunsoft created a video prototype to demonstrate their ideas for the new interface design, and the resulting overall user experience, offered by the next generation Starfire computer. This video became an influential vision to the computer of the future. Today many companies such as Microsoft use videos to describe their interaction ideas and visions. Wendy Mackay, Ratzer, and Janecek [26] have proposed that video can be added to a design brainstorming session. In this "video brainstorming" method, participants of the brainstorming session select a few ideas and demonstrate them in front of the camera, creating "video sketches"—outcomes of the session that are easier to understand and remember than text notes.

Video prototyping is also becoming a common practice for design students. Some interaction design education programs have included a special course in video prototyping in their curriculum,[1] exploring the different levels of fidelity that can be used to communicate a concept through video. The ease of editing and sharing video has made video prototypes feasible and useful not only for selling ideas to management but for sharing ideas at many stages of the design process—from very sketchy, low-fi videos shot with a simple camera and no editing to more planned and designed videos with various camera angles, edited effects, and the like.

In their 1990 CHI tutorial on "Storyboards and Sketch Prototypes for Rapid Interface Visualization" [27], Curtis and Vertelney described the idea of using "special effects" to prototype interaction ideas—in their case, printing out screen visuals and using camera effects such as zoom, pan, and the like to simulate interactive screen elements. Somewhat similarly, in our sonic overlay video prototypes, students shoot video and overlay the sounds and effects over it at a later time.

The following attributes of sonic overlay make it a useful tool for the early stages of designing sonic interactions:

Sound-centered   Students are instructed to shoot simple videos and focus their efforts on the sonic part of the video. The method is aimed at allowing the student to get an impression of different sound options over a fixed interaction: different sound options can be easily compared, thus sensitizing students to the impact of sound and giving them a tool to test ideas with. Filming an interaction makes the continuous aspects of interaction prominent and can push students to develop sonic interactions that are tightly coupled to actions.

Diverse   Sonic overlaying gives the designer the best possible conditions for creating the desired sound. At the editing table, the variety of sonic materials available to the designer can be found or created, be they voice, everyday objects, music, downloaded sound samples, and the like. The sounds can be overlaid with temporal precision because even simple video editing programs provide audio tracks on the timeline. Also useful are the options to easily record over video using the built-in microphone as well as the ability to create multiple sound tracks and thus to easily layer a number of different sounds.

Good communication tool   With Web video-sharing platforms such as YouTube, students can bounce ideas back and forth between themselves and tutors with ease; they can even annotate the videos directly. The language of video is highly communicative and easy to understand, and thus, video prototypes can be shown to different stakeholders, and opinions received, during the design process.

Sonic Overlay has some disadvantages, described below:

Nonenactive   The greatest disadvantage of this method is the passive experience it entails. Both the creator and viewer do not directly experience the interaction but rather view it secondhand.

Non–real-time   Sonic overlay cannot be performed spontaneously as part of a brainstorm or group design session but rather requires the designer to go "to the drawing board." As video editing tools become simple, it may be possible to find methods to use them in more integrated fashions within the design session.

No "reality check"   Like vocal sketching, sonic overlay enables high degrees of creativity in thinking about sound. The designers are not limited by their technical skill set;

they are limited mainly by their imagination. This promotes interesting and original solutions. The obvious challenge here are the unknown feasibility of the design solutions and the fact that the difficulty of implementation is not a factor in the design process.

4.3.3   Example: Sound and Pepper—A Project about Adding Information through Sound

This section describes a project developed using the sketching and prototyping methodologies described above as well as a working prototype.

The Context

The Sound and Pepper project was created within a class called "Interaction Design Hands On," at Holon Institute of Technology in Israel. The class (4 hours weekly for one semester) combines students from various disciplines, predominantly design and engineering. The project spanned over 2 weeks and was the work of two students: an industrial designer and a graphic designer.

The Brief

The design brief was to use sound in order to add information to an everyday object. Students were instructed to produce a sonic overlay in the first week and a working demo in the second week.

The Concept

During the initial brainstorming, the students reviewed daily activities of an imaginary person, moving from the bedroom to the bathroom and on, going over mundane actions and the information that might enhance them. This process brought them to the kitchen and to the activity of cooking. Using the boiling kettle as an example, they sought to add information to the spice containers. They identified two opportunities: giving each spice a sound, such that the right container can be identified without looking at it, and giving an indication of the amount of spice poured into a dish, to enhance the feedback and prevent overspicing.

The Process

A video prototype was produced to explore and communicate the concept. One of the students was filmed in her kitchen, stirring a dish and shaking different spice containers before selecting one and pouring spice into the dish (figure 4.1). This video was overlaid with a new soundtrack in which each shaking and each pouring action was

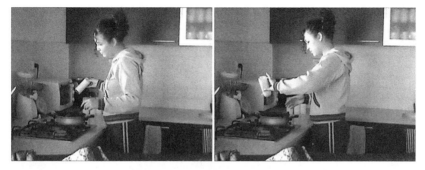

Figure 4.1
A video prototype of the Sound and Pepper project.

accompanied by a sound. Although the sounds were not precisely placed and were not yet chosen with coherence (one spice gave a liquid sound, another a musical sound, another an everyday grating sound), the video was very successful in communicating the concept and convincing the designers, their teachers, and co-students of its value.

A discussion of the sounds of different spices was the next step, with the students analyzing the character and characteristics of each spice. Two main methods were used. The first was an association game in which each student gave keywords related to the spice (e.g., for cinnamon: belly dancers, orientalism, bells; for salt: crystals, glass). These associations were often visual, sometimes conceptual, and sometimes auditory. The second method was vocal sketching, predominantly to portray the relationship between the shaking action and the sounds. For example, for pepper, the sound associated with the grinding action in a pepper grinder was performed vocally. The keywords generated from this process were used to find sound files on the web. A collection of sounds was made for each spice, and the best was selected and trimmed.

The Final Deliverable
A working prototype was produced in a 1-week process. Five spice containers (salt, pepper, cinnamon, chili, and paprika; figure 4.2) were embedded with mercury switches for detecting pouring motions, and light sensors for detecting a hand passed over the container. These were connected via an input-output board to the PC. When a hand was detected over the spice, the sound of the spice was played once. When the container was shaken, the same sound was generated with every shake, making the sounds play over each other. This most simple solution, in implementation terms, proved effective. The spice sound texture became more dense with every shake of the

Figure 4.2
A prototype of the Sound and Pepper project.

container. An epiphenomenon was that when the shaking was stopped, the sound slowly faded, as though in parallel to the small cloud of spice dispersing in the air. Although this prototype was limited in many ways (by cables coming out of the containers, by the feedback related not to the actual amount of spice but to the shaking behavior), it worked, with real spices in the containers enhancing the experience.

Evaluation

A large number of people tried the prototype in an educational setting: an "open house" in the lab and an exhibition of the final course. This enabled a meaningful amount of feedback on the concept and the protoype, which was largely enthusiastic. The prototype was also filmed in a kitchen setting, and this video was posted to YouTube (http://www.youtube.com/watch?v=zF_3ZlpxiZk). The project also attracted media attention and appeared on TV and radio. However, the prototype was not evaluated in the real setting of a kitchen due to lack of time. In this sense, the project can be seen more as a "sensitizer" and primer toward SID than as an attempt to build a real product. The use of sonic overlaying on video proved an extremely effective way to work in this project in that it got the designers focused and committed to the concept early enough to enable the creation of a working prototype.

4.4 Problem-Based Learning and Sonic Interaction Design

The problem-based learning (PBL) approach is a pedagogical method adopted at Aalborg University to introduce students to projects in sonic interaction design that last a full semester.

Historically, PBL started in the early 1970s at the medical school of McMaster University in Canada and was slowly adopted by different faculties worldwide [28]. PBL can also be traced back to the problem-solving approach described in section 4.1.

In PBL students are active learners and collaboratively solve problems while reflecting on their experiences. The instructor in this approach is considered mostly as a facilitator who helps students solving the problem. PBL becomes interesting when a variety of disciplines need to be incorporated to address a problem, which is clearly the case in sonic interaction design.

Problems are chosen by the students themselves and structured in such a way to be able to integrate and apply knowledge from different disciplines. This also allows students to realize connections among disciplines and promote carryover of knowledge from one discipline to another. In this way, PBL is a method that facilitates transdisciplinarity, defined as the ability to start from a problem and, using problem solving, bring the knowledge of those disciplines that contribute to the solution.

Most of the problems addressed by students are transdisciplinary by nature in that they start from a given question and use several disciplines to address it and solve it. PBL projects require a high level of social, communication, and cooperative skills among students. These skills are in high demand in professional work. Given the high amount of workload a project requires, usually the final results are very satisfactory, and learning can be effectively measured. PBL has proven to be particularly suitable for education dealing with design of interactive systems [29] and multidisciplinary settings [30].

As also observed by Schultz and Christensen [29], PBL is a valid methodology for approaching interaction design projects, especially for the possibility to explore, analyze, and define the problem space, the importance of teamwork and team development, and, eventually, to find a solution to the given problem. The problem space, domain, and context have to be analyzed, and problem definition and requirements need to be defined. Team members have different roles, which must be clear to all. In teamwork both interpersonal and intrapersonal skills [31] are important.

A related pedagogy of sonic interaction design, albeit based on workshops taking a few days of collective practice, has been developed by Hug [32]. In particular, he uses filmic materials to extract narrative metatopics or abstractions of narrative fragments. Each metatopic is assigned to a group, which develops a project around it. Narrative and performative elements are shown to emerge and combine both at the analysis and at the prototyping/demonstration stage.

4.4.1 Example: The Soundgrabber—Combining Sonic Interaction Design and PBL

The Context

The Soundgrabber installation (see figure 4.3) was created during the fourth semester of the medialogy education at Aalborg University in Copenhagen. The Soundgrabber represented the final project of the semester, in which courses in audio design, physical interface design, measurement of user experiences, and sensors technology were offered.

The Brief

The Soundgrabber installation investigated the following problem: Is it possible to make sound tangible by means of an intangible user interface?

The Concept

The aspiration of this project was to challenge the physical impossibility of designing an intangible installation that creates the illusion that people are tangibly interacting with sounds.

The Process

A group of six students worked on this project. Students were all enrolled in the medialogy program, but their main interests ranged from programming, graphical interface design, interaction design, animation, and audiovisual effects.

The design of the Soundgrabber went through several iterations before reaching the shape shown in figure 4.3. The first prototype was made of carton boxes, and was built with the mere purpose of testing the possibility of grabbing sounds and moving them around in space. All prototypes were created using the Max/MSP software platform for the auditory feedback. One of the last prototypes had also visual feedback in order to help the user to locate the position of the sounds. However, the visual feedback distracted the user from focusing on the audio and pseudohaptic experience, so it was not used in the last prototype.

Figure 4.3
The Soundgrabber in use at Sound Days in Copenhagen, 2008.

The Final Deliverable

The Soundgrabber is a physical interface designed as a semicircle. At the top of the semicircle, four columns are placed. At the bottom of each column a speaker is installed. Each column is embedded with light sensors that allow it to detect the position of the hand of the user moving vertically parallel to the column. Moreover, a bucket is placed in the center of the semicircle. The user interacts with the Soundgrabber using a glove embedded with a bend sensor. By bending the hand inside the bucket, the user is able to grab a sound, listen to it (thanks to the speaker embedded inside the glove) and release it in one of the columns.

Evaluation

In order to evaluate if the installation answered the problem formulated, the Soundgrabber was evaluated by allowing users to play with it and then answer a questionnaire inspired by the sensory substitution presence questionnaire [33]. In such a questionnaire, statements such as "I felt that I was able to grab a sound" or "I felt that I was able to relocate the individual sounds" were made, and subjects were asked to answer in a scale from 1 to 5 if they agreed or not with the statement. Results showed that the sensory substitution between audition and touch worked because there was a statistically significant number of subjects who felt they were able to move sounds around and grab them.

4.5 Physical Prototyping with Interactive Sound

How do product designers approach the design process? They start by sketching with paper and pencil. They produce many sketches and compare them. They use sketches as generators of thoughts. Then they build mockups that can give a physical impression of the product and even allow a limited form of experience in use. Mockups must be developed quickly, and they must be cheap and easy to abandon. Then there are prototypes that allow a full experience and a more complete evaluation. Prototypes could be evolved into products.

Nowadays, products can include visual, haptic, or auditory displays. Via an auditory display we can actually mold the acoustic behavior of objects to be passive (responding to actions), active (stimulating actions), or continuously coupled with actions.

In a pedagogy of sonic interaction design, the connection between gesture and sound is further investigated by testing basic sonic interactions in physical realizations. This goal is achieved either by extending everyday objects with sensors or by creating novel sonic objects. Students are provided with a palette of basic sensors together with a microcontroller to acquire the sensors' data. As basic sensors, students have the possibility to use buttons, pressure sensors, tilt sensors, and accelerometers. They are also introduced to tools that make it possible to perform basic sound synthesis and processing in real time. An exercise of this kind takes usually 3 or 4 days, divided into an introduction to the technology used, both for the sensors and the sound part, development and implementation of ideas, and presentations of final results. By the end of the 3- or 4-day workshop, students usually acquire a basic understanding of how to design novel objects embedded with sensors.

4.5.1 Sound Models

As emerging from the exercises described in section 4.2, the sounds of everyday objects are immediately associated with events or processes. We may distinguish between basic acoustic events and temporal organizations of continuous or discrete signals. If such organizations have some temporal regularity, we call them textures. If a designer wants to augment an object with a sonic indicator of action (or sonic affordance), events, processes, and textures are the classes of sound that should be considered. Methods and tools inherited from the field of sound and music computing [34] are readily available for the designer. However, further research is needed in interaction-centered sound modeling for the goal of designing better sounding objects and environments [35, 36].

The palette of sound synthesis methods is quite large [37], although they are not equally suitable for prototyping interactive artifacts. Abstract synthesis methods, such as frequency modulation, were introduced as an economical means to produce rich musical spectra but are not very suitable for contemporary product sound design. They would produce, in most cases, abstract sounds that are difficult to relate to events, processes, forces, and dynamics. Some general principles of frequency modulation and nonlinear distortion are, however, still useful, especially at the stage of dynamic processing of sound. For example, making a sound spectrally thicker by modulation is an easily conceived thing to do.

Additive synthesis is certainly rich and easy to think of, but it goes against the goal of economic representations of sound, as each sinusoidal component must be specified in its temporal behavior. Still, sinusoidal modeling is most easily coupled with sound analysis, so that resynthesis with modifications becomes one of the most effective ways to approach auditory displays. A designer could start from recordings, even of vocal sketches or other forms of imitation, derive a noise plus sines plus transients model [38], and process this material in its components. The challenge with this kind of processing is to have transformations that are meaningful to the designer. For example, if I record a water drop in a sink, I would like to make the drop bigger, or the liquid denser, and these transformations can be nontrivial if expressed through a sinusoidal model.

The subtractive synthesis model, including linear predictive coding, is useful for timbral transformation and sound hybridization. It makes it possible to preserve the temporal articulation of sound processes as it is captured by a recording, for example of a vocal imitation.

Sometimes the designer is faced with the problem of devising a sound process that does not sound repetitive and has relatively constant long-term characteristics. Many natural phenomena such as fire or water flow have such textural character, and they can afford sustained listening without inducing fatigue. Special techniques are required to synthesize convincing sound textures without using overly long recordings. Again, linear predictive coding is one of the enabling techniques for texture generation, together with wavelet decomposition and tiling and stitching of samples [39].

Thinking of sound as a side effect of physical interactions allows the organization of basic events, processes, and textures into hierarchies that also have a strong perceptual basis [40]. Sound synthesis by physical modeling is the natural framework to exploit such organization, from elementary events such as impacts or drops to complex processes such as rolling.

4.5.2 Software Tools

The repertoire of software tools that can be used to make sounds with numbers is very large [41]. Most of them, however, have been designed for musical use. So we have software applications for recording, composition, and performance.

In sonic interaction design, the focus is mainly on what has been called procedural audio [42]. This is the possibility of generating sound algorithmically, using some sound synthesis method, and of relating such generation to the events and processes as they are captured by a set of sensors. Among the software available for procedural audio, which are the programs that are so rapidly accessible as to become sketching tools in the hand of the designer? There is no definite answer to this question, as it heavily depends on the designer's background. Some languages and environments, such as SuperCollider, are enablers of performing practices such as "live coding," which can be considered as dynamic production of code sketches that have immediate audible effect. Actually, if the designer becomes proficient with an audio-oriented programming language, the production of sonic sketches can become faster than what is achievable with any other means. Nevertheless, most interaction designers show their legacy with visual design in preferring visual languages and environments such as Puredata or Max/MSP. The latter is the only software for audiovisual interaction that is mentioned in Buxton's book on sketching user experiences [43]. Several sensor boxes are available on the market that come with software modules that are ready to be used in Puredata or Max/MSP patches, and this gives the designer an effective toolbox to produce interactive prototypes relatively quickly.

Although the aforementioned tools embed general-purpose languages and give the freedom to produce virtually any sound, in many cases it makes sense to use specialized tools. For example, sound for interaction sometimes requires the composition of scenes or sonic tapestries. For this specific purpose, Misra and colleagues developed software for texture generation and transformation [44].

Between the specific and the general are those systems that make it possible to represent a wide range of phenomena within a consistent frame. An example is the Sound Design Toolkit [45], a set of software modules for Max/MSP that rely on the accurate physical modeling of basic physical phenomena (impact, friction, bubble) to construct a hierarchy of events, processes, and textures that are easily described in terms of interaction with everyday objects.

4.5.3 Example: The Sonified Moka—An Exercise in Basic Sonic Interaction Design

A basic design exercise on the theme of screw connections, quite compelling in terms of interaction, was developed at IUAV in Venice [46]. It shows how physical prototyping can be combined with sound modeling while maintaining a focus on the direct experience of interaction.

The Context

As part of design education at the graduate level at IUAV, the exercise is part of a series of workshops aimed at extending the basic design methods to interactive contexts, as explained in section 4.1.

The Brief

The exercise was formalized into three components:

Theme   Continuous feedback for mechanical connections.

Objective   Design the feedback for a screw connection, such as found in the moka, in such a way that the right degree of tightness in coupling can be easily reached.

Constraints   The feedback should be continuous, nonsymbolic, immediately apparent (or preattentional), and yet divisible into three clear stages.

The Concept

The purpose of the exercise was to explore the effectiveness of sound in guiding and conditioning continuous manipulations of objects.

The Process

A solution to this exercise was found by adapting a physics-based sound model of friction, which gives rise to a wide palette of timbres. Depending on the vertical force exerted by an object sliding on a surface, the sound can range from a gentle noise to a harmonic squeak to a harsh and irregular grind. The transition can be made gradual, yet the three different qualities of coupling (loose, tight, too tight) can be clearly perceived. The effectiveness of this gesture sonification largely depends on sound design choices, such as parameter mapping, range of parameter values, and temporal articulation of sonic events.

The Final Deliverable

The chosen sound model was applied to a moka augmented by a force sensor that was giving a continuous measure of the tightness in coupling.[2]

Evaluation

Evaluation is an intrinsic part of the basic design process. The various design solutions were readily compared through direct object manipulation, and group discussion made it possible to develop a consensus on reported phenomena and to highlight possible problems [46]. For example, the degree of expressiveness afforded by the sonic object was such that interactions turned playful, performative, and even extreme, thus challenging the robustness of the prototype.

4.6   Conclusions

We have presented a collection of methods and tools ranging from general to specific, wide to narrow, easy to laborious, and so on. These methods can obviously be used together, and in fact in our teaching practice this is usually the case.

For example, the strengths and challenges of the vocal sketching and sonic overlay show that they are to some extent complementary: vocal sketching is quick, low-fi, and transient; sonic overlay is slower to make and higher in finesse. A good option is to combine them. In our teaching, students use vocal sketching in a group brainstorming session to reach initial ideas for their sonic projects. They then present their initial design ideas in class using their voices. This performance is captured by one of the group members on camera. After this activity, students are asked to use the vocal sketch and captured video as a base for a more elaborate sonic design to be presented in the next lesson. The students overlay the video with new sound options, creating more elaborate sound designs to be presented and discussed in class. In this process, the experience and sensibility acquired during training sessions such as those described in section 4.2 are highly valuable. A natural next step is to use the video prototype as a reference and guide for the implementation of a working prototype.

One of the big challenges in creating a framework for the pedagogy of sonic interaction design is the breadth of contexts in which this topic is taught and applied. General design programs may want to introduce sound design in a short workshop; graphic design programs may strive to equip students with tools for introducing responsive sound to visual interfaces; product design programs may search for ways to make students aware of the potential of sonic feedback in digital products; and interaction design programs may require a more complete set of skills for the sonic domain. From the "other side" of the educational map, sound design programs may search for ways to help students move from the design of fixed sounds to the design of responsive sound; composition and computer music programs may want to introduce students to this potential field of application for their skills; computer science programs with a focus on human-computer interaction may want to create awareness to this design topic, and the list goes on. Each of these disciplines has a different set of terms, different needs, different skills, a different angle. It is our hope that this attempt to group together an initial collection of teaching methods can contribute to a sharing of knowledge between educators and to the further development of awareness of sonic interaction design.

Notes

1. Video Prototyping, Copenhagen Institute of Interaction Design, http://ciid.dk/education/portfolio/idp11/courses/video-prototyping/overview/.

2. Basic sonic interaction design Web site, http://soundobject.org/BasicSID/.

References

1. Albers, J. (2006). *Interaction of color*. New Haven, CT: Yale University Press. (Original edition published 1963).

2. Anceschi, G. (2006). Basic design, fondamenta del design. In G. Anceschi, M. Botta, & M. A. Garito (Eds.), *L'ambiente dell'apprendimento—Web design e processi cognitivi* (pp. 57–67). Milan: McGraw-Hill.

3. Findeli, A. (2001). Rethinking design education for the 21st century: Theoretical, methodological, and ethical discussion. *Design Issues, 17*(1), 5–17.

4. Lupton, E., & Phillips, J. (2008). *Graphic design: The new basics*. New York: Princeton Architectural Press.

5. Svanaes, D. (2000). *Understanding interactivity: Steps to a phenomenology of human-computer interaction*. PhD thesis, NTNU, Computer Science Department, Trondheim, Norway.

6. Hallnäs, L., & Redström, J. (2002). From use to presence: On the expressions and aesthetics of everyday computational things [TOCHI]. *ACM Transactions on Computer-Human Interaction, 9*(2), 106–124.

7. Schaal, S., Ijspeert, A., & Billard, A. (2003). Computational approaches to motor learning by imitation. *Philosophical Transactions of the Royal Society of London. Series B, Biological Sciences, 358*(1431), 537–547.

8. Franinović, K. (2009). Toward basic interaction design. *Elisava Temes de Disseny Journal* (special issue).

9. Essl, G., & O'Modhrain, S. (2006). An enactive approach to the design of new tangible musical instruments. *Organised Sound, 11*(3), 285–296.

10. Sinico, M. (2008). Demonstration in experimental phenomenology: How to bring out perceptual laws. *Theory & Psychology*, *18*(6), 853.

11. Vicario, G. B. (1993). On experimental phenomenology. In S. C. Masin (Ed.), *Foundations of perceptual theory* (pp. 197–219). Amsterdam: Elsevier Science Publishers.

12. Bozzi, P. (1978). L'interosservazione come metodo per la fenomenologia sperimentale. *Giornale Italiano di Psicologia*, *5*, 229–239.

13. Press, P. A., Lupton, E., & Miller, J. (Eds.). (1996). *ABC's of the Bauhaus: The Bauhaus and design theory*. New York: Princeton Architectural Press.

14. Franinović, K., Gaye, L., & Behrendt, F. (2008). Exploring sonic interaction with artifacts in everyday contexts, In *Proceedings of the 14th international conference on auditory display*.

15. Schafer, R. M. (1977). *The tuning of the world*. New York: Random House.

16. Beck, A. (1997). *Radio acting*. London: A & C Black.

17. Pauletto, S., Hug, D., & Luckhurst, S. B. M. (2009). Integrating theatrical strategies into sonic interaction design. In *Proceedings of the audio mostly conference*, Glasgow, UK.

18. Lederman, S., & Klatzky, R. (1990). Haptic classification of common objects: Knowledge-driven exploration. *Cognitive Psychology*, *22*(4), 421–459.

19. Houde, S., & Hill, C. (1997). What do prototypes prototype. In M. G. Helander, T. K. Landauer, & P. V. Prabhu (Eds.), *Handbook of human-computer interaction* (vol. 2, pp. 367–381). Amsterdam: Elsevier/North-Holland.

20. Newman, F. (2004). *Mouth sounds*. New York: Workman Publishing.

21. Ekman, I., & Rinott, M. (2010). Using vocal sketching for designing sonic interactions. In *Proceedings of the 8th ACM conference on designing interactive systems* (pp. 123–131). New York: ACM.

22. Burns, C., Dishman, E., Johnson, B., & Verplank, B. (1995). *Min(d)ing future contexts for scenario-based interaction design*. Palo Alto, CA: BayCHI/Informance.

23. Buchenau, M., & Suri, J. (2000). Experience prototyping. In *Proceedings of the 3rd conference on designing interactive systems: Processes, practices, methods, and techniques* (pp. 424–433). New York: ACM.

24. Dessein, A., & Lemaitre, G. (2009). Free classification of vocal imitations of everyday sounds. In *Sound and music computing* (pp. 213–218). Porto, Portugal: SMC.

25. Spence, R., & Apperley, M. (1982). Data base navigation: An office environment for the professional. *Behaviour & Information Technology*, *1*(1), 43–54.

26. Mackay, W., Ratzer, A., & Janecek, P. (2000). Video artifacts for design: Bridging the gap between abstraction and detail. In *Proceedings of the 3rd conference on designing interactive systems: Processes, practices, methods, and techniques* (p. 82). New York: ACM.

27. Curtis, G., & Vertelney, L. (1990). Storyboards and sketch prototypes for rapid interface visualization. *Tutorial notes from CHI conference on human factors in computing systems*, Seattle, WA.

28. Kolmos, A., Krogh, L., & Fink, F. (2004). *The Aalborg PBL model: Progress, diversity and challenges*. Aalborg, DK: Aalborg University Press.

29. Schultz, N., & Christensen, H. (2004). Seven-step problem-based learning in an interaction design course. *European Journal of Engineering Education, 29*(4), 533–541.

30. Kimmons, J., & Spruiell, P. (2005). Using problem-based learning in a multidisciplinary setting. *Clothing & Textiles Research Journal, 23*(4), 385.

31. Larsen, L., Andersen, S., Fink, F., & Granum, E. (2009). Teaching HCI to engineering students using problem based learning. In *Interact workshop of IFIP WG*, vol. 13.

32. Hug, D. (2010). Investigating narrative and performative sound design strategies for interactive commodities. In S. Ystad, M. Aramaki, R. Kronland-Martinet, & K. Jensen (Eds.), *Auditory display*, vol. 5954 of *Lecture notes in computer science* (pp. 12–40). Berlin: Springer.

33. Biocca, F., Harms, C., & Burgoon, J. K. (2003). Toward a more robust theory and measure of social presence: Review and suggested criteria. *Presence: Teleoperators and Virtual Environments, 12*(5), 456–480.

34. Polotti, P., & Rocchesso, D. (Eds.). (2008). *Sound to sense—sense to sound: A state of the art in sound and music computing*. Berlin: Logos Verlag.

35. Serra, X., Bresin, R., & Camurri, A. (2007). Sound and music computing: Challenges and strategies. *Journal of New Music Research, 36*(3), 185–190.

36. Widmer, G., Rocchesso, D., Välimäki, V., Erkut, C., Gouyon, F., Pressnitzer, D., et al. (2007). Sound and music computing: Research trends and some key issues. *Journal of New Music Research, 36*(3), 169–184.

37. Roads, C. (1996). *The computer music tutorial*. Cambridge, MA: MIT Press.

38. Verma, T., & Meng, T. (2000). Extending spectral modeling synthesis with transient modeling synthesis. *Computer Music Journal, 24*(2), 47–59.

39. Strobl, G., Eckel, G., Rocchesso, D., & le Grazie, S. (2006). Sound texture modeling: A survey. In *Proceedings of the sound and music computing conference*, Marseille, France.

40. Houix, O., Lemaitre, G., Misdariis, N., Susini, P., Franinović, K., Hug, D., et al. (2007). *Everyday sound classification: Sound perception, interaction and synthesis: Deliverable of project closed*. Paris: IRCAM.

41. Bernardini, N., & Rocchesso, D. (2002). Making sounds with numbers: A tutorial on music software dedicated to digital audio. *Journal of New Music Research, 31*, 141–151.

42. Farnell, A. (2010). *Designing sound*. Cambridge, MA: MIT Press.

43. Buxton, B. (2007). *Sketching user experiences: Getting the design right and the right design.* Waltham, MA: Morgan Kaufmann.

44. Misra, A., Wang, G., & Cook, P. (2009). TAPESTREA: A new way to design sound. In *Proceedings of the seventeenth ACM international conference on multimedia* (pp. 1033–1036). New York: ACM.

45. Delle Monache, S., Polotti, P., & Rocchesso, D. (2010). A toolkit for explorations in sonic interaction design. In *AM '10: Proceedings of the 5th audio mostly conference* (pp. 1–7). New York: ACM.

46. Rocchesso, D., Polotti, P., & Delle Monache, S. (2009). Designing continuous sonic interaction. *International Journal of Design, 3,* 55–65.

5 Perceptual Evaluation of Sound-Producing Objects

Bruno L. Giordano, Patrick Susini, and Roberto Bresin

The experimental study of sonic interactions can elucidate one of the most important aspects of the design process: "How should the sonic interaction be structured to produce a target perceptual result or to induce a specific motor behavior of the user?" Very similar questions are the object of basic research on the human processing of sensory events (e.g., "What is the perceptual effect of this sound stimulus?"), and have spurred the development of a large number of experimental methods. This chapter is meant as an introductory guide to the behavioral methods for the experimental study of complex sound events and sonic interactions. Table 5.1 reports a list of design questions that can be answered with each of the presented methods. Throughout this chapter, we reference a number of empirical studies of complex and naturalistic sounds based on the described methods. Interested readers can find in these studies more detailed descriptions of the various behavioral paradigms.

Studying sonic interactions in the laboratory implies focusing on conditions where the action or motor behavior of the participant influences the properties of the presented sounds. This generic definition encompasses a large number of everyday events: a sonic interaction can indeed be as simple as the playback of a sound following a button press (e.g., touch tones of a mobile phone). Sonic interactions can be tentatively organized along a continuum of complexity according to the number of sound properties that can be modified by a change in the motor behavior of the user (see figure 5.1). The generation of touch tones lies at one extreme of this continuum because the only property of a sound that can be modified by the user is its presence or absence (the same sound will be played back independently of large variations in the force exerted on the key). Examples of complex sonic interactions, in order of increasing complexity are the turning of a volume knob in a sound amplification system, the striking of an object with a hammer (e.g., Giordano, Avanzini, Wanderley, & McAdams [37]) and the crunching of potato chips (Zampini & Spence [154]), up to the perhaps most complex type of sonic interaction—conducting a symphonic

Table 5.1
Examples of questions answered with the methods described in this chapter

Section and method	Examples of answered questions
5.1: Psychophysical methods	Can the user perceive each of the configurations of a sonic interaction?
	Can the user differentiate between configurations?
5.2: Identification and categorization	What naturalistic object is recognized in each of the configurations?
	What emotional category is recognized in a sonic artifact?
5.3: Scaling and rating	How does perceived effort vary between sonic feedbacks for robotic surgery applications?
	How should the user-controlled gain for sound level vary so as to produce a linear increase in perceived loudness?
5.4: Dissimilarity estimation	Which properties of a complex sonic interaction are most relevant to the user?
	Do different individuals focus on different attributes of the sensory events?
5.5: Sorting	How many categories of perceived materials can a sound synthesis algorithm reproduce?
	What is the most typical configuration for each of the material categories?
5.6: Verbalization	Which words capture the semantic correlates of a sonic interaction?
	What are the individual interactive strategies? Are there problems in the prototype design?
5.7: Semantic differential	Which configuration has the highest aesthetic and functional value?
	How do preference, perceived sound brightness, and perceived efficiency covary for these particular sonic interactions?
5.8: Preference estimation	Which configuration has the highest aesthetic and functional value?
	Which configuration is the least annoying?
5.9: Continuous evaluation	Do users' gestures map onto changes in the perceptual attributes of the sonic events?
	How does the emotional response to a complex sound vary in time?
5.10: Multisensory contexts	What influences most strongly preference for cars? The sound of its doors closing or their felt weight?
	Do sonic feedbacks significantly shorten the time it takes to park a car?
5.11: Measurement of acoustical information	What sound properties should be manipulated to induce a target perceptual result (e.g., maximize preference)?
5.12: Motion capture	How do we use our body in interaction with a sonic artifact?
	How do gestures and artifacts mutually influence a sonic interaction?

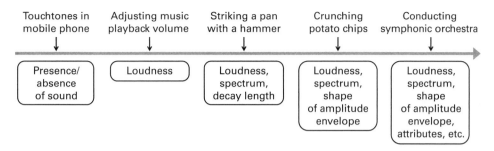

Figure 5.1
Variable complexity of everyday sonic interactions. With more complex interactions, changes in the motor behavior of the user lead to changes in a higher number of properties of the sound signal (from left to right).

orchestra (e.g., Kolesnik & Wanderley [74]). Importantly, a large number of experimental methods can be adopted to study both simple and complex sonic interactions. For example, participants in the study by Giordano, Rocchesso, and McAdams [41] triggered with the click of a mouse the playback of sounds recorded by striking objects of different hardness, whereas participants in the study by Lederman [85] actively generated sounds by scraping a rough surface with their fingers. Nonetheless, both studies adopted the same method, ratings, to measure the perceived properties of the sound-generating objects (hardness and surface roughness, respectively).

Another important distinction between studies of sonic interactions is the type of variable relevant to the experimenter. To make this distinction clear, it is helpful to summarize the various stages involved in the interactive production of sounds and in their perception (see figure 5.2). In general, a sonic interaction begins with a motor behavior or action carried out on a mechanical system, a sound-generating object (e.g., slapping the membrane of a bongo drum with the hand). The action-induced displacements of the components of the mechanical system will ultimately result in the production of a sound, which constitutes a source of acoustical information for the listener. At the same time, the motor behavior itself and the sound-generating object will produce information for additional sensory systems: kinesthetic, tactile, and visual. All these types of sensory information will then trigger various physiological and neural processes, resulting in conscious sensations, perceptions, and cognitions. Eventually, the processing of sensory information will feed back into the planning and control of further sound-generating actions.

The experimental study of sonic interactions can thus focus on four different types of variables: (1) quantitative measures of the motor behavior, as frequently measured

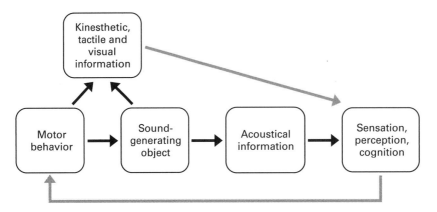

Figure 5.2
Chain of events and processing stages involved in the perception and production of interactive
sonic events. Arrows symbolize causal connections.

with motion capture systems (e.g., [73]); (2) measures of the properties of the sound-
generating system (e.g., the movement of the hammers and keys of a piano). Section
5.12 reviews a number of studies based on the measurement of the motor behavior
and of the behavior of sound-generating systems in interactive contexts. (3) A third
type of variable measures the properties of the stimuli impinging on the sensory
systems (e.g., properties of the sound signals). Section 5.11 illustrates the main
approaches that can be used to establish what properties of the stimulation, specifically of sound stimuli, affect the perceptual responses of experiment participants.
Section 5.10 illustrates a number of experimental paradigms for comparing the perceptual effects of information from different sensory modalities. (4) A final category
of variables aims at quantifying the sensory, perceptual, and cognitive responses to
the sensory stimulation. In contrast with the previously described components of the
sonic interaction chain, these variables are not measured directly, but are commonly
inferred from the responses of experiment participants in a variety of behavioral tasks
(e.g., "rate the hardness of the object you are manipulating," see section 5.3). The
majority of the methods presented in this chapter are designed just for this purpose
(table 5.1).

5.1 Psychophysical Methods: Detection, Discrimination, and Equivalence

The psychophysical methods presented in this section make it possible to answer
basic questions concerning the user experience: Is a particular attribute of the sonic

interaction perceivable? Are two different settings of a designed sonic interaction perceptually equivalent? Throughout the section, we focus on a simple attribute of the sonic interaction: the loudness of the sound. This example could be easily translated to a variety of sonic interactions such as the perceived loudness of sound effects in a videogaming context.

Psychophysics is the study of the mapping from physical attributes of the stimuli (e.g., sound level), to attributes of the corresponding sensations (e.g., loudness; see Gescheider [35] for an excellent handbook of psychophysics methodology). Classical psychophysical methods are often concerned with the measurement of two sensory quantities: the absolute threshold, which is the smallest or highest detectable value of a stimulus attribute (e.g., the lowest detectable sound level), and the differential threshold, which is the smallest discriminable difference in a stimulus attribute (e.g., the smallest discriminable difference in level).[1]

The *method of constant stimuli* has been widely used for measuring both absolute and differential thresholds. When measuring an absolute threshold (e.g., absolute threshold for sound level), participants are repeatedly presented with a small set of stimuli ranging from hardly perceivable (e.g., very low level) to clearly perceivable. Participants are asked if they detect the stimulus or not. When measuring a differential threshold, on each trial participants are presented with two stimuli: a standard stimulus whose properties remain constant across all trials (e.g., a 60 dB SPL sound) and a comparison stimulus that varies from trial to trial (e.g., one sound from a set ranging from barely weaker to barely louder than the comparison stimulus). Participants indicate for which of the paired stimuli the target attribute has the largest or lowest value (e.g., which of the two stimuli is louder). Figure 5.3 shows the likely outcome of a constant stimuli experiment: the function relating response probabilities to stimulus values is called the *psychometric function*. The absolute threshold can thus be defined as the stimulus intensity perceived 50 percent of the time; the differential threshold can instead be measured as the average of the stimulus values judged greater than the standard 25 percent and 75 percent of the time. More often than not, none of the presented stimuli is associated with the exact response probabilities used to calculate the thresholds. In part for this reason, and in part for the need to integrate experimental data across all the investigated stimuli, a psychometric function is usually fit to the observed response probabilities for all stimuli (e.g., a cumulative normal distribution [152, 153]), and the thresholds are estimated from the parameters of the fitted function.

The method of constant stimuli provides another important measure of sensation: the *point of subjective equality* (PSE). The paradigm used for this purpose is the same as

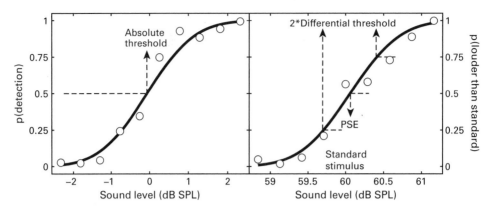

Figure 5.3

Hypothetical outcome of a constant-stimuli experiment for the measurement of the absolute and differential threshold for sound level (left and right panel, respectively). Circles show the response probabilities measured for each of the stimuli; solid lines show the psychometric function fitted to the experimental data and used to calculate the threshold measures. PSE, point of subjective equality. The differential threshold corresponds to half the distance between the stimuli associated to a response probability of 0.25 and 0.75, respectively. Note the slight misalignment of the PSE relative to the standard stimulus (60 dB SPL).

that used for measuring the differential threshold. In the level discrimination experiment described above, the PSE equals the value of the comparison stimulus judged louder than the standard 50 percent of the time (see figure 5.3). It should be noted that when the standard and comparison stimuli differ only in intensity, a PSE that significantly deviates from the intensity of the standard stimulus is a sign of imperfections in the experimental design [35, pp. 52–53]. A more interesting case of PSE measurement is the study by Robinson and Dadson [119]. In this experiment, the comparison stimulus was a 1-kilohertz tone of variable intensity, and the standard stimulus was a fixed-intensity sound of different frequency (e.g., 2 kilohertz). Participants judged whether the comparison or the standard stimulus was louder. The PSE was defined as the intensity of the comparison stimulus judged louder than the standard 50 percent of the time and measured the change in loudness brought by the frequency difference between the standard and comparison stimulus.

One shortcoming of the method of constant stimuli is its susceptibility to response biases. In an experiment designed to measure the absolute threshold for sound level, a very cautious participant might, for example, answer "I do not hear a sound" 80 percent of the time even though level is above threshold for only 50 percent of the trials. The framework of *signal detection theory* (SDT [48]) remedies this problem by computing bias-independent measures of sensitivity. The reader is referred to the works of McNicol [104] and of MacMillan and Creelman [95] for recent handbooks on SDT.

A second potential shortcoming of the method of constant stimuli is a low engagement of the experimental participant, who might quickly grow bored with the repetitive task. The *adjustment method* is a less tedious alternative that measures absolute and differential thresholds and PSEs. Accordingly, participants actively adjust the value of a stimulus property until a desired sensory result is achieved. For example, the participant might use a volume knob to adjust the intensity of a 1-kilohertz comparison stimulus so that it is perceived as loud as a 200-hertz standard stimulus of fixed intensity, thus producing an estimate of the PSE. The price of the adjustment method is an increase in the noise of the experimental data. In the PSE measurement experiment, imperfections in the manual control of the volume knob can, for example, produce a reduced accuracy of the adjustment response.

A final shortcoming of the method of constant stimuli is its low efficiency, which is the fact that many answers are required for each of the stimuli, even for the least informative ones, to yield reliable estimates of the parameters of the psychometric function. For example, when measuring the absolute threshold, the experimenter might be interested in a single point of the psychometric function associated with a response probability of 50 percent. However, the method of constant stimuli would also require collecting many responses for stimuli that are far from the absolute threshold. *Adaptive methods* are a more efficient alternative to the method of constant stimuli because the presented stimuli are concentrated around the point of interest on the psychometric function [34, 88, 143]. This is achieved by determining the level of the stimulus presented at a given trial based on the responses given at the preceding trials. The simplest and earliest example of adaptive method is the staircase or von Békésy tracking method [18, 146]. In an experiment for the measurement of the absolute threshold for sound intensity, the participant is asked to tell whether he hears the presented sound or not. Importantly, if the participant reports a detection at one trial, the intensity at the succeeding trial is decreased, whereas the intensity at the succeeding trial is increased if no detection has been reported (see figure 5.4). This simple rule for determining the value of the presented stimuli allows the experimenter to present sound intensities that are close to the absolute threshold. Alternative methods for determining the stimulus values provide target points on the psychometric function other than the 50 percent [88].

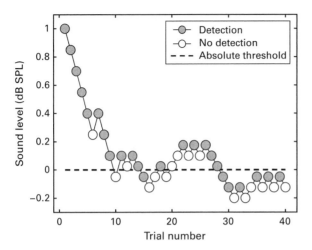

Figure 5.4

Hypothetical data from an experiment for the measurement of the absolute threshold for sound level based on the von Békésy tracking method. The dashed line shows the hypothetical absolute threshold. The level is decreased or increased by a constant step size after a detection or a no-detection response, respectively. Note the slightly larger step size for the initial trials meant to accelerate the convergence around the absolute threshold.

5.2 Identification and Categorization

Within a sonic interaction design context, identification and categorization methods can be adopted to assess the ecological interpretation of the displays: "Does this sonic interaction correspond to the scraping of a metallic surface or of a piece of styrofoam?" For this reason, these methods are among the most frequently adopted in the study of the perception of naturalistic sound events [39, 55, 89, 112, 118, 148]. In general, identification and categorization experiments allow the designer to assess the mapping from a set of display configurations to a set of meaningful verbal labels (e.g., identi-fication of naturalistic events but also mapping of a stimulus set onto emotion-related categories such as "sad" or "happy").

During an identification/categorization experiment, participants are asked to assign each of the stimuli to one among a set of prespecified verbal labels (e.g., "Is this sound a violin, a guitar, or a flute tone?"). Whereas in an identification experiment the number of response categories equals the number of stimuli (e.g., "violin," "guitar" and "flute" in an experiment with three stimuli, one violin, one guitar, and one flute tone), with categorization the response categories are fewer than the number of stimuli (e.g., multiple violin, guitar, and flute tones of different pitch). Whereas in a classical psychophysical experiment stimuli are often highly controlled and vary along a very low number of dimensions (e.g., only sound level in the absolute-threshold example reported in section 5.1), identification and categorization experiments can be carried out with complex stimuli that differ along a large number of properties.

Data analysis can focus either on measures of performance (e.g., "Which among these tones has been identified correctly most often?") or on the raw probabilities of assigning each stimulus to each of the response categories (e.g., "Which among these tones has been most frequently identified as a violin tone?"). A third analysis option is to adopt SDT methods to compute measures of the sensory distance between response categories (e.g., "Is the sensory distance between violin and guitar tones shorter than that between violin and flute tones?") independent of response biases (e.g., in a categorization experiment with an equal number of violin, guitar, and flute tones, the tendency to use the response category "violin" more often than any other). This latter analysis approach was adopted by Giordano et al. [42] in an experiment investigating the effects of multisensory information (auditory, tactile, kinesthetic) on the identification of walked-upon materials. Measures of bias-independent sensory distance among walked-upon materials were computed within the framework of *general recognition theory* (GRT, see figure 5.5 [1, 2]). Unlike classical SDT methods, GRT makes it possible to deal with experiments in which stimuli vary along multiple properties, a frequent case when one is investigating naturalistic stimuli, and in which participants are allowed more than two response categories. Another advantage of GRT is that it considers within a single theoretical and analytical framework data from a variety of methods: identification, categorization but also dissimilarity (section 5.4), and preference (section 5.8).

5.3   Scaling and Rating

Perceptions can be organized in several different ways. In section 5.2 we saw that perceptions can be mapped onto discrete categories, each described by a verbal label. In this section we present a number of methods for measuring ordered relations among perceptions (e.g., the touch sensation of silk is smoother than that of wool). The concept of *sensory continuum* is central to these methods. Sensory continua are the result of a mental computation that allows us to order stimuli relative to a specific attribute and can be conceptualized as directional lines in a cognitive space (e.g., the sound of a flying bee is closer to the origin of the sensory continuum for loudness than the sound of a jet plane). Within a sonic-interaction design context, the methods presented in this section allow the answering of questions such as: "How does

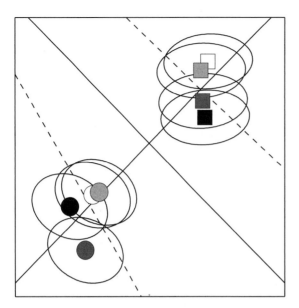

Figure 5.5
General recognition theory analysis of data for the multisensory identification of walked-upon materials [42]. The identification responses modeled in this example were collected after participants walked blindfolded on one of eight different materials: four solids (e.g., ceramic; circle symbols) and four aggregates (e.g., small gravel; square symbols). The sensory representation of each material is modeled as a normal distribution in a two-dimensional space of sensory effects. The average of each normal distribution (filled symbols) denotes the most frequent sensory effects for a particular walking ground; the oval surrounding each filled symbol (0.05 equal-likelihood contour) gives an approximate representation of the extent to which the sensory effects for a given material vary across repeated presentations. Four lines (decision boundaries) divide the two-dimensional space into eight regions: sensory effects that fall within the same region receive the same identification response (e.g., sensory effects above the top-left-to-bottom-right diagonal line are identified as aggregate materials; sensory effects below the same diagonal are identified as solid materials). The measure of the ability to differentiate between two walking grounds independent of response biases is approximated in this figure by the area of overlap between their respective normal distributions.

perceived effort vary between sonic feedbacks for robotic surgery applications?" and "How should the user-controlled gain for sound level vary to produce a linear increase in perceived loudness?"

Scaling methods make it possible to measure the *psychophysical function*, that is, the function that relates physical attributes of the stimuli to sensory continua (e.g., sound level and perceived loudness).[2] We can distinguish between two families of scaling methods: partition and ratio scaling (see Gescheider [35] for a thorough presentation).

In a *partition scaling* experiment, participants are asked to divide a target physical dimension (e.g., sound level) into perceptually equivalent intervals. For example, participants are presented with two stimuli that bracket the range of variation of the physical dimension of interest (e.g., a low- and a high-level sound) and are asked to divide the sensory continuum of interest (e.g., loudness) into a prespecified number of equal sensory intervals (e.g., to adjust the level of this sound so that the difference between its loudness and that of the low-level sound is equal to the difference between its loudness and that of the high-level sound). Similar information can be collected using the *category scaling* method. Accordingly, participants are presented with a set of stimuli that differ along the physical dimension of interest and are asked to assign them to one of a set of prespecified categories, each representing a different level of the sensory quantity and each bracketing an equal range of sensory magnitudes (e.g., to assign each of these fifty differently loud sounds to five classes of progressively increasing loudness, with each class bracketing an equal range of loudness variation).

The methods of *ratio scaling* and *magnitude scaling* produce a mapping from the sensory continuum to a numeric continuum. Both methods have one *estimation* variant and one *production* variant. In a ratio estimation task, participants estimate numerically the ratio between the sensory magnitude of two stimuli (e.g., "What is the ratio between the loudness of sound A and that of sound B?"). In a ratio production task, participants adjust the physical properties of a stimulus so that the ratio of its sensory magnitude to that of a reference stimulus equals a prespecified number (e.g., "Adjust the level of sound A so that its loudness is one-fourth of that of the reference sound B"). With magnitude estimation, participants assign a number to the sensory magnitude of the first presented stimulus and estimate numerically the sensory magnitude of subsequent stimuli based on the number assigned to the first stimulus (e.g., "Given that you estimated the loudness of sound A to equal 100, what number quantifies the loudness of sound B?"). With magnitude production, the sensory magnitude for a reference stimulus is assigned a numerical value, and the participant is asked to manipulate the physical properties of a new stimulus so that its sensory magnitude equals a given number (e.g., "The loudness of this reference sound equals 20; adjust the level of sound B so that its loudness equals 50"). Results from estimation and production scaling methods are known to diverge systematically because participants tend to avoid extreme values along the response continuum (the numerical and physical continuum, for estimation and production methods, respectively). This systematic divergence is termed regression bias [136]. The regression bias can be dealt with by estimating an unbiased psychophysical function defined as the "average" of the functions obtained with estimation and production methods.

The *rating method* can be conceived as a variant of the magnitude estimation method. Accordingly, participants estimate the sensory magnitude by choosing an integer number within a prespecified range (e.g., "Rate the hardness of a hammer used to generate this sound by using the integer numbers from 1 to 7" [33]). Before any experimental data are collected, it is good practice to allow participants to establish a mapping between the range of variation of the target sensory property within the set of stimuli and the response scale. To illustrate the need of this additional step, we can hypothesize that participants are asked to rate loudness on a scale from 1 to 10, and that participants are not familiarized with the experimental set of stimuli before rating each of them. A participant at the beginning of this experiment might, for example, rate the loudness of one stimulus using the highest allowed response (10), and find out that subsequent stimuli are louder than the previously heard ones. In such a case, the ratings of the participant will not accurately measure the perceived loudness. A variant of the rating task uses a nonnumerical continuous response scale. For example, in an experiment on the estimation of the hardness of struck sounding objects [41], participants rated hardness by moving a slider along a continuous scale marked "very soft" and "very hard" at the endpoints. This approach circumvents eventual response biases originating from the tendency to use particular numbers more frequently than others (e.g., multiples of 5 in a 1 to 100 scale) and maximizes the amount of experimental information (e.g., the number of different rating answers that a participant can give is usually much larger with a slider than with a numerical scale including only integer numbers).

The method of *cross-modality matching* finally relies on the comparison of sensory magnitudes from different modalities. With this method, participants adjust the properties of a comparison stimulus so that the target sensory magnitude it evokes matches the magnitude of a target sensory attribute of the reference stimulus presented in a different modality. This method was, for example, used by Grassi [47] to investigate the estimation of the size of a ball from the sound it makes when bouncing on a plate.

On each trial, the reference stimulus was a bouncing sound. Participants were asked to estimate the size of the bouncing ball by manipulating the diameter of a circle presented on a computer screen. Focusing on the design of sonic interactions, cross-modality matching could, for example, be adopted to calibrate the sensory properties of simultaneous auditory and tactile displays.

5.4   Dissimilarity Ratings

Many naturalistic sonic interactions generate rich sensory signals from various modalities (e.g., walking on various gravels or grass; crumpling a sheet of paper). Similarly, sonic interactions designed in the laboratory can involve many different parameters that control the sensory information delivered to the user. Within such rich domains it is often unclear what sensory properties dominate perceptions (e.g., which sensory properties of a naturalistic event; which synthesis parameters of a designed sonic interaction). The method of *dissimilarity ratings*, also known as paired comparisons method, sheds light on this issue and, in combination with particular data-analysis methods, makes it possible to answer questions such as: "Which properties of a complex sonic interaction are most relevant for the user?" and "Do different individuals focus on different properties of the sensory event?" Dissimilarity ratings have been frequently adopted to characterize the perception of complex auditory stimuli such as musical sounds [51, 75, 102] or environmental sounds [57, 105].

The goal of a dissimilarity ratings experiment is to measure the perceptual distance between stimuli: very similar/dissimilar stimuli are separated by a short/large perceptual distance. The structure of a dissimilarity ratings experiment is very similar to that of a standard ratings experiment. On each, participants are presented with two stimuli and are asked to rate how dissimilar they are on a "very similar" to "very dissimilar" scale. Throughout the experiment, participants rate the dissimilarity of each possible pair of stimuli.[3] so that each of the participants yield a matrix of between-stimulus perceptual distances.

The most common strategy for the analysis of dissimilarity ratings relies on the mathematical model of multidimensional scaling (MDS [9, 20])[4]. In general, MDS represents the dissimilarity ratings as the distance between the stimuli in a Euclidean space with a given number of dimensions (see figure 5.6).[5,6] Pairs of stimuli rated as very dissimilar are also far apart in the MDS space; stimuli rated as very similar are very close in the MDS space.

A notable limitation of classic MDS arises from the fact that it yields *rotationally invariant* solutions. A rotationally invariant solution can be rotated arbitrarily without

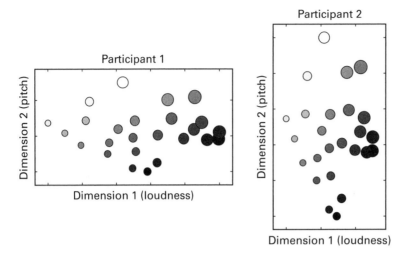

Figure 5.6

Weighted Euclidean model for a hypothetical dissimilarity ratings experiment carried out with 25 sounds differing in loudness (larger symbols denote louder sounds) and pitch (brighter symbols denote higher-pitched sounds). The MDS model in this example has two dimensions: dimension 1 is strongly correlated with the loudness of the stimuli; dimension 2 is strongly correlated with pitch. Participant 1 (left) carries out the task by focusing on loudness differences, hence the wider spread of the stimuli along the loudness-related dimension. Participant 2 (right) focuses on pitch differences, hence the larger spread of the stimuli along the pitch-related dimension.

affecting the extent to which the MDS distances accurately reproduce the input dissimilarities (e.g., a geographic map can be arbitrarily rotated without affecting the extent to which it accurately represents the distances between cities). In this sense, the dimensions of classic MDS spaces, that is, the location of the stimuli along each of the dimensions, do not provide meaningful information about the perceptual structure of the stimuli. A second limitation of classic MDS algorithms is that they accept in input only one matrix of dissimilarities and are thus not appropriate for modeling interindividual differences. The weighted Euclidean MDS model (wMDS) is free of these limitations (e.g., INDSCAL [141]): it is capable of analyzing simultaneously dissimilarity matrices from multiple experiment participants, and yields psychologically meaningful dimensions.

As in classical MDS, wMDS represents dissimilarities as the distance between stimuli in a Euclidean space. Notably, wMDS assumes that each of the participants weights differently the dimensions of a common group space. The individual weights are multiplicative terms specific to each of the participants that modify the spread of the stimuli along the dimensions of the group space, and allow her to reconstruct a space of mental distances specific to the participant that better accounts for her perceptions (see figure 5.6). Importantly, the wMDS model is not rotationally invariant. For this reason, the dimensions of the group space can be taken as a model of the response criteria followed to carry out the dissimilarity ratings task. For example, stimulus features that strongly correlate with the location of the stimuli along a specific dimension of the wMDS model are likely to have been used by participants to estimate the between-stimulus dissimilarity (note that the correlation between a dimension of the wMDS model and a stimulus feature is not affected by the multiplication of the dimension by a participant-specific weight). Further, within the wMDS model the range of variation of a given dimension for each of the participants can be taken as a measure of the perceptual relevance of the dimension to the participant herself, that is, if for one participant the first dimension has a larger range of variation than the second dimension, she likely carried out the dissimilarity ratings task by focusing more on the feature that correlates with the first dimension than on the feature that correlates with the second dimension (see figure 5.6).

A very important aspect of the dissimilarity ratings method is that it does not constrain participants to focus on a specific property of the stimuli. Indeed, such a generic task as "rate the dissimilarity between these two stimuli" leaves the participant free to decide which stimulus properties most strongly affect dissimilarity. In this sense, dissimilarity rating is the most liberal among the experimental methods presented up to this point. Dissimilarity rating thus frees the experimenter from assumptions about the perceptual structure of the stimulus set: asking participants to scale a particular property of the stimuli (e.g., the size of a bouncing ball from its sound [47]) indeed comes with the hidden assumption that perceptions are organized along the rated property. Such an assumption can prove hard to be tested empirically (e.g., What is the object of perception when hearing the sound of a bouncing ball? Size, weight, or the acceleration at the moment of impact with a sound-generating object?). The liberal nature of dissimilarity ratings thus makes it a particularly good method for exploratory studies of complex perceptual domains (e.g., environmental sounds [57]), for which it might just not be possible to formulate a working hypothesis on what stimulus attributes are perceptually relevant (e.g., "Which stimulus attributes capture the perceptual difference between the sound of an airplane engine and that of a dog howling?"). For all these reasons, the quantification of perceptual distances can be considered the method of choice for the exploration of novel, complex perceptual domains (see also Borg & Groenen [9, pp. 9–11]).

We have seen that dissimilarity ratings, in conjunction with wMDS modeling, makes it possible to (1) discover perceptually relevant attributes of the stimuli; (2) establish hierarchies of perceptual relevance for stimulus attributes (e.g., loudness is more perceptually relevant than pitch); and (3) analyze interindividual differences for the perceptual relevance of stimulus attributes. From the applied perspective, the designer of sonic interactions might focus the modeling efforts on those properties of the interactive events that dominate the perceptions of the majority of the individuals.

5.5 Sorting

One of the goals for a designer of sonic interactions might be to establish a perceptually meaningful palette of presets for the parameters of complex audio-haptic synthesis algorithms. Palettes of presets are, for example, common in the control of digital color spaces (e.g., the user is able to select from various color presets rather than having to specify the RGB value corresponding to the desired color). Within the context of sonic interactions, the designer might, for example, plan to create a palette of presets for the control of the perceived material of virtual audio-haptic objects. Faced with this problem the designer needs to decide how many presets should be included in the palette and what parameter values should be assigned to each of them: "How many categories of perceived materials can the synthesis algorithm reproduce, and what is the most typical configuration for each of them?" Sorting methods make it possible to answer these questions and, in general, can be adopted to define meaningful categories of events within a perceptual domain. Further, sorting methods can be adopted as a more efficient although less reliable and accurate alternative to dissimilarity ratings (see section 5.4) [38].

With sorting experiments, participants are quite simply asked to create groups of stimuli. In an image-sorting experiment, each image might be printed on a card, and participants might be asked to create groups of cards. In a sound-sorting experiment, each sound might be represented by an icon on a computer screen, and participants might be asked to group the icons. One of the first decisions to be made when designing a sorting experiment how many groups of stimuli the participant should create. In the *free sorting* variant [121], the experimenter does not specify the number of groups and leaves the decision up to the participant. A free sorting task could, for example, be used to answer the materials-palette problem described at the beginning of this section (e.g., "group together stimuli generated with the same material; create as many groups you think are necessary"). With *constrained sorting* the number of groups is instead specified by the experimenter. Gingras, Lagrandeur-Ponce, Giordano, and McAdams [36] used constrained sorting to test the recognition of the individuality of music performers. In this experiment participants were presented with excerpts of organ-performance recordings from six different performers and were asked to create six different groups of excerpts played by the same individual.

A second important decision to be made when designing a sorting experiment concerns the criteria that participants should follow when creating the groups of stimuli. In the above examples participants receive specific instructions about the criteria they should follow to create the groups. When this is the case, sorting methods bear a strong resemblance to the categorization task described in section 5.2. The only major differences are indeed that categories are not labeled verbally by the experimenter (free and constrained sorting) and that the number of categories may not be specified (free sorting). Most often, however, participants are given the more generic instructions to create groups based on the similarity of the stimuli (e.g., for studies carried out with complex naturalistic sounds [8, 38, 53, 57]). When this is the case, the sorting task is the "categorization analogue" of dissimilarity ratings and shares with this method the ability to uncover the structure of perceptual domains within an assumption-free framework (see section 5.4). Similarity-based free sorting further allows one to measure the so-called basic level of categorization in a stimulus domain [120]. Sorting data, coding for the group membership of each of the stimuli, are frequently converted into co-occurrence data (see Coxon [21] for various data-analytic strategies for sorting data). For each possible pair of stimuli, the square co-occurrence matrix uses a binary variable to code whether stimuli have been assigned to the same group or not. Co-occurrences collected with a similarity-focused sorting are often considered as binary measures of dissimilarity: dissimilar stimuli do not belong to the same group (co-occurrence = 0), whereas similar stimuli do (co-occurrence = 1).

Sorting makes it possible to measure a full dissimilarity matrix in a shorter time than dissimilarity ratings [4, 38, 121]. For this reason sorting is often adopted to measure the dissimilarity of large sets of stimuli, or the dissimilarity of stimuli that might easily produce adaptation effects (e.g., taste stimuli [84]). The higher efficiency of sorting comes, however, with a price: dissimilarity data are less accurate, that is, less likely to reflect stimulus properties, and less reliable, that is, less likely to be replicated with a different group of participants [38]. Another drawback of sorting methods concerns the data-modeling aspect, in particular the extent to which MDS algorithms accurately model sorting dissimilarities. As explained above, each of the participants in a sorting experiment yields a binary dissimilarity matrix, the co-occurrence matrix.

Notably, MDS models of binary dissimilarities are known to present strong artifacts that prevent an accurate account of the input data [44]. As such, interindividual-differences MDS models of sorting data (e.g., wMDS, see section 5.4) are prone to significant errors. For this reason sorting data are less than ideal for measuring and modeling interindividual differences in perceived dissimilarity. It is thus advisable to carry out MDS analyses of sorting data by focusing on the co-occurrence matrix pooled across participants.

The *hierarchical sorting method* finally yields richer participant-specific data than free sorting (e.g., Rao & Katz [117]; see Giordano et al. [38] for various comparisons between sorting methods and dissimilarity ratings). In the agglomerative variant of this method, participants start from a condition where each of the stimuli is in a different group and, at each stage of the procedure, merge together the two most similar stimuli or groups of stimuli. The procedure is iterated until all stimuli are merged in the same group. Between-stimulus dissimilarity is estimated by the step of the merging procedure when two stimuli are first grouped (e.g., dissimilarity = 1 and 10 if stimuli A and B have been merged together at the first or tenth stage of the sorting procedure). With this method, each of the participants thus yields a square dissimilarity matrix with as many different values as the number of merging steps. Figure 5.7 shows the hierarchical sorting for one of the participants in an experiment carried out with environmental sounds [40] (also see other studies of sound stimuli based on the hierarchical sorting method [38, 61]). Note that in the first step of this hierarchical sorting experiment participants created 15 groups of similar stimuli. In other words, the hierarchical sorting started with a constrained sorting step (*truncated hierarchical sorting*; see Giordano et al. [38]).

5.6 Verbalization

Verbalization tasks can be used to explore qualitatively the perceptually relevant attributes of sound-interactive events: "Do users focus on the characteristic of the sound signal (e.g., this sound is very bright), on the characteristic of the source (e.g., this is the sound of a breaking glass), or on more abstract symbolic contents associated with the sound event (e.g., harmfulness for pieces of shattered glass)"? Focusing on interactive events, verbalizations might be collected during an interview where participants are shown the video recording of their own interaction with a sonic prototype (auto-confrontation interview). Analysis of the verbalizations will, for example, provide useful feedback on individual interactive strategies or on problems in the design of the prototype.

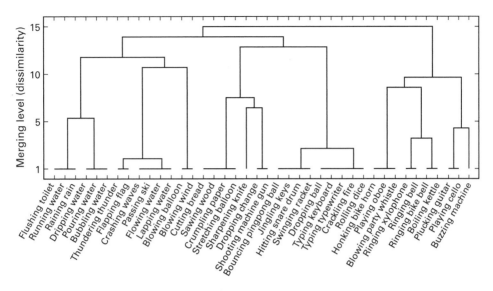

Figure 5.7
Truncated hierarchical sorting data for nonliving environmental sounds (data for one participant in the unbiased condition in Giordano et al. [40]). Participants initially created fifteen groups of similar sounds. At each subsequent step, participants merged together the two most similar sounds or groups of sounds. Similar sounds are assumed to be merged earlier than dissimilar sounds: the sorting step where two sounds are first merged yields an estimate of their dissimilarity.

Free verbalization is the simplest method for the collection of verbal data. Accordingly, participants are asked to describe in words the presented stimuli, with no restriction being imposed on the format of the response (e.g., "describe the sound you heard"). The high discovery potential of a free verbalization experiment comes at the price of a high complexity of data that can differ greatly across experiment participants. Simpler verbal data that require less sophisticated analysis strategies can be obtained using *constrained verbalization* methods that impose limits on the format of the verbal responses (e.g., "use one verb and one/two nouns to describe the sound-generating event" [40]). Verbalization methods can be used either to characterize stimuli presented in the laboratory [145] or as a method to sample the environment (verbalization of sounds as they are experienced in outside the laboratory [3]). Verbalization can be used as a method of its own or as a complement to a preceding task (e.g., free verbal description of the groups of sounds created in a sorting task, section 5.5) or as a strategy for designing further experiments (e.g., verbalization data can be used to design semantic-differential experiments, section 5.7).

Various software routines are available for the analysis of verbalization data. The software LEXICO [16] can be used to tabulate the terms used by participants according to their frequency of occurrence in the verbalization responses. The software STONE [115] can be used to carry out a lexical analysis that organizes the semantic fields among the different verbal descriptions. Methods of natural language processing can be used to carry out various analyses of the semantic content of the verbal descriptions. For example, latent semantic analysis techniques [83] can be used to compute measures of the semantic similarity of the verbal descriptions used to identify a sound-generating event [40]. Specific to research on complex sounds, methods for the scoring of the accuracy of verbal identifications of a sound-generating event have been detailed [3, 40, 98]. Further methods for the analysis of verbal response data have been developed by Nosulenko and Samoylenko [108]. Notably, the approach developed by Nosulenko and colleagues makes it possible to organize and encode verbal units at different levels: logical, perceptual, and semantic.

Verbalization tasks have been frequently used in the study of complex sounds. Peters [114] investigated the verbalization of synthetic sounds (e.g., pure tones, white noise), speech, music, and environmental sounds. Overall, terms describing the sensory properties of the sounds (e.g., high, low, soft, loud) were used more often than words related to objects, actions, and events involved in the generation of a sound. Nonetheless, only one-third of the participants used terms describing the sensory quality of the stimuli for real sounds. Similarly, in a study on the free identification of environmental sounds, Vanderveer [145] observed that listeners most often described the sound-generating event and, more specifically, the actions and objects involved, , and the context where the sound was generated. Faure [32] asked participants to freely describe pairs of musical sounds presented during a dissimilarity estimation experiment. Verbal responses belonged to one of three semantic categories: descriptors of the sound source (e.g., material, action), descriptors of the temporal evolution of a sound (e.g., attack, progression, resonance), and descriptors of sensory aspects of a sound (e.g., sharp, light, bright). Kyncl and Jiricek [82] investigated the free verbalization of vacuum cleaner sounds. Thirty-three pairs of semantic opposites were derived from the verbal responses, five of which were consistently judged as relevant to describe the vacuum cleaner sounds: fuzziness, atypicality, inefficiency, loudness, and pleasantness.

5.7 Semantic Differential

The *semantic differential* method can be conceived as a multidimensional extension of the ratings method (section 5.3). The main difference between the two methods indeed stands in the number of psychological or perceptual attributes simultaneously evaluated by the experiment participant: one for the latter, more than one for the former. The semantic differential method can, for example, be adopted to optimize a sound-interactive system according to multiple psychological attributes (e.g., "Which configuration has the highest aesthetic and functional attributes?") and to analyze the interdependence among multiple properties of sonic interactions (e.g., "How do preference, perceived sound brightness, and perceived efficiency covary?"). Various studies adopted the semantic differential to assess the multidimensional character of complex sound stimuli (e.g., Solomon [133] and von Bismarck [147] for early applications; and various sources for studies on musical timbre [71, 116, 134, 151]; for studies on environmental sounds [6, 72, 81, 155]; for studies of sounds generated with various human-made objects such as cars, vacuum cleaners, air conditioning systems, and refrigerators [5, 15, 65, 67, 70, 82, 131, 139]).

In the most popular variant of the semantic differential method [109, 110], participants rate each stimulus along several bipolar scales defined by opposing semantic descriptors (e.g., "soft" and "loud" or "pure" and "rich"). Each bipolar scale is usually divided into an odd number of intervals (e.g., seven). The task of the participant is thus to choose which of these intervals most appropriately describes the location of the stimulus along the continuum defined by the opposing semantic descriptors (e.g., this sound has a loudness of 5 along a soft-to-loud scale with seven intervals). As described for the ratings method, the rating scale does not need to be clearly divided into a predefined number of intervals (e.g., ratings along each bipolar scale can be collected using on-screen sliders or by marking with a pen a position along a line connecting the two opposing descriptors). The method of *verbal attribute magnitude estimation* (VAME) is a variant of the semantic differential method that might improve the interpretability of the results [59, p. 259]. With the VAME method, bipolar scales are not defined by opposing semantic descriptors but by one semantic descriptor and its negation (e.g., "not loud" and "loud").

Semantic differential studies are not restricted to particular attributes: they can be adopted to evaluate various properties of the stimuli such as sensory attributes (e.g., sound roughness), higher-level psychological attributes (e.g., pleasantness), or emotional attributes (e.g., dominance). As a result, this method can be a good choice for the study of previously unexplored perceptual domains. Two considerations are important concerning the design of a semantic differential experiment. First, in the absence of previous literature on the investigated stimuli, the choice of what semantic attributes should be considered can be based on a preliminary verbalization experiment (see section 5.6). For example, participants in the preliminary experiment might be presented with each of the stimuli and asked to describe verbally their most salient attributes. The semantic differential can thus be designed by considering those semantic attributes used by the majority of the participants in the verbalization experiments. A second design consideration concerns the number of attributes included in the semantic differential. It is likely that as the number of semantic attributes grows, participants are more likely to answer using response strategies meant to reduce the difficulty of the task. In particular, with an excessive number of semantic attributes, participants might start using response scales in a correlated manner (e.g., sounds that are rated as brighter are also rated as more pleasant) not because of a genuine association between the semantic attributes within the stimulus set (e.g., all feminine voices have a high pitch) but simply in order to minimize fatigue. To minimize these effects it is advisable to limit the number of semantic attributes to the minimum necessary. Independently of the origin of the correlation between different response scales, statistical methods such as factor analysis or principal components analysis can be adopted to reduce the raw semantic differential data into a set of independent dimensions of evaluation.

5.8 Preference

A sound designer often aims to improve the overall quality of sonic interaction experiences. To this purpose she might seek an answer to questions such as "Which configuration of a sonic feedback system do users prefer?" or, almost equivalently, "Which configuration is the least annoying?" Preference judgments have been often adopted in the applied sector to improve the design of a variety of products [54] (see Ellermeier & Daniel [27, 28] for studies of tire sounds and environmental sounds; Susini and colleagues, [139, 140] for studies of car sounds and air-conditioning noises; Lemaitre, Susini, Winsberg, McAdams, & Letinturier [87] for a study of car horns).

We can distinguish between two types of preference data: revealed and stated preferences [7]. Revealed preferences are usually derived from choice data collected in ecological conditions (e.g., product sales). One of the main disadvantages of revealed preferences is that the complete set of alternative choices is often unknown (e.g., data on the sales of Gibson Les Paul and of Fender Stratocaster guitars lack information about all of the brands and makes of electric guitars considered by the costumers of a musical instrument shop). In the following, we focus on stated preferences which are directly elicited from the participants and are not characterized by this important drawback of revealed preferences.

Preference data can be collected with various experimental methods [50], some of which have been described in previous sections. In a *paired preference comparison* task, participants are presented with all the possible pairs of stimuli, one at a time, and are asked to choose which of the two they prefer. In a *preference ranking* task, participants are presented with all of the stimuli at once and are asked to arrange them from the least to the most preferred. With *preference ratings*, participants rate their preference for each of the stimuli on a categorical or continuous scale (see section 5.3).

The most peculiar aspect of the study of preference is not the behavioral method used to collect the data, but strategy adopted for their analysis. Here, we briefly describe several models of preference judgments and list various references for the interested reader.

The law of comparative judgment by Thurstone [142] can be considered as one of the earliest models developed for the analysis of paired preference comparison data. Within this framework, the probabilities of preferring one stimulus over the other are used to determine the position of each of the stimuli along a preference continuum. A similar representation can be computed based on the Bradley-Terry-Luce (BTL) model [10, 90]. More complex preference models locate stimuli in a preference space with a given number of dimensions. We can distinguish between two classes of such models (see figure 5.8): ideal point and ideal vector models. Ideal point models represent experiment participants as points in the same preference space where stimuli are positioned. The location of the participant within the space models the hypothetical stimulus he prefers the most, that is, the ideal point, and his preference for each of the experimental stimuli is represented by their distance from the ideal point (stimuli farther from the ideal point are less preferred). Ideal point models can, for example, be computed by using unfolding algorithms [9, 17, 20, 24, 25]. With ideal vector models, stimuli are also represented as points in a space with a given number of dimensions, whereas participants are represented as vectors oriented toward a maximum preference point located at infinity. The preference for the experimental stimuli is thus modeled as the location of their projections onto the ideal vector, with preferred stimuli located further along the direction of the ideal vector. Ideal vector models can be fitted using the MDPREF algorithm [14].

The dimensions of various multidimensional models of preference can, in general, be interpreted similarly as the dimensions of the weighted Euclidean model (wMDS). For example, with the MDPREF model the dimensions can be interpreted as corresponding to the stimulus attributes that most strongly influence preference in the population of participants. Further, within the same model the length of the projection of a participant-specific ideal vector onto a dimension measures the relevance of that dimension, that is, stimulus attribute for the participant.

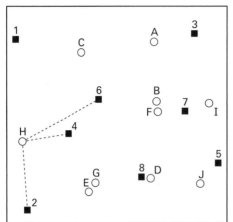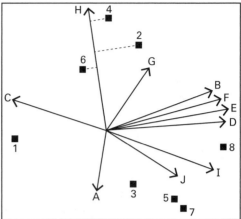

Figure 5.8

Multidimensional models of preference (hypothetical data). In both panels stimuli are represented as numbers, and participants are represented as letters. In the ideal point model (left),
preferred stimuli are located closer to the ideal point (one for each of the participants). For
example, the three most preferred stimuli for participant H are stimuli numbers 4, 2, and 6. In
the ideal vector model (right), participants are represented as directional vectors of preference.
The preferences of a given participant can be visualized by projecting the stimuli onto his ideal
vector (dashed lines): preferred stimuli are located farther along the direction of the ideal
vector.

All of the abovementioned models of preference operate on behavioral data and
do not require a knowledge of the stimulus properties (e.g., price of a guitar). For this
reason the study of what properties of the stimuli influence preference is usually
carried out after the preference model has been created (see section 5.11). The quantification and prediction of choice and preference based on measured properties
of the stimuli are instead an integral part of a family of experimental and analysis
methods that flourished within the field of marketing research: conjoint measurement [49]. The reader interested in conjoint measurement has several sources [13,
50, 54].

In summary, the mathematical modeling of preference can be used by the designer
to discover those configurations of a sonic feedback system that are most preferred
(or least annoying) for the majority of the participants. In the case of meaningful
dimension of a multidimensional preference space, the designer might thus decide
to focus the modeling efforts on those stimulus properties/synthesis parameters that
most strongly correlate with the most relevant dimensions of the preference space (see
also section 5.11).

5.9   Continuous Evaluation

Sonic interactions are time-varying events: posture, sound, touch, and visual properties all change in time as a consequence of the motor behavior of the user. Importantly, the temporal variation in sensory information is likely to result in a temporal variation of the experienced properties of the sonic interaction itself. The behavioral methods presented in the previous sections are not able to account for this level of complexity of a sonic interaction because, in general, they produce one single data point for each of the presented stimuli. Continuous evaluation methods instead permit the measurement of the temporal dynamics of the perceptions of a user (e.g., "How does the emotional response to a complex sound vary in time?") because they continuously sample perceptions throughout the entire duration of the stimulus.

Participants in a continuous evaluation experiment are asked to repeatedly judge some perceptual/cognitive attribute as the stimulus unfolds in time (e.g., a musical composition or a soundscape). No limitation is imposed on the type of task carried out by the listener: we can thus have a continuous ratings task (e.g., rate continuously along a pleasant/unpleasant scale), a continuous categorization task (e.g., "What emotions are you currently feeling? Sadness, happiness, fear, or anger?"), or a continuous preference estimation task (e.g., "Rate your current preference for the sound material on a least preferred to most preferred scale"). Table 5.2 provides a classification of the type of judgments and stimuli investigated in a number of continuous evaluation studies.

Table 5.2
Summary of behavioral methods used in previous continuous evaluation studies

Behavioral method	Sound stimuli	Study
Ratings (categorical scale)	Road traffic	[78, 79]
	Trains	[106]
	Helicopters	[80]
	Car accelerations	[77]
Ratings (continuous scale)	Road traffic	[30, 31, 45]
	Music	[107, 97, 125, 101, 144]
	Speech	[19]
Ratings (continuous and categorical scales)	Road traffic	[150, 64, 111]
	Synthetic sounds	[62, 137, 138]
	Speech	[60, 52]
Cross-modality matching	Pure tones	[137]

The design of a continuous evaluation experiment should take into account the maximum temporal resolution of the responses that an experiment participant can reliably give. The temporal resolution of the response is indeed limited by various factors such as the constants of temporal integration governing a particular perceptual or sensory process (e.g., loudness integration) or the speed of the motor responses of a participant (e.g., how rapidly a participant can move a slider). For example, whereas it is unlikely that a listener can rate the time-varying loudness of a stimulus with a temporal resolution higher than 100 milliseconds (ten ratings per second), it is plausible that a reliable rating can be collected at slower temporal rates such as two ratings per second. Because of these limitations, continuous evaluation methods are better suited for quantifying the perceptual dynamics for relatively long stimuli (e.g., several minutes) rather than for very short ones (e.g., sounds shorter than 1 second).

In regard to sonic-interaction contexts, logistics make it hard if not impossible to ask a participant to interactively generate a sound while evaluating simultaneously his time-varying perceptions. For this reason, it might be advisable to carry out continuous evaluations on offline recordings of sonic interactions (e.g., videos, sound recordings). Despite this limitation, this method will permits to address important sonic-interaction issues, such as the temporal correlation between the gestures of a user and the perceptions of a passive listener. The implementation of continuous evaluation must finally meet a number of requirements: judgments must be made easily, rapidly, and without discontinuity. Several methods combined with specific devices for response collection have been proposed. The interested reader is referred to the work of Schubert [126, 127] for a description of the methodological issues associated with the continuous evaluation of complex musical materials.

Continuous evaluation methods have been used for the study of speech quality [52, 60], for the quantification of the effects of running bus sounds on comfort [111], to estimate auditory brightness [62], and to measure emotional responses to music [76, 101, 132, 144]. Various studies have assessed the relationship between the instantaneous judgment of a given perceptual property of the time-varying stimulus and the global judgment of the overall level of the same property for the entire stimulus. Kuwano and Namba [79] and Fastl [31], for example, showed that judgments of the global loudness of a complex sound are strongly related to the peaks of the time-varying loudness (see also [45, 63, 137]).

5.10 Multisensory Contexts

By definition a sonic interaction involves multiple sensory modalities: audition, touch and kinesthesia (haptics), and possibly vision.[7] Within the context of multisensory interactions, a relevant issue might, for example, be whether each of the involved sensory modalities affects a behavioral variable of interest (e.g., "Does the sound of car doors influence preference?") or which of the sensory modalities has the strongest behavioral effect (e.g., "Which has the stronger impact on preference: the sound of a car's doors closing or the felt weight of the doors?"). Alternatively, the goal of the designer might be to improve some aspect of the performance of the user based on multiple sensory inputs: "Do sonic feedbacks shorten the time required for parking a car?" Answering most of these questions does not usually require experimental tasks different from those described in previous sections. As detailed in this section, the study of multisensory contexts might nonetheless pose specific challenges and require specialized experimental paradigms.

The study of multimodal perception and performance typically involves a comparison of data collected when only one modality is stimulated with data collected in multisensory contexts (e.g., when only touch is available vs. when both sound and touch are available). Designing an experimental condition where information from one single modality is presented might not always be trivial, especially when participants interact with naturalistic objects.

A recent study by Giordano et al. [42] illustrates this issue. The goals of this study were to measure the effects of various sensory modalities and of their combination on the identification and discrimination of walked-on grounds: audition, kinesthesia (e.g., perception of limbs movement), and touch. Participants carried out a simple identification task (see section 5.2). The ideal experiment would have included seven experimental conditions: three conditions for each of the sensory modalities in isolation, three conditions where only one of the three modalities was suppressed, and one condition that combined information from all modalities. The general approach adopted to suppress information from one modality was based on the phenomenon of masking: a signal cannot be perceived if presented together with a sufficiently intense masking stimulus (e.g., random signal or noise). Auditory information was thus suppressed by presenting walking participants with an intense noise over headphones, and leaving both touch and kinesthetic information intact. Similarly, touch information was disrupted by presenting walking participants with a random mechanical vibration at the foot. The acoustical and tactile noises were presented simultaneously in a condition where only kinesthetic information was left intact. Notably, the mechanical actuator used to vibrate the foot also generated an audible noise (the range of frequencies that can be perceived with touch partially overlaps with the range of auditory frequencies). For this reason, an auditory-kinesthetic condition could not be investigated. Further limitations in the experimental design arose from the impossibility of suppressing kinesthetic information while participants walked. Apart from the logistic challenge, such an experimental condition would have likely produced abnormal, and most importantly, unstable locomotion. For this reason the auditory-tactile condition could not be investigated, and the auditory-only condition could be carried out only with nonwalking participants who were presented with recordings of the sound they had generated while walking on the grounds. The study by Giordano et al. [42] thus exemplifies the compromises and challenges faced in the study of multimodal naturalistic sonic interactions and shows the usefulness of noise-based disruption of sensory information in both the auditory and tactile domains.

A frequent goal of studies on multisensory contexts is to establish modality dominance hierarchies, that is, to assess which modality affects most strongly the perception of multimodal events. This question is often answered by investigating *multisensory conflicts*. Conflicting multisensory events combine contrasting information from different modalities, that is, modality-specific stimuli that induce different perceptual results when presented in isolation (e.g., a light located slightly to the left of the point of fixation and a sound presented slightly to the right of the same point). The experimental paradigm relies on the comparison of perceptions for the unimodal stimuli with perceptions of multisensory conflicts (e.g., "Provided that this light is perceived to be located on the left, and that this sound is perceived to be located on the right, what will be the perceived location of the light-sound event?"). In general, the perceptions for a multisensory conflict will be most similar to those for the dominant-modality stimulus presented in isolation. This basic paradigm can be used to reveal the dominance of vision on audition in the estimation of the spatial position of an event (ventriloquist effect [66]), the effect of seeing the movements of the lips of a speaker on the speech perception [103], or the influence of the number of sound beeps on the perceived numerosity of visual flashes ([129]; see Shams, Kamitani, & Shimojo [130] for an overview of the effects of auditory information on visual perception and Lederman & Klatzky [86] for an overview of research on the multisensory perception of audio, haptic, and visual textures). From a design perspective, it would be reasonable to focus modeling efforts on the modalities that dominate the perceptual responses of a user.

The work of Ernst and Banks [29] combines the study of multisensory conflicts with an interesting experimental manipulation of sensory noise. In simple words, the ML model predicts that in a multimodal context modality-specific information that is better discriminated when presented in isolation affects more strongly the multisensory perceptions. To test this prediction, Ernst and Banks investigated the perception of the height of virtual visual-haptic ridges. The method of constant stimuli was used to measure the perceived height of ridges in visual-haptic conflicting stimuli where haptic and visual height differed. The main goal of the experiment was to measure how the conflicting visual and haptic heights were combined to yield an estimate of visual-haptic height. The experiment included various visual noise conditions where the visual ridge stimulus was perturbed randomly with a noise of variable intensity. Notably, increased noise levels produced a more pronounced impairment in the ability to discriminate ridge height based on visual information alone. In these conditions the ML model predicted that the perceived height of the visual-haptic ridge progressively approached the haptic height as the level of the visual noise increased. The experimental results were in strong agreement with this prediction. From the methodological point of view, the work by Ernst and Banks shows how noise can be used not only to occlude completely the information from one sensory modality but also to modulate its discriminability. The possibility of predicting multisensory dominance based on measures of the discrimination of unimodal information might also be particularly useful to the design process. Indeed, with this paradigm the designer will be able to estimate which modality will dominate a multisensory interaction based on unimodal displays even before further efforts are spent on combining each of the modalities in a single multimodal display.

The paradigm developed by Giordano et al. [37] adopts a different approach for the study of multisensory dominance in interactive contexts, based on measures of motor activity rather than based on non-action responses (e.g., "Is this light presented to your right or left?"). In this study participants are asked to strike an audio-haptic virtual object with a target velocity. Each trial is divided in three phases (see figure 5.9). During a first training phase, participants receive feedback on whether their striking velocity is within the target range. During a subsequent adaptation phase, feedback is eliminated, and the virtual object is kept unchanged. Participants are asked

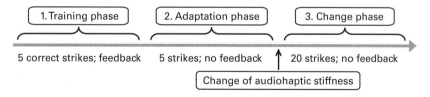

Figure 5.9
Structure of experiment trials in the study by Giordano et al. (audio-haptic condition [37]). The training phase ended after five consecutive correct strikes within the target velocity range. The adaptation and change phases ended after five and twenty strikes, respectively, independently of whether they were within the target velocity range or not. Participants received correctness feedback only during the training phase.

to continue striking the object with the same velocity. In a final change phase, the properties of the virtual object are changed (auditory or haptic stiffness). Multisensory dominance is inferred from the strength of the effects of modality-specific changes on striking velocity, where a change in the dominant modality (haptics, in this experiment) leads to larger changes in striking velocity. The assessment of multisensory perceptions based on the motor activity of the user might be particularly useful for those contexts where a continuous interaction of the user with the display precludes the possibility of easily collecting non-action judgments such as verbally communicated responses.

A further topic of interest in the study of multisensory contexts focuses on the effects of the congruence of different modalities on perceptions [68]. For example, Giordano et al. [37] measured the extent to which the congruence of the auditory and haptic stiffness of the virtual object influenced velocity-tracking abilities. On each audio-haptic trial the change in audio-haptic stiffness could be congruent (e.g., both the auditory and haptic stiffnesses increase above the initial level) or incongruent (e.g., an increase in auditory stiffness and a decrease in haptic stiffness). Only the behavioral effects of the change in auditory stiffness were modulated by congruence. In particular, when auditory stiffness changed in the opposite direction from haptic hardness, auditory information appeared to have no effect on striking velocity. Further paradigms in the study of multimodal congruence are illustrated by Marks [99]. From a design point of view, measurements of the effects of multimodal congruence can, for example, be used to predict more accurately the perceptions of users or can inform the process of linking together modality-specific displays into a single multisensory interactive object.

5.11 Measurement of Acoustical Information

One of the frequent goals of studies on the perceptual processing of complex sounds is to pinpoint what sound features influence the answers of the experimental participant. From the scientific perspective, this analysis stage aims at quantifying the acoustical information relevant to the perception and sensory processing of complex sounds. From the applied point of view, this analysis makes it possible to discover those sound properties that should receive the attention of the sound designer in order to achieve a target perceptual result (e.g., maximize preference, simulate variations in the perceived hardness of struck objects; see Gygi & Shafiro [58] for various applications of research on complex naturalistic sounds).

With most of the approaches to the characterization of the perceptually relevant acoustical information, a priori hypotheses about what properties of the sound might influence the judgments of experiment participants are necessary. To this purpose, each of the experimental stimuli is often characterized by a short sequence of numbers, each of the numbers measuring one attribute of the sound or acoustical feature (e.g., attack time, approximately the speed of level increase at sound onset [102]). It is important to note that our knowledge of the psychophysics of complex everyday sounds is still far from complete. Thus, the problem of defining sound features is largely unconstrained and can be practically endless: a large number of mathematical operations can indeed be adopted to describe a sound with a single number [41]. For these reasons, we omit from this presentation a definition of the various sound features previously used to study the perception of complex sounds. We nonetheless refer the interested reader to the work of Peeters, Giordano, Susini, Misdariis, and McAdams [113], for a recent and extensive system for the extraction of acoustical features.

In general, two different strategies can be adopted to discover perceptually relevant sound features. The first operates on unmodified recordings of the sound events. We can term this approach *correlational* because it often relies on tests of the statistical association between the sound features on the one hand and some behavioral variable on the other. The behavioral variables can be either the responses given by the experiment participant (e.g., probability of choosing the response "female" in gender categorization of walking sounds [89]) or the parameters of a mathematical model of the behavioral data (e.g., the dimensions of an MDS model of dissimilarity ratings [102], see section 5.4; the parameters of a spatial model of preferences, see section 5.8). One of the main issues associated with the correlational approach is that different sound properties can be statistically associated (e.g., natural impact sounds whose level decays more slowly also have a longer duration [39]). Because it is not possible to tell which of two strongly correlated features has the strongest influence on a target behavioral variable, a reasonable answer to this issue has been to recognize this ambiguity in the data-analysis process and to eliminate the need to choose arbitrarily among strongly correlated features by reducing them to one single variable [41].

The second approach adopted to assess perceptually relevant sound features is based either on the manipulation of sound recordings or on purely synthetic sounds. Notably, this approach somehow mitigates the features-correlation problem that characterizes studies of unmodified sound stimuli. The sound-manipulation approach was for example adopted by Li, Logan, and Pastore [89] to investigate the identification of the gender of a walker in sounds where the spectral mode (i.e., the most prominent spectral frequency) and the spectral slope were actively manipulated. Based on these stimuli, Li et al. [89] confirmed an influence of both acoustical factors, as previously revealed by the correlational analysis of the data collected for unmodified sounds. Grassi [46] investigated the extent to which various modifications of the sounds of a ball bouncing on a dish (e.g., removal of bounces) alter the ability to correctly estimate the size of the ball. Another interesting sound-manipulation paradigm has been adopted by Gygi et al. [55, 56] and Shafiro [128] to assess the relevance of temporal information for the identification of environmental sounds. In these studies the spectrum of the sound signals was progressively smeared while keeping the amplitude envelope untouched. After spectral smearing, sounds whose identification relies on the temporal variation of amplitude are comparatively better recognized than sounds whose identification relies on spectral factors. A final sound-manipulation strategy relies on the synthesis of novel sound stimuli. This approach was adopted by Caclin [11] to investigate the sound properties affecting timbre judgments. Importantly, the dimensions of acoustical variations of the synthetic stimuli corresponded to the acoustical attributes associated with the perception of unmodified musical stimuli in a previous correlational study [102]. The study by Caclin thus exemplifies the potential confirmatory function of sound-manipulation studies: whereas correlational studies can be used to generate hypotheses about what features influence the perception of complex sounds, sound-manipulation studies can be carried out to explicitly test these hypotheses.

A final interesting sound-manipulation approach is based on the method of *perturbation analysis* [149], frequently adopted by Lutfi to investigate the perception of the properties of sound sources [91–94]. In short, perturbation analysis relies on a trial-by-trial random perturbation of acoustical parameters of interest: if an acoustical parameter is relevant to the perceptual task, statistical analyses will reveal a significant association between the trial-specific responses and the trial-specific value of the perturbed acoustical parameters. In Lutfi and Oh [93] participants were presented with synthetic impact sounds and were asked to identify the material of the struck object. Sounds were generated by perturbing independently from trial to trial three parameters that characterize the spectral components of an impact sound: frequency, starting amplitude, and decay modulus, a measure of the temporal velocity of amplitude decay. A statistical analysis of the association between trial-specific responses and synthesis parameters made it possible to measure the extent to which each of the participants identified materials by focusing on frequency, amplitude, or decay modulus. It should be noted that whereas the majority of the studies mentioned in the previous paragraphs focus on global acoustical properties that characterize the entire sound signal (e.g., measures of the spectral distribution of energy), the molecular approach of Lutfi focuses on the attributes of the single spectral components that constitute a sound. The method of perturbation analysis has indeed never been applied to investigate the perceptual processing of global sound features and should represent an interesting methodological tool for future studies on this topic.

5.12 Motion Capture

Motion-capture methods are a natural choice for the analysis and evaluation of sonic interactions because they allow direct measurement of the motor interaction of the experiment participant with the sound-producing system. Being in general nonintrusive (no need of wires, and in some cases only very light sensors are placed on objects or on the body of the experiment participant), these methods allow an accurate measurement of the motor interaction of the user with the sound-producing system (e.g., "How do we use our body when interacting with a sonic artifact?" or "How do gestures and artifacts mutually influence a sonic interaction?"). Motion-capture techniques are often versatile and simple, and can also be used in real time to modify the structure of the sound signal reaching the user.

Motion-capture methods are frequently used in the study of music performance. Playing a musical instrument is indeed a complex form of sonic interaction in which the gestures of a user, the musician, ultimately trigger the generation of a musical sound and often influence the properties of ongoing sound events. Methods developed in this field can thus be easily generalized to the study of any type of sonic interaction. Within this research domain, sensors of different nature (e.g., accelerometers, cameras) have been used to collect high-temporal-resolution data on the movements of objects involved in the sound-generation process (e.g., hammer movements in a piano) and on the motor activity of the performer (e.g., motion of the arms and fingers of a pianist). In its most frequent application, motion capture data are collected for an off-line analysis. Goebl and Bresin [43] measured the effect of different touches and of striking speed on the acceleration of the keys and hammers in a piano. Dahl [22] used motion capture to uncover different percussion strategies in drumming performance. Schelleng [122] used these techniques to measure the parameters of a bowing gesture necessary to optimize the quality of violin sounds. Dahl and Friberg [23] used motion-capture methods to analyze the quality of movements interacting with sounding objects. Cameras of various kinds have been used to track user gestures in studies of music performance. For example, Schoonderwaldt

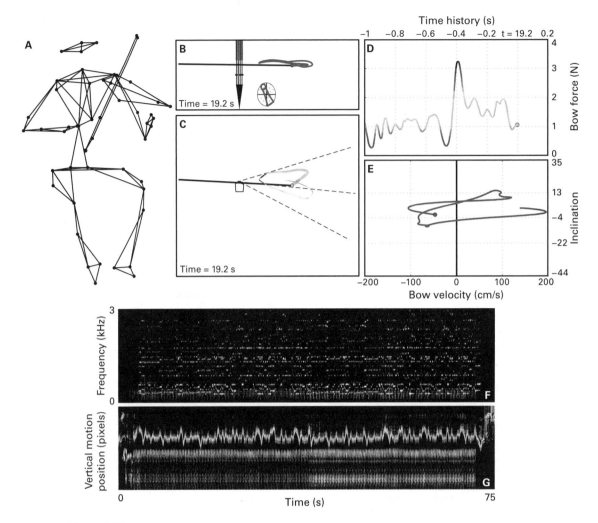

Figure 5.10

Motion capture analysis of violin performance (panels A–E, courtesy of Erwin Schoonderwaldt), and analysis of audio and video recordings (panels F–G, courtesy of Alexander Refsum Jensenius). An extended video of this violin performance can be seen at http://youtu.be/Y2zkrT76CvI. (A) Reflective markers (black dots) attached to the violinist's body and to the violin. (B) Top view of the motion of the bow (horizontal line) on the violin strings (vertical lines). The blue line displays the past bow trajectory. The bottom polar plot displays the rotation of the bow frog (keyhole shape), that is, the tilt of the bow. (C) Same as B, seen from the player perspective (bow trajectory = colored line). Each of the colored fanshaped areas correspond to one of the four violin strings, selected by changing the bow inclination. (D) Time-varying bow force. (E) Plot of time-varying bow inclination as a function of bow velocity, useful for assessing the coordination of string crossings and bow reversals. (F) Spectrogram of the sound signal resulting from the motions measured in panels B–E. (G) Motiongram of the violin performance analyzed in panels B–E. The vertical axis shows the vertical motion of the player. The grayscale measures the activity level (black = no activity; white = peak activity).

and Wanderly [124] used cameras to measure violin bowing gestures and to develop visual feedback methods for students in violin performance. Schoonderwaldt and Demoucron [123] designed a nonintrusive system combining optical motion capture with sensors. This system allows accurately measurement of all bowing parameters in bowed-string instrument performance. Video data have been captured by Dahl [23] to investigate the influence of visual information on the perception of emotional expression and gesture quality in music performance. Further methods for the study of body movement in music performance based on the real-time analysis of video data have been developed by Jensenius [69]. These methods are based on a collapsed visualization of vertical and horizontal motion of musicians (motiongram) as registered in video recordings and on the synchronization of the motiongram with the corresponding audio spectrogram (see figure 5.10 for further explanation).

Promising and largely unexploited applications of motion-capture data, particularly for the study of sonic interactions, are based on their on-line use. Indeed, motion-capture data can, for example, be used to control in real time the parameters of a sound-generating algorithm based on measures of motor activity (i.e., the spatial position, speed, and acceleration of a hand hitting and grabbing a sounding object). For example, in a study on percussion performance Giordano et al. [37] used measures of the velocity with which participants struck a virtual object to control the parameters of a real-time model for the generation of synthetic impact sounds. In another study DeWitt and Bresin [26] used the real-time gestures (speed and pressure) of a pen on a tablet to investigate the control of sound models mimicking the sound of a pen scribbling and its emotional content.

Acknowledgments

The authors wish to thank Richard E. Pastore, Carl Gaspar, and Philip McAleer for comments on earlier versions of this chapter and Erwin Schoonderwaldt and Alexander Refsum Jensenius for providing the materials displayed in figure 5.10.

Notes

1. The absolute threshold can be considered as a special case of the differential threshold. For example, the absolute threshold for sound level can be conceived as the lowest sound level that can be discriminated from the lowest possible level: silence.

2. In this section we focus on so-called direct scaling methods which measure scales of sensation directly, based on judgments of sensory magnitudes.. In contrast, indirect scaling methods derive scales of sensory magnitude from measures of the discrimination of the stimuli [35, p. 185].

Direct and indirect scaling methods are largely based on the work of S. S. Stevens and L. L. Thurstone, respectively [135,142].

3. In common practice the paired stimuli are always different (i.e., participants never rate the dissimilarity between one stimulus and itself) because the mathematical models most frequently used to analyze the dissimilarity ratings (e.g., multidimensional scaling models, see below) do not take into account same-stimulus dissimilarities. It should be nonetheless noted that self-dissimilarity data are useful for assessing whether participants correctly understood the use of the response scale, in which case same-stimulus dissimilarities should be on average lower than different-stimulus dissimilarities (see McAdams, Roussarie, Chaigne, & Giordano [100] for an example).

4. Although the most popular, MDS is one of several different distance models available for analysing dissimilarity data; see the appendix of Giordano et al. [38].

5. In a Euclidean space, the distance between two points A and B equals

$$\sqrt{\sum_{i=1}^{N}(A_i - B_i)^2} \, ,$$

where N equals the number of dimensions of the space. The Euclidean distance is a special case of the Minkowsky distance:

$$\left(\sum_{i=1}^{N}|A_i - B_i|^p\right)^{1/p} ,$$

where p is the power of the Minkowski metric, and the Euclidean distance equals the Minkowski distance for $p = 2$. Not all MDS algorithms assume a Euclidean distance: measuring the power of the Minkowsky metric that best accounts for the dissimilarity ratings has indeed been a research topic in several previous studies [96, p. 149].

6. We can distinguish between metric and nonmetric MDS depending on whether model distances approximate a linear or monotonic transformation of the input dissimilarities, respectively.

7. See Calvert, Spence, and Stein [12] for an excellent overview of research on multisensory processes.

References

1. Ashby, F. G. (1992). *Multidimensional models of perception and cognition.* Hillsdale, NJ: Lawrence Erlbaum Associates.

2. Ashby, F. G., & Perrin, N. A. (1988). Toward a unified theory of similarity and recognition. *Psychological Review, 95*(1), 124–150.

3. Ballas, J. A. (1993). Common factors in the identification of an assortment of brief everyday sounds. *Journal of Experimental Psychology. Human Perception and Performance, 19*(2), 250–267.

4. Bijmolt, T. H. A., & Wedel, M. (1995). The effects of alternative methods of collecting similarity data for multidimensional scaling. *International Journal of Research in Marketing, 12*(4), 363–371.

5. R. Bisping. (1997). Car interior sound quality: Experimental analysis by synthesis. *Acta Acustica united with Acustica, 83*(5), 813–818.

6. Björk, E. A. (1985). The perceived quality of natural sounds. *Acustica, 57*(3), 185–188.

7. Böckenholt, U. (2005). Scaling of preferential choice. In B. S. Everitt & D. C. Howell (Eds.), *Encyclopedia of statistics in behavioral science* (pp. 1790–1794). Chichester, UK: John Wiley & Sons.

8. Bonebright, T. L. (1996). An investigation of data collection methods for auditory stimuli: Paired comparisons versus a computer sorting task. *Behavior Research Methods, Instruments, & Computers, 28*(2), 275–278.

9. Borg, I., & Groenen, P. (1997). *Modern multidimensional scaling.* New York: Springer-Verlag.

10. Bradley, R. A., & Terry, M. E. (1952). Rank analysis of incomplete block designs. *Biometrika, 39*, 324–345.

11. Caclin, A., McAdams, S., Smith, B. K., & Winsberg, S. (2005). Acoustic correlates of timbre space dimensions: A confirmatory study using synthetic tones. *Journal of the Acoustical Society of America, 118*(1), 471–482.

12. Calvert, G., Spence, C., & Stein, B. (2004). *Handbook of multisensory processes.* Cambridge, MA: MIT Press.

13. Carroll, J. D., & Green, P. E. (1995). Psychometric methods in marketing research: Part I, conjoint analysis. *JMR, Journal of Marketing Research, 32*(4), 385–391.

14. Chang, J. J., & Carroll, J. D. (1968). *How to use MDPREF, a computer program for multidimensional analysis of preference data. Technical report.* Murray Hill, NJ: Bell Telephone Laboratories.

15. Chouard, N., & Hempel, T. (1999). A semantic differential design especially developed for the evaluation of interior car sounds. *Journal of the Acoustical Society of America, 105*(2), 1280.

16. CLA2T research group, UPRES SYLED. (n.d.). Lexico3. URL: http://www.tal.univ-paris3.fr/lexico/index-gb.htm [retrieved June 21 2012]. Université de la Sorbonne Nouvelle, Paris 3.

17. Coombs, C. H. (1964). *A theory of data.* New York: John Wiley & Sons.

18. Cornsweet, T. N. (1962). The staircase-method in psychophysics. *American Journal of Psychology, 75*(3), 485–491.

19. Cowie, R., Douglas-Cowie, E., Savvidou, S., Mcmahon, E., Sawey, M., & Schröder, M. (2000). "feeltrace": An instrument for recording perceived emotion in real time. In *ISCA tutorial and research workshop (ITRW) on speech and emotion* (pp. 19–24). Newcastle, Northern Ireland.

20. Cox, T. F., & Cox, A. A. (1997). *Multidimensional scaling* (2nd ed.). New York: Springer-Verlag.

21. Coxon, A. P. M. (1999). *Sorting data: Collection and analysis. Sage university papers on quantitative applications in the social sciences*. Thousand Oaks, CA: Sage Publications.

22. Dahl, S. (2004). Playing the accent-comparing striking velocity and timing in an ostinato rhythm performed by four drummers. *Acta Acustica united with Acustica, 90*, 762–776.

23. Dahl, S., & Friberg, A. (2007). Visual perception of expressiveness in musicians' body movements. *Music Perception, 24*(5), 433–454.

24. de Leeuw, J. (2005). Multidimensional unfolding. In B. S. Everitt & D. C. Howell (Eds.), *Encyclopedia of statistics in behavioral science*. Chichester, UK: John Wiley & Sons.

25. De Soete, G., & Carroll, J. D. (1992). Probabilistic multidimensional models of pairwuse choice data. In F. G. Ashby (Ed.), *Multidimensional models of perception and cognition* (pp. 61–88). Hillsdale, NJ: Lawrence Erlbaum Associates.

26. DeWitt, A., & Bresin, R. (2007). *Sound design for affective interaction* (vol. 4738, pp. 523–533). Berlin/Heidelberg: Springer.

27. Ellermeier, W., & Daniel, P. (2002). Tonal components in tire sounds: Refined subjective and computational procedures. In G. Ebbitt & P. Davies (Eds.), *Proceedings of the sound quality symposium (SQS2002) at Inter-noise 2002*, Ames, IA: Institute of Noise Control Engineering (INCE-USA), Iowa State University.

28. Ellermeier, W., Mader, M., & Daniel, P. (2004). Scaling the unpleasantness of sounds according to the BTL model: Ratio-scale representation and psychoacoustical analysis. *Acta Acustica united with Acustica, 90*(1), 101–107.

29. Ernst, M. O., & Banks, M. S. (2002). Humans integrate visual and haptic information in a statistically optimal fashion. *Nature, 415*(24), 429–433.

30. Fastl, H. (1989). Average loudness of road traffic noise. In *Proceedings of the international congress of noise control engineering* (Vol. II, pp. 815–820). Newport Beach, CA.

31. Fastl, H. (1991). Loudness versus level of aircraft noise. In A. Lawrence (Ed.), *Proceedings of Inter-noise '91* (Vol. 1, pp. 33–36). Sydney, Australia.

32. Faure, A. (2000). *Des sons aux mots, comment parle-t-on du timbre musical?* PhD thesis, Ecole des Hautes Etudes en Sciences Sociales (EHESS), Paris, France.

33. Freed, D. J. (1990). Auditory correlates of perceived mallet hardness for a set of recorded percussive events. *Journal of the Acoustical Society of America, 87*(1), 311–322.

34. García-Pérez, M. A. (1998). Forced-choice staircases with fixed step sizes: Asymptotic and small-sample properties. *Vision Research, 38*(12), 1861–1881.

35. Gescheider, G. A. (1997). *Psychophysics: The fundamentals* (3rd ed.). Mahwah, NJ: Lawrence Erlbaum Associates.

36. Gingras, B. T., Lagrandeur-Ponce, T., Giordano, B. L., & McAdams, S. (2011). Perceiving musical identity: Performer identification is dependent on performer expertise and expressiveness, but not on listener expertise. *Perception, 40*(10), 1206–1220.

37. Giordano, B. L., Avanzini, F., Wanderley, M., & McAdams, S. (2010). Multisensory integration in percussion performance. In *Actes du 10eme congrès francais d'acoustique, Lyon* [CD–ROM], Paris, France: Société Française d'Acoustique.

38 Giordano, B. L., Guastavino, C., Murphy, E., Ogg, M., Smith, B. K., & McAdams, S. (2010). Comparison of methods for collecting and modeling dissimilarity data: applications to complex sound stimuli. *Multivariate Behavioral Research, 46*(5), 779–811.

39. Giordano, B. L., & McAdams, S. (2006). Material identification of real impact sounds: Effects of size variation in steel, glass, wood and plexiglass plates. *Journal of the Acoustical Society of America, 119*(2), 1171–1181.

40. Giordano, B. L., McDonnell, J., & McAdams, S. (2010). Hearing living symbols and nonliving icons: Category-specificities in the cognitive processing of environmental sounds. *Brain and Cognition, 73*, 7–19.

41. Giordano, B. L., Rocchesso, D., & McAdams, S. (2010). Integration of acoustical information in the perception of impacted sound sources: The role of information accuracy and exploitability. *Journal of Experimental Psychology. Human Perception and Performance, 36*(2), 462–479.

42. Giordano, B. L., Visell, Y., Yao, H. Y., Hayward, V., Cooperstock, J., & McAdams, S. (2012). Identification of walked-upon materials in auditory, kinesthetic, haptic and audio-haptic conditions. *Journal of the Acoustical Society of America, 131*(5), 4002–4012.

43. Goebl, W., & Bresin, R. (2003). Measurement and reproduction accuracy of computer-controlled grand pianos. *Journal of the Acoustical Society of America, 114*, 2273.

44. Goodhill, G. J., Simmen, M. W., & Willshaw, D. J. (1995). An evaluation of the use of multidimensional scaling for understanding brain connectivity. *Philosophical Transactions of the Royal Society B. Biological Sciences, 348*(1325), 265–280.

45. Gottschling, G. (1999). On the relations of instantaneous and overall loudness. *Acta Acustica united with Acustica, 85*(3), 427–429.

46. Grassi, M. (2002). Recognising the size of objects from sounds with manipulated acoustical parameters. In J. A. Da Silva, E. H. Matsushima, & N. P. Ribeiro-Filho (Eds.), *Proceedings of the 18th annual meeting of the international society for psychophysics (Fechner day 2002)* (pp. 392–397). Rio de Janeiro, Brasil.

47. Grassi, M. (2005). Do we hear size or sound? Balls dropped on plates. *Perception & Psychophysics, 67*(2), 274–284.

48. Green, D. M., & Swets, J. A. (1966). *Signal detection theory and psychophysics*. New York: John Wiley & Sons.

49. Green, P. E., & Rao, V. R. (1971). Conjoint measurement for quantifying judgmental data. *JMR, Journal of Marketing Research, 8*(3), 355–363.

50. Green, P. E., & Srinivasan, V. (1978). Conjoint analysis in consumer research: Issues and outlook. *Journal of Consumer Research, 5*(2), 103.

51. Grey, J. M. (1977). Multidimensional perceptual scaling of musical timbres. *Journal of the Acoustical Society of America, 61*(5), 1270–1277.

52. Gros, L., & Chateau, N. (2001). Instantaneous and overall judgements for time-varying speech quality: Assessments and relationships. *Acta Acustica united with Acustica, 87*(3), 367–377.

53. Guastavino, C. (2007). Categorization of environmental sounds. *Canadian Journal of Experimental Psychology, 61*(1), 54–63.

54. Gustaffson, A., Herrmann, A., & Huber, F. (2007). Conjoint analysis as an instrument of marketing research practice. In A. Gustaffson, A. Herrmann, & F. Huber (Eds.), *Conjoint measurement: Methods and applications* (4th ed., pp. 3–30). Berlin: Springer-Verlag.

55. Gygi, B. (2001). *Factors in the identification of environmental sounds.* PhD thesis, Indiana University, Department of Psychology.

56. Gygi, B., Kidd, G. R., & Watson, C. S. (2004). Spectral-temporal factors in the identification of environmental sounds. *Journal of the Acoustical Society of America, 115*(3), 1252–1265.

57. Gygi, B., Kidd, G. R., & Watson, C. S. (2007). Similarity and categorization of environmental sounds. *Perception & Psychophysics, 69*(6), 839–855.

58. Gygi, B., & Shafiro, V. (2007). General functions and specific applications of environmental sound research. *Frontiers in Bioscience, 12*, 3152–3166.

59. Hajda, J. M., Kendall, R. A., Carterette, E. C., & Harshberger, M. L. (1997). Methodological issues in timbre research. In I. Deliege & J. Sloboda (Eds.), *The perception and cognition of music* (pp. 253–306). London: Lawrence Erlbaum Associates.

60. Hansen, M., & Kollmeier, B. (1999). Continuous assessment of time-varying speech quality. *Journal of the Acoustical Society of America, 106*(5), 2888–2899.

61. Harbke, C. R. (2003). *Evaluation of data collection techniques for multidimensional scaling with large stimulus sets.* Master's thesis, Washington State University, Department of Psychology.

62. Hedberg, D., & Jansson, C. (1998). *Continuous rating of sound quality. Technical report.* Technical Audiology, Karolinska Institutet.

63. Hellbrück, J. (2000). Memory effects in loudness scaling of traffic noise: How overall loudness of short-term and long-term sounds depends on memory. *Acoustical Science and Technology, 21*(6), 329–332.

64. Hellbrück, J., & Zeitler, A. (1999). Evaluating sequences of environmental noise using the method of absolute judgment in laboratory and outdoor situations. Some methodological considerations. *Journal of the Acoustical Society of America*, *105*, 1083.

65. Hempel, T., & Chouard, N. (1999). Evaluation of interior car sound with a new specific semantic differential design. *Journal of the Acoustical Society of America*, *105*(2), 1280.

66. Howard, I. P., & Templeton, W. B. (1966). *Human spatial orientation*. New York: John Wiley & Sons.

67. Ih, J. G., Lim, D. H., Shin, S. H., & Park, Y. (2003). Experimental design and assessment of product sound quality: Application to a vacuum cleaner. *Noise Control Engineering Journal*, *51*(4), 244–252.

68. Jacobs, R. A. (2002). What determines visual cue reliability? *Trends in Cognitive Sciences*, *6*(8), 345–350.

69. Jensenius, A. R. (2007). *Action-sound: Developing methods and tools to study music-related body movement*. PhD thesis, Department of Musicology, University of Oslo.

70. Jeon, J. Y., You, J., & Chang, H. Y. (2007). Sound radiation and sound quality characteristics of refrigerator noise in real living environments. *Applied Acoustics*, *68*(10), 1118–1134.

71. Kendall, R. A., & Carterette, E. C. E. C. (1993). Verbal attributes of simultaneous wind instrument timbres: I. Von Bismarck's adjectives. *Music Perception*, *10*(4), 445–468.

72. Kidd, G. R., & Watson, C. S. (2003). The perceptual dimensionality of environmental sounds. *Noise Control Engineering Journal*, *51*(4), 216–231.

73. Kitagawa, M., & Windsor, B. (2008). *MoCap for artists: Workflow and techniques for motion capture*. Burlington, MA: Focal Press.

74. Kolesnik, P., & Wanderley, M. (2004). Recognition, analysis and performance with expressive conducting gestures. In *Proceedings of the international computer music conference (ICMC '04)*. Miami, FL: ICMA.

75. Krumhansl, C. L. (1989). Why is musical timbre so hard to understand? In S. Nielzén & O. Olsson (Eds.), *Structure and perception of electroacoustic sound and music* (pp. 43–51). Amsterdam: Excerpta Medica.

76. Krumhansl, C. L. (1998). Topic in music: An empirical study of memorability, openness, and emotion in Mozart's String Quintet in C Major and Beethoven's String Quartet in A Minor. *Music Perception*, *16*, 119–134.

77. Kuwano, S., Hayakawa, Y., & Namba, S. (1993). Psychological evaluation of noise in passenger cars using the method of continuous judgment by category: Analysis in different noise sources. In Chapelle, P., Vermier, G. (Eds.) *Proceedings of Inter-Noise 93* (Vol. II, pp. 919–922). Leuven, Belgium.

78. Kuwano, S., & Namba, S. (1978). On the loudness of road traffic noise of longer duration (20 min.) in relation to instantaneous judgment. *Journal of the Acoustical Society of America, 64,* S127–S128.

79. Kuwano, S., & Namba, S. (1985). Continuous judgment of level-fluctuating sounds and the relationship between overall loudness and instantaneous loudness. *Psychological Research, 47*(1), 27–37.

80. Kuwano, S., & Namba, S. (1990). Temporal change of timbre of helicopter noise. *Noise Control Engineering Journal, 2,* 1177–1180.

81. Kuwano, S., & Namba, S. (2001). Dimension of sound quality and their measurement. In *Proceedings of the 17th international congress on acoustics (ICA),* Rome, Italy.

82. Kyncl, L., & Jiricek, O. (2001). Psychoacoustic product sound quality evaluation. In *Proceedings of the ICA* (p. 90). Rome, Italy.

83. Landauer, T. K., McNamara, D. S., Dennis, S., & Kintsch, W. (2007). *Handbook of latent semantic analysis.* Mahwah, NJ: Lawrence Erlbaum Associates.

84. Lawless, H. T., Sheng, N., & Knoops, S. S. C. P. (1995). Multidimensional scaling of sorting data applied to cheese perception. *Food Quality and Preference, 6*(2), 91–98.

85. Lederman, S. J. (1979). Auditory texture perception. *Perception, 8,* 93–103.

86. Lederman, S. J., & Klatzky, R. L. (2004). Multisensory texture perception. In G. Calvert, C. Spence, & B. Stein (Eds.), *Handbook of multisensory processes* (pp. 107–122). Cambridge, MA: MIT Press.

87. Lemaitre, G., Susini, P., Winsberg, S., McAdams, S., & Letinturier, B. (2007). The sound quality of car horns: A psychoacoustical study of timbre. *Acta Acustica united with Acustica, 93*(3), 457–468.

88. Levitt, H. (1971). Transformed up-down methods in psychoacoustics. *Journal of the Acoustical Society of America, 49*(2), 467–477.

89. Li, X., Logan, R. J., & Pastore, R. E. (1991). Perception of acoustic source characteristics: Walking sounds. *Journal of the Acoustical Society of America, 90*(6), 3036–3049.

90. Luce, R. D. (1959). *Individual choice behavior: A theoretical analysis.* New York: John Wiley & Sons.

91. Lutfi, R. A. (2001). Auditory detection of hollowness. *Journal of the Acoustical Society of America, 110*(2), 1010–1019.

92. Lutfi, R. A., & Liu, C. J. (2007). Individual differences in source identification from synthesized impact sounds. *Journal of the Acoustical Society of America, 122,* 1017–1028.

93. Lutfi, R. A., & Oh, E. L. (1997). Auditory discrimination of material changes in a struck-clamped bar. *Journal of the Acoustical Society of America, 102*(6), 3647–3656.

94. Lutfi, R. A., & Wang, W. (1999). Correlational analysis of acoustic cues for the discrimination of auditory motion. *Journal of the Acoustical Society of America, 106,* 919.

95. Macmillan, N. S., & Creelman, C. D. (1991). *Detection theory: A user's guide* (2nd ed.). Cambridge, MA: Cambridge University Press.

96. Maddox, W. T. (1992). Perceptual and decisional separability. In G. Ashby (Ed.), *Multidimensional models of perception and cognition* (pp. 147–180). Hillsdale, NJ: Lawrence Erlbaum Associates.

97. Madsen, C. K. (1996). Empirical investigation of the "aesthetic response" to music: Musicians and nonmusicians. In *Proceedings of the 4th international conference for music perception and cognition* (pp. 103–110). Montréal, Canada: Society for Music Perception and Cognition.

98. Marcell, M. E., Borella, D., Greene, M., Kerr, E., & Rogers, S. (2000). Confrontation naming of environmental sounds. *Journal of Clinical and Experimental Neuropsychology, 22*(6), 830–864.

99. Marks, L. E. (2004). Cross-modal interactions in speeded classification. In G. Calvert, C. Spence, & B. Stein (Eds.), *Handbook of multisensory processes* (pp. 85–105). Cambridge, MA: MIT Press.

100. McAdams, S., Roussarie, V., Chaigne, A., & Giordano, B. L. (2010). The psychomechanics of simulated sound sources: Material properties of impacted thin plates. *Journal of the Acoustical Society of America, 128*(3), 1401–1413.

101. McAdams, S., Vines, B. W., Vieillard, S., Smith, B. K., & Reynolds, R. (2004). Influences of large-scale form on continuous ratings in response to a contemporary piece in a live concert setting. *Music Perception, 22*(2), 297–350.

102. McAdams, S., Winsberg, S., Donnadieu, S., De Soete, G., & Krimphoff, J. (1995). Perceptual scaling of synthesized musical timbres: Common dimensions, specificities, and latent subject classes. *Psychological Research, 58,* 177–192.

103. McGurk, H., & MacDonald, J. (1976). Hearing lips and seeing voices. *Nature, 264,* 746–748.

104. McNicol, D. (2004). *A primer of signal detection theory.* Mahwah, NJ: Lawrence Erlbaum Associates.

105. Misdariis, N., Minard, A., Susini, P., Lemaitre, G., McAdams, S., & Parizet, E. (2010). Environmental sound perception: Metadescription and modeling based on independent primary studies. *EURASIP Journal on Audio. Speech and Music Processing.*

106. Namba, S., Kato, T., & Kuwano, S. (1995). Long-term evaluation of the loudness of train noise in laboratory situation. In M. Newman (Ed.), *Proceedings of the international congress on acoustics* (pp. 215–218). Trondheim, Norway.

107. Namba, S., Kuwano, S., Hatoh, T., & Kato, M. (1991). Assessment of musical performance by using the method of continuous judgement by selected description. *Music Perception, 8*(3), 251–276.

124. Schoonderwaldt, E., & Wanderley, M. M. (2007). Visualization of bowing gestures for feedback: The Hodgson plot. In *Proceedings of the 3rd international conference on automated production of cross media content for multi-channel distribution (AXMEDIS07)* (Vol. 2, pp. 65–70). Barcelona, Spain.

125. Schubert, E. (1996). Measuring temporal emotional response to music using the two dimensional emotion space. In *Proceedings of the 4th international conference for music perception and cognition* (pp. 263–268). Montréal, Canada.

126. Schubert, E. (2001). Continuous measurement of self-report emotional response to music. In P. N. Juslin & J. A. Sloboda (Eds.), *Music and emotion: Theory and research* (pp. 393–414). New York: Oxford University Press.

127. Schubert, E. (2010). Continuous self-report methods. In P. N. Juslin & J. A. Sloboda (Eds.), *Handbook of music and emotion: Theory, research, applications* (pp. 223–253). New York: Oxford University Press.

128. Shafiro, V. (2004). *Perceiving the sources of environmental sounds with a varying number of spectral channels*. PhD thesis, City University of New York.

129. Shams, L., Kamitani, Y., & Shimojo, S. (2000). What you see is what you hear. *Nature, 408,* 788.

130. Shams, L., Kamitani, Y., & Shimojo, S. (2004). Modulations of visual perception by sound. In G. Calvert, C. Spence, & B. Stein (Eds.), *Handbook of multisensory processes* (pp. 27–33). Cambridge, MA: MIT Press.

131. Siekierski, E., Derquenne, C., & Martin, N. (2001). Sensory evaluation of air-conditioning noise: Sensory profiles and hedonic tests. In *Proceedings of the 17th international congress on acoustics (ICA)* (p. 326). Rome, Italy.

132. Sloboda, J. A., & Lehmann, A. C. (2001). Tracking performance correlates of changes in perceived intensity of emotion during different interpretations of a Chopin piano prelude. *Music Perception, 19*(1), 87–120.

133. Solomon, L. N. (1958). Semantic approach to the perception of complex sounds. *Journal of the Acoustical Society of America, 30,* 421–425.

134. Stepánek, J. (2006). Musical sound timbre: Verbal description and dimensions. In *Proceedings of the international conference on digital audio effects (DAFx-06)* (pp. 121–126). Montréal, Canada.

135. Stevens, S. S. (1956). The direct estimation of sensory magnitude—loudness. *American Journal of Psychology, 69,* 1–25.

136. Stevens, S. S., & Guirao, M. (1962). Loudness, reciprocality, and partition scales. *Journal of the Acoustical Society of America, 34,* 1466.

137. Susini, P., McAdams, S., & Smith, B. K. (2002). Global and continuous loudness estimation of time-varying levels. *Acta Acustica united with Acustica, 88*(4), 536–548.

138. P. Susini, S. McAdams, and B. K. Smith. (2007). Loudness asymmetries for tones with increasing and decreasing levels using continuous and global ratings. *Acta Acustica united with Acustica, 93*(4), 623–631.

139. Susini, P., McAdams, S., & Winsberg, S. (1999). A multidimensional technique for sound quality assessment. *Acta Acustica united with Acustica, 85*(5), 650–656.

140. Susini, P., McAdams, S., Winsberg, S., Perry, I., Vieillard, S., & Rodet, X. (2004). Characterizing the sound quality of air-conditioning noise. *Applied Acoustics, 65*(8), 763–790.

141. Takane, Y., Young, F. W., & De Leeuw, J. (1977). Nonmetric individual differences multidimensional scaling: An alternating least squares method with optimal scaling features. *Psychometrika, 42*(1), 7–67.

142. Thurstone, L. L. (1927). A law of comparative judgment. *Psychological Review, 34*(4), 273–286.

143. Treutwein, B. (1995). Adaptive psychophysical procedures. *Vision Research, 35*(17), 2503–2522.

144. Upham, F., & McAdams, S. (2010). Approaching an audience model of listening experience. In S. M. Demorest, S. J. Morrison, & P. S. Campbell (Eds.), *Proceedings of the 11th international conference on music perception and cognition (ICMPC11)* (pp. 489–490). Seattle, WA.

145. Vanderveer, N. J. (1979). *Ecological acoustics: Human perception of environmental sounds.* PhD thesis, Cornell University. *Dissertation Abstracts International, 40,* 4543B. (University Microfilms No. 80–04–002).

146. von Békésy, G. (1947). A new audiometer. *Acta Oto-Laryngologica, 35*(5–6), 411–422.

147. Von Bismarck, G. (1974). Timbre of steady sounds: A factorial investigation of its verbal attributes. *Acustica, 30*(3), 146–159.

148. Warren, W. H., & Verbrugge, R. R. (1984). Auditory perception of breaking and bouncing events: A case study in ecological acoustics. *Journal of Experimental Psychology. Human Perception and Performance, 10*(5), 704–712.

149. Watson, C. S. (1992). *Signal detection and certain physical characteristics of the stimulus during the observation interval.* PhD thesis, Indiana University, Bloomington, IN.

150. Weber, R. (1991). The continuous loudness judgement of temporally variable sounds with an "analog" category procedure. In A. Schick, J. Hellbrück, & R. Weber, (Eds.). *Fifth Oldenburg symposium on psychological acoustics* (pp. 267–294). Oldenburg, Germany: BIS.

151. Wedin, L., & Goude, G. (1972). Dimension analysis of the perception of instrumental timbre. *Scandinavian Journal of Psychology, 13*(3), 228–240.

152. Wichmann, F. A., & Hill, N. J. (2001). The psychometric function: I. Fitting, sampling, and goodness of fit. *Perception & Psychophysics, 63*(8), 1293–1313.

153. Wichmann, F. A., & Hill, N. J. (2001). The psychometric function: II. Bootstrap-based confidence intervals and sampling. *Perception & Psychophysics*, *63*(8), 1314–1329.

154. Zampini, M., & Spence, C. (2004). The role of auditory cues in modulating the perceived crispness and staleness of potato chips. *Journal of Sensory Studies*, *19*(5), 347–363.

155. Zeitler, A., & Hellbrück, J. (2001). Semantic attributes of environmental sounds and their correlations with psychoacoustic magnitudes. In *Proceedings of the 17th international congress on acoustics* (Vol. 28), Rome, Italy.

II Case Studies

Audio and Touch

As discussed in the previous chapters, sonic interaction design in the broadest sense is concerned with how to design the sonic aspect of an interaction as well as how to bring sonic aspects into the design process per se. A critical question to ask is: How do the sonic aspects of the interaction relate to other aspects such as tangible and visual design decisions? The purpose of the first five case studies is to discuss the role of the interaction of sound and touch using the design, study, and experience of PebbleBox, CrumbleBag, Scrubber, Sound of Touch, Gamelunch, Zizi, Sonic Texting, and A20 as illustrative examples. All these interfaces link auditory and tactile experience through active exploration. PebbleBox, CrumbleBag, and Scrubber were initially designed to be musical instruments that provided control for various kinds of sound synthesis methods including granular synthesis and physical models of friction. However, they have in numerous ways transcended these initial design ideas, as discussed in the first case study. Sound of Touch is an interactive installation in which the sound produced by different physical textures is explored. The Gamelunch experience is an installation based around a dining table where the loop among interaction, sound, and emotion is explored. Zizi is a couch that provides both physical and emotional support. Sonic Texting is a gesture-based text entry system that uses tactile input and auditory output. In A20, methods from human-centered design are combined with techniques from the field of new interfaces for musical expression (NIME).

6 Perceptual Integration of Audio and Touch: A Case Study of PebbleBox

Sile O'Modhrain and Georg Essl

The design philosophy of PebbleBox, CrumbleBag, and Scrubber [5, 13] was strongly inspired by both Gibson's theory of *direct perception* [8] and, more importantly, by that of *enaction* [2, 4, 6, 14]. Gibson [8] and later Noe [12] both emphasize that the process of obtaining knowledge through action requires that there exists a well-defined relationship between the actions of the perceiver and the results of these actions in the environment. If one drops an object, for example, our experience concerning the law of gravity dictates that the object will fall until it comes into contact with a surface. If this defined relationship is violated, our past experience is no longer valid, and we seek an alternative law to govern the object's behavior. In building interfaces that seek to promote the formation of enactive knowledge, such relationships must either be borrowed from the real world or constructed within a novel interaction space. The question then becomes to what extent the notion of lawfulness can be stretched and under what conditions such relationships are nonnegotiable. One can thus view the designs we discuss here as examples of artifacts to explore perceptually guided actions and the extent to which the notions embedded in the theory of enaction can be stretched within constructed interaction spaces that flexibly combine tactile, sonic, and to some extent visual cues.

These designs belong to the broader field of new interfaces for musical expression. Musical instruments are a particularly interesting case for exploring enactive principles because action and response are physically coupled through the sound production mechanism of the instrument. A key problem in instrument design is the *mapping problem*, which is the challenge of mapping sensed gestures to sound synthesis algorithms [9, 11].

Our approach, which circumvents the mapping problem, is to start with instances where the sonic and tangible response of an interaction are coupled through some shared, physically informed relationship with its associated gestures or actions. The goal is to design an interface that allows and retains such actions but also offers new variability within the range of available perceptual outcomes. The actions are grounded in familiar physical circumstances; hence, we restore crucial elements of the instrumental gesture approach [6]. However, we go beyond the notion of what might be considered instrumental gestures by allowing for much weaker couplings between sonic and tangible responses of the interface. It is in these more loosely defined notions of action-response coupling that the power of our design approach can be found. Although the designs we introduce here are not instruments per se, the principle of flexible coupling between gesture and sound suggests that there may be more latitude in the notion of the instrumental gesture than we have hitherto considered. The presence of friction in some form, for example, may be sufficient to support the control of the bow-string interaction, even if the bow is a ski and the string is a mountain.

6.1   Design for Perceptually Guided Actions

Given that the goal of PebbleBox, CrumbleBag, and Scrubber 1.1 (figure 6.1) was to create a context where the theory of enaction could be instantiated, certain design challenges had to be considered.

First, the object must provide a context within which the user can explore, through his or her actions, the opportunities within the constructed environment. In this way it should be possible to support the building up of the kind of bodily knowledge central to the enactive approach. The challenge in meeting these design criteria was to find a situation that was sufficiently constrained. We decided to build an object that resembled a musical instrument in that it would tightly couple the actions of the user to some sound response from the object. As noted above, prior work on the role of touch in musical instrument design had suggested that the haptic or tangible response of a musical instrument played a crucial role in playability. For this reason we looked for examples of real-world interactions where the feel and sound resulting

Figure 6.1
PebbleBox, CrumbleBag, and Scrubber.

Table 6.1
The relation of physical law, auditory response, and tactile response in the design of PebbleBox, CrumbleBag, and Scrubber

Physical law	Heard behavior	Felt behavior	Artifact name
Collision	Granular synthesis of natural sound recordings	Coarse-grain materials (typically decorative flowerpot stones)	PebbleBox
Crushing	Granular synthesis of natural sound recordings	Fine-grain brittle materials (styrofoam filling material, broken coral shells, cornflakes)	CrumbleBag
Dry contact friction	Granular and looped wavetable synthesis of natural sound recordings	Contact friction between held object and surface	Scrubber

from an action were clearly related through the physics of the interaction context. This decision, in turn, led us to construct a design space and populate it with examples of interactions with objects that could be felt as well as heard. Some of these are outlined in table 6.1.

In order to truly explore the instantiation of the enactive approach, it was necessary for us to go beyond the recreation of a real-world scenario and create a constructed environment that the user would never encounter in the real world. Only in this way could we learn whether it was possible for someone approaching an interactive system for the first time to acquire new "enactive" knowledge.

The first of our ideas to be instantiated in an artifact was collision, and the artifact was the PebbleBox [13, 15]. PebbleBox is a musical instrument designed for granular sound synthesis. The granular synthesis is made tangible by providing the performer with physical grains—in the form of pebbles—that can be manipulated. The sound of the instrument is generated by a granular synthesis algorithm that takes parameters derived from the sound of the colliding pebbles as its input. These parameters are event amplitude, event timing, pitch centroid of colliding pebbles, and so on. Although the performer can choose the sound grains to be used, the main performative control is in the temporal structure and the dynamic content of the grains.

The PebbleBox itself consists of a foam-lined wooden box that contains a layer of polished stones. The motion of the stones is sensed using a small microphone embedded in the foam lining. The microphone picks up sounds within the box, which by nature of the padding and the filling material are dominated by collision sounds of the coarse material within the box. Audio processing allows a time series of dynamic parameters to be extracted. This information is then used to drive arbitrary parametric synthesis algorithms.

Typically, Pebblebox uses decorative flower-pot stones, which have a smooth surface and produce distinct collision sounds. Initially, two synthesis algorithms were used. One is a wavetable of short natural sound recordings. When it is played we modify the intensity and playback speed based on the detected intensity and spectral centroid of collision sounds. The other synthesis method is a parametric modal approach called Physically Informed Sonic Modeling (or PhiSM) [3].

The resulting design makes the relationship between the physical dynamics of pebbles colliding in the box and the sound output that is driven by these collisions very immediate. Natural aspects driving the dynamics, such as the frequency of events, the collision intensity, and spectral variation in collisions retain their original semantic meaning when used to drive the synthesis of new sounds. Hence, although the new sounds can be completely different from the actual collision noises, the broader physical properties of the interaction are retained. Abrupt hand movements through the pebbles, for example, result in rapid, high-amplitude sequences of output sound events. Hence, the sound output, although different from the sound of colliding pebbles by virtue of how it is mapped through the granular synthesis algorithm, retains the expressive contour of the original hand gestures that caused the pebbles to move. This is very much a design intention. This design decouples action and tactile response from the sonic response while it retains the core physical dynamics of the original link. Hence, PebbleBox allows for the exploration of the decoupling of tangible and sonic responses in a physical sound interaction environment [7].

CrumbleBag [13] and Scrubber [6] followed very much the same design principles for other classes of environmental sound mechanisms. CrumbleBag is concerned with the relationship between the sound and tactile feel resulting from crumbling actions performed on fine-grain brittle materials. It is very much inspired by the Foley artists' methods for emulating footstep sounds on gravel, leaves, or other surfaces where matter is crushed underfoot. Often such methods involve filling a bag with some material—cornflakes for crunchy gravel, cotton wool for snow—and then squeezing the bag to produce individual footstep sounds. In the case of CrumbleBag, a hand-size bag made of rubber material was sewn into a pocket shape with interior fleece padding, and a microphone was then embedded into the bottom of the bag. Plastic bags containing brittle filler materials appropriate for different crushing effects could then be inserted into the pockets. Again the physical dynamics and feel of acting on fine brittle material is retained and used as the input to a granular synthesis algorithm, and the sounding response is made flexible by virtue of the sound grains used for synthesis.

As with PebbleBox, this design ensures that the original expressive gestures performed by the hands are retained and directly inform the sound output.

Scrubber is an interface for dry friction sounds. It uses a gutted whiteboard eraser containing a custom silicon filling to encase two microphones. It also uses a force-sensing resistor to detect contact with a surface. As its name suggests, the gestures encouraged by the interface are scrubbing gestures such as wiping, rubbing, or bowing. By using two microphones, it is possible to derive the direction of motion of the Scrubber. The force-sensing resister conveys how hard the Scrubber is pressed against a surface. The force-sensing resistor also acts to detect if there is contact at all and hence allows the microphone signal to be silenced when there is none, thereby filtering out unwanted environmental noise.

One of the most important design decisions governing all of these interfaces was the decision to make the physical form and feel of the Scrubber as it is moved across a surface the source of tactile and haptic responses. In other words, we do not attempt to artificially generate the feel of friction but design the device in such a way that friction between it and the surface being scrubbed is transmitted to the hand; at the same time, the sound arising from scrubbing is picked up and used to alter the sound that is synthesized. The same is true for collision of pebbles in PebbleBox and the crushing of materials in CrumbleBag. In this way responsibility for generating the feel of the interaction remains in the real world and does not have to be constructed through computer-generated haptic feedback.

By employing the design approach described here, we deliberately retain considerable flexibility and control over the relationship between the sound and the feel of the system. PebbleBox can be filled with a variety of materials, as can CrumbleBag. Scrubber can be used on any dry flat surface. This also holds for sound. Due to the decoupling of sound one can resynthesize new sonic responses to the action at hand. Alternatively, one can take advantage of the properties of the different materials used in the physical interface to shape the sonic response in different ways; filling PebbleBox with highly regular objects such as ball bearings, for example, produces very uniform collisions that result in an output sound with a comb filter effect. Hence, the choice of the material to interact with already influences characteristics of the sound.

In accordance with our original design goal, we can say that we have indeed designed interfaces that provide a flexible coupling between their tangible and auditory response. The next steps in this process are a more detailed exploration of the relationship between actions and sounds and then to ask how certain combinations of action, tactile percept, and sonic response are perceived as a whole.

6.2   How Actions Guide the Cognitive Grouping of Sounds

It is still very much an open question as to how complex sonic events are grouped into a cohesive cognitive experience. Certain aspects of the problem are, however, understood. Temporal patterns, in particular, give rise to important perceptual cues [15]. Warren and Verbrugge showed that temporal events generated by bouncing are well separated, grouped, and recognized compared to breaking and shattering events. For example, whereas the first is deterministic, the latter has a strong stochastic nature.

The sounds used in the original instantiation of PebbleBox can be grouped by the temporal characteristics of the typical contexts in which they occur. This is very much by design because an important parameter in the control of sound events is the temporal characteristics of onsets generated as the user moves his or her hands within the PebbleBox.

The sounds chosen for PebbleBox were intended to cover a wide range of actions and temporal patterns. Some of the sounds used are included in table 6.2.

The boundaries between these categories are often difficult to define: sounds may not neatly fit into any category, or they may even belong to just one category. In fact, often sounds were chosen because they appeared to belong to a given category but, when implemented in PebbleBox, turned out to belong to another. For example, a collision sound such as a coin dropping worked well for both quasiperiodic timings and stochastic timings. A coin can indeed be dropped or can be part of a pile of coins

Table 6.2
Sounds, their generating actions, and their ecologically likely temporal patterns

Apple	Chewing	Quasiperiodic
Bell	Striking	Quasiperiodic
Run	Steps	Quasiperiodic
Walk	Steps	Quasiperiodic
Tap-dancing	Steps	Quasiperiodic
Paper	Crumbling/shuffling	Mostly stochastic
Knuckles	Stretch fingers	Stochastic
Breaking glass	Dropping	Stochastic
Coins	Dropping	Quasiperiodic/stochastic
Hammer	Striking	Quasiperiodic/stochastic
Dishes	Collisions	Quasiperiodic/stochastic
Fencing	Fencing	Quasiperiodic/stochastic
Tools	Shuffle	Stochastic/quasiperiodic
Ice cubes	Collisions	Stochastic/quasiperiodic

that are shuffled around or dropped together. The individual impact sound is essentially the same in all these cases. This explains why coins and other impact sounds are easily recognized even within diverse temporal patterns.

Some sounds resist being forced across category boundaries because they fail to make sense when triggered in different ways. Periodic sounds such as footsteps, for example, lose their coherence when they are triggered by aperiodic gestures such as random pebble collisions. And yet other sounds work unexpectedly well given different temporal triggering patterns. This happens, for example, with the sound of fencing. If played stochastically the collision of fencing blades sounds like the shuffling of blades or tools on a floor or inside a tool chest. There are a number of these cases that, in hindsight, could be explained by plausible if unrelated ecologically valid relationships between action and response.

In summary, what we obtain are broad groups in which we can replace sounds that have similar temporal patterns, even if the relationship between action and response for a given sound is not inherently meaningful. Some sounds can belong to multiple groups, whereas others cannot. This observation can be seen as a design principle. In supporting the performance of certain actions, an interface will be perceived differently depending on the ecologically relevant groups that a sound belongs to. If this design principle is violated, the interaction may become unconvincing or unrecognizable. In PebbleBox, for example, footsteps work well when they are performed as regular, temporally organized events, but they lose their identity when they are performed randomly because, in ecological terms, bipedal beings cannot move in this way, and we therefore do not expect to hear temporally disorganized footsteps.

This observation relates to the known phenomenon of human sensitivity to biological motion [10]. Studying visual observation of biological motion, Johansson noticed that, even if observers are given very sparse representations of moving objects (in the form of correlated moving dots), ecologically based motions, such as walking and running, can be easily recognized and distinguished from random movements of dots, only some of which are correlated. There is also physiological evidence suggesting that the breakdown of the auditory perception of footsteps in nonecological contexts similar to the case we observed in PebbleBox may have similar neurological bases that relate to the distinction between nonbiological and biological motion in vision. In both cases the posterior superior temporal sulcus activates in response to the presentation of biological motion [1], suggesting that the recognition of such movements is multimodal in nature. Even though in our designs we may manipulate the causal link between action and response by making footsteps with our hands, it appears that forcing sounds from a context where we have a well-established expectation of their temporal organization into a context where this is removed forces such sounds across a boundary of plausibility, and we can no longer recognize them for what they are.

6.3 Conclusions

One of the primary observations from enactive and ecologically motivated designs such as PebbleBox, CrumbleBag, and Scrubber is that there is considerable malleability in the relationship between contributions of different modalities in an ecologically inspired interaction. This space of malleability is rather difficult to delineate precisely. Emotion and memory play as much a role in this as the complex nature of the multisensory interaction. We are as yet only beginning to understand the relationships among motor action, sounding response, tactile feedback, emotion, and prior experience in designing interfaces such as these, and many questions are as yet unanswered. Designs of interactions that allow for flexibility and variation along the dimensions of interest such as those described here provide us with the means to probe this space. For example an early subjective study showed independent variation of likability and believability of touch-to-sound relations for such interfaces [6].

Although this work is as yet in its infancy, we believe that we have demonstrated that the concept underlying the instrumental gesture, namely that a physical interaction that has an associated tangible and sonic response, can be extended beyond the realm of musical instrument design to encompass a whole class of interfaces where coupling of sound and touch is required. Moreover, we have illustrated that such a coupling is somewhat more malleable than might be predicted, provided that there is a somewhat defined relationship between the actions performed by the user and the response of the system. As we have shown, certain expectations such as those related to biological motion appear to resist manipulation. What will be of interest in future work is to determine what are the dimensions along which such relationships are malleable and where in this new design space other points of resistance to manipulation lie.

References

1. Bidet-Caulet, A., Voisin, J., Bertrand, O., & Fonlupt, P. (2005). Listening to a walking human activates the temporal biological motion area. *NeuroImage, 28*(1), 132–139.

2. Bruner, J. (1968). *Processes of cognitive growth: Infancy*. Worcester, MA: Clark University Press.

3. Cook, P. R. (1997). Physically informed sonic modeling (PhISM): Synthesis of percussive sounds. *Computer Music Journal, 21*(3), 38–49.

4. Dourish, P. (2000). *Where the action is*. Cambridge, MA: MIT Press.

5. Essl, G., & O'Modhrain, S. (2005). Scrubber: An interface for friction-induced sounds. In *Proceedings of the international conference on new interfaces for musical expression (NIME)*, Vancouver, Canada, May 26–28 (pp. 70–75).

6. Essl, G., & O'Modhrain, S. (2006). An enactive approach to the design of new tangible musical instruments. *Organised Sound*, *11*(3), 285–296.

7. Essl, G., Magnusson, C., Eriksson, J., & O'Modhrain, S. (2005). Towards evaluation of performance, control of preference in physical and virtual sensorimotor integration. In *Proceedings of ENACTIVE 2005, the 2nd international conference on enactive interfaces*, Genoa, Italy, November 17–18.

8. Gibson, J. J. (1979). *The ecological approach to visual perception*. Boston: Houghton Mifflin.

9. Hunt, A., & Wanderley, M. M. (2002). Mapping performer parameters to synthesis engines. *Organised Sound*, *7*(2), 97–108.

10. Johansson, G. (1973). Visual perception of biological motion and a model for its analysis. *Perception & Psychophysics*, *14*(2), 201–211.

11. Miranda, E. R., & Wanderley, M. M. (2006). *New digital musical instruments: Control and interaction beyond the keyboard*. Middleton, WI: A-R Editions.

12. Noe, A. (2005). *Action in perception*. Cambridge, MA: MIT Press.

13. O'Modhrain, S., & Essl, G. (2004). PebbleBox and CrumbleBag: Tactile interfaces for granular synthesis. In *Proceedings of the conference for new interfaces for musical expression*, Hamamatsu, Japan (pp. 74–79).

14. Varela, F., Thompson, E., & Rosch, E. (1991). *The embodied mind*. Cambridge, MA: MIT Press.

15. Warren, W. H., & Verbrugge, R. R. (1984). Auditory perception of breaking and bouncing events: A case study in ecological acoustics. *Journal of Experimental Psychology*, *10*(5), 704–712.

7 Semiacoustic Sound Exploration with the Sound of Touch

David Merrill and Hayes Raffle

People have a great deal of experience hearing the sounds produced when they touch and manipulate different materials. We know even without executing the action what it will sound like to bang our fist against a wooden door, or to crumple a piece of newspaper. We can imagine what a coffee mug will sound like if it is dropped onto a concrete floor. These intuitions are useful when we want to create a particular sound with objects in our environment, and they can guide exploration if we just want to play and experiment with new sounds. However, they do not always prove as useful for creating sounds with electronics or computers.

7.1 The Problem of Mapping

Modern electronic instruments afford musicians new and compelling modes of sound synthesis and control. However, these tools have been received less enthusiastically in performance contexts. Although the explosion of sampler/sequencers and digital audio production software has allowed musicians to easily compose music by arranging prerecorded or synthesized sounds for later playback, the expressive performance-oriented affordances of electronic instruments have been criticized. Some musicians report that electronic synthesizers and digital communication standards interfere with expression, lacking the continuous and organic feel that is found in acoustic instruments.

We propose that musicians' dissatisfaction with electronic music instruments stems at least in part from the fact that electronic instruments lack the subtle affordances and potential for acoustic spontaneity featured by acoustic instruments. The decoupling of performer gesture and output sound inherent in electronic instruments comes at a cost: most musical affordances must be deliberately created by the instrument designer. In order for a device to respond to pressure on its body, a pressure sensor must be incorporated in the appropriate location. To respond to bending, or blowing, or changes in tilt or temperature, each of these affordances must be engineered with sensing elements, and the data must be mapped to synthesis parameters. Furthermore, the most straightforward configuration for an electronic music instrument makes the sensing parameters independent, each capable of being actuated separately by a performer without affecting the others. If the control parameters are to feature interconnected behavior like physical objects, this too must be an explicitly designed feature.

This requirement for explicit design of the devices' operation stands in contrast to acoustic instruments, which feature many unplanned or serendipitous affordances that need not have been foreseen completely by the instrument's designer. These ways to play derive from the materials an instrument is built of and its mechanical construction. For instance, plucked notes on an acoustic guitar can be detuned by physically pushing on the guitar neck, causing it to bend and change the distance between the endpoints of the string. These additional affordances are learned partially through experience with an instrument, but they are also informed by our life experience manipulating physical materials. Since most people have handled and bent wooden objects before, the fact that a guitar neck is bendable can be intuited directly by looking at the instrument.

Our wealth of experience handling physical materials does not typically provide much intuition for operating a new electronic instrument given its inherently arbitrary mapping from gesture to sound. A primary design goal with the Sound of Touch has been to build an interactive system for manipulating digital sound that will allow people to utilize their already-existing intuitions about the sonic results of striking, scraping, and other physical gestures on well-known materials.

7.2   The Sound of Touch

We adapted Aimi's methods [1] for real-time percussion instruments, which allow a stored digital sound sample to be stimulated continuously by the signal from a piezo-electric vibration sensor attached to a drum brush. Aimi's work develops a number of semiacoustic percussion instruments that convolve prerecorded samples with the signal from piezoelectric sensors manipulated in real time to provide greater realism and intuitiveness to digital percussion. The underlying mechanism of this stimulation is a continuously running digital convolution [2] of the stored sound sample and the digitized incoming signal from the piezoelectric element. In the Sound of Touch

Figure 7.1
One of the authors playing the Sound of Touch using one of two texture palettes that were built.

(figure 7.1) [3], every nuance of the physical contact between the wand and a texture elicits sound that incorporates the common frequencies of the physical interaction and the recorded audio sample. For example, recording the word *hello* (figure 7.2) and then tapping the wand against a piece of felt produces a soft, muffled version of the *hello* wherein the recording incorporates the acoustic quality of the wand's interaction with the felt.

7.2.1   User Interface Advance

With its direct convolution of two user-created acoustic signals, The Sound of Touch sidesteps the common and often central challenge of designing an effective mapping from sensor input to synthesizer parameters [4]. Digital convolution can create

Figure 7.2
Singing into the microphone of the brass blade.

acoustically rich cross-filtering of audio samples, but traditional methods of using the algorithm are abstracted (typically via graphical waveform editors) and do not promote real-time experimentation and improvisation convolving a variety of audio samples. The Sound of Touch puts convolution into a person's hands quite literally, pairing a *low-latency* (less than 20 milliseconds) and *continuous* convolution algorithm with an acoustically sensitive tangible interface (wand). The physicality of the wand and textures makes manipulating digital sounds with the Sound of Touch akin to manipulating ordinary physical objects, but the system allows a user to work with a larger variety of sounds than physical objects alone would permit.

Our interaction design is inspired by work in Tangible Interfaces such as IO Brush [5], which presents a single brush-like tool with embedded video camera to record and then manipulate a visual recording on a computer display. With IO Brush, users paint with digital ink, whereas with the Sound of Touch, users are painting with sound. Additionally, work has been done to leverage users' experience with physical objects and/or haptics to inform their expressive manipulation of sound or music [6]. There is also a related history of kinesthetic explorations into musical composition using contact microphones to amplify small sounds such as Cage and Tudor's *Cartridge Music* [7].

Figure 7.3
From left to right: Brass Blade, Drum Brush, Plastic Wand, Painting Knife.

7.2.2 Wands

The wand in our system affords both sound recording and manipulation. We built four distinct wands, each using different materials for the handle and tip, and we found that these materials greatly impacted the user experience and expressive potential of the instrument (figure 7.3). All the wands featured a pushbutton to initiate recording, an electret microphone to capture sound, and a piezo element to sense the vibrations induced from contact with physical objects and textures.

Brass Blade

We fitted a wooden dowel handle with a tip made from thin brass. A flat disk piezo sensor was affixed directly to the blade with epoxy, and a button was taped to the handle. Rigid, disc-style piezo sensors capture a relatively limited frequency range, which reduced the sonic potential of this wand. Our experiments with flexible piezo sensors showed a much greater range of frequency sensitivity, and we used them in all successive designs with good results. The flexibility of the Brass Blade also allowed it to kink to a degree that can cause undesirable sonic artifacts. We used stiffer materials for subsequent wands.

Drum Brush

We fitted a flexible piezo sensor into the end of the rubber housing of a drummer's steel brush. The sensor was positioned such that its end would rest against the base of the bristles, sensing their movements and vibrations. The Drum Brush's numerous steel bristles created a large effective surface area for sensing compared to the Brass Blade. This was sometimes an advantage when the brush was scraped against uniform textures such as stone tiles or fur because many points of parallel contact with the surface produced a rich, dense chorusing effect. However, this high contact density became problematic when we tried to hear the effect of a particular surface feature, such as individual aquarium pebbles or slats of a wooden window blind. The effect of the brush's bristles themselves became the dominant acoustic impression, obscuring these surface patterns and making many of the textures produce very similar effects. In response to this observation, we built two follow-up wands that featured single stiff blades.

Plastic Wand

We instrumented the body of a felt-tip permanent marker with a nylon guitar pick mounted in a slot cut into its tip. The marker's body provided a handle that was lightweight and appropriately sized for a typical user's hand. The guitar pick was pliable but durable enough to survive many repeated uses. We laminated a piezo sensor directly to the guitar pick with tape. The Plastic Wand's single tip proved to be more satisfying than the Drum Brush for exploring textures. Acoustic characteristics of different textures were more perceptible when the Plastic Wand scraped across them, and its fine edge allowed more precise sonic exploration of surface features. Furthermore, the flexibility of the body and the tip and the pliable coupling between them resulted in an absorption of higher frequencies, making the sound from this tool more muffled than than from the metal-tipped wands.

Painting Knife

We embedded a microphone, pushbutton, and wiring directly into the wooden handle of a carbon steel painter's knife. The piezo sensor is laminated to the knife under a layer of durable metallic tape. The Painting Knife proved to be the most versatile wand. Like the Plastic Wand, its single tip makes it better for articulately exploring a wide range of textures. However, the stiffer steel and wooden handle allow vibrations to propagate to the sensor with less absorption into the tool itself. The result is a richer, more full-spectrum sound. The density of the materials gives the Painting Knife the feel of a serious, high-quality tool.

7.2.3 Textures

We paired the record-and-playback wands of the Sound of Touch with a diverse set of textures on which to sculpt live-recorded or prerecorded sounds, affording a great range of sonic possibilities. A recorded digital sample can be stimulated in extremely diverse ways depending on the tool, the texture, and how the two are used together.

Two *texture kits* rest on flat, free-standing tables that measured approximately 1 × 1 meter. The surface of each table was divided into a series of rectangular regions with a different material in each. In order to leverage visitors' lifetime of experience hearing the sounds of physical objects, we sought to offer a wide range of materials that were acoustically and texturally diverse, yet familiar. These included hard uniform bathroom tile, sheep's wool, broom bristles with varying stiffness, artificial turf, aquarium pebbles, shag carpeting, metal screen, and wicker curtain pieces. The patterns of holes in the tables were modeled after paintings by Piet Mondrian for aesthetic interest.

7.3 Installation and Feedback

The Sound of Touch system has been installed in a series of art and technology exhibitions in North America including the Boston Cyberarts Festival and SIGGRAPH Emerging Technologies 2007 (part of the SIGGRAPH international computer graphics and interaction conference). At SIGGRAPH it was used by thousands of visitors over the course of 5 days. During that time, we received a vast amount of observational data and feedback from users with backgrounds ranging from user interface research to Foley and cinematic sound design.

In exhibition contexts the two texture tables featured two different spatial layouts of similar textures. One table featured our aforementioned record-and-play wand, which allowed for the exploration of sounds created by the user. This setup allowed for quick *sketching* in sound, where a user would record a new sound, experiment with the various textures for a minute or two, then record a new sound and iterate. The other table had two wands that each allowed certain preselected samples to be selected using buttons mounted in the table (figure 7.4). The available preselected sounds were mostly percussive in nature, ranging from cymbal crashes to a piano impulse response to a "laser gun" sample.

7.3.1 Observations

People engaged in many different styles of use. One axis of variation was the degree to which people used the wands as percussive versus sculpting instruments. At one end some users played with the system as if it were an electronic drum kit, striking the wand sharply against the textures. In this mode of use the surface features of the textures were not as important as their overall density and pliability. This style of use was more common at the table featuring preselected samples. Even though the convolution algorithm was identical on both tables, it is likely that our choice of percussion-oriented samples contributed to this usage pattern. Other users made very slight, deliberate scraping and brushing gestures, sensitively exploring the sonic variations that came from the details of the materials' surfaces (figure 7.5). In both classes many

Figure 7.4
People interacting with the Sound of Touch at SIGGRAPH 2007.

users seemed determined to try every texture, methodically working their way across the entire table.

Some visitors noticed that different size scales of surface features provided interesting variation in granularity. For instance, one mosaic-like tile arrangement contained a repeating pattern of large and small tiles. When a wand is moved across this pattern, it produces a repeating rhythmic tempo. This large-scale low-frequency periodicity contrasts with the tiny high-frequency stick-and-slip vibrations possible across a sand-blasted slab of marble when the knife scrapes the material quickly at a glancing angle.

7.3.2   Feedback

Feedback from Installations
Musicians and sound engineers were particularly interested in the Sound of Touch. Musicians reported that they would like to use the tool in performance settings,

Figure 7.5
A little girl tries all the textures.

particularly in improvisational ones. They were intrigued by the question of how notation could be written for the system and how the spatial layout of the textures might be reconfigured to support specific compositions or playing styles. The record-and-playback table was most appealing to those with experimental interests, whereas those with more traditional tendencies (particularly percussionists) liked the table with preselected samples.

Sound engineers and Foley artists reported that the Sound of Touch could be a new way for them to create sound effects and adapt them to different contexts. The possibilities that the Sound of Touch offers for real-time exploration and quick iteration on ideas could allow them to get more use from their vast and underutilized libraries of samples or to manipulate synthesized or newly recorded digital sounds in more intuitive ways. We were told that the record-and-playback wand could be useful for quick sketching and exploration of sonic ideas, whereas the table featuring preselected sounds might be used for final renderings.

Figure 7.6
Student performance at MIT with the Sound of Touch.

Feedback from MIT Composition Class

Students in a music composition course at MIT used the Sound of Touch for a week-long assignment focusing on the compositional possibilities of electronic tools (figure 7.6). Groups of four or five students were each given a simple version of the palette knife wand with just the piezo sensor (no microphone) and were asked to assemble a collection of objects and digital sound samples. Students were enthusiastic about the sonic possibilities of the tool and offered a great deal of feedback. Suggestions included adding the ability to use longer sounds and to change the length or material of the blade to tune the instrument.

7.4 Conclusion

The Sound of Touch enables a flexible capture and manipulation of audio that is characteristic of digital tools—but in a direct and physical manner that approaches the continuous experience of manipulating acoustic musical instruments and found objects. The system advances user interface research by giving people physical access to an extremely responsive and low-latency convolution algorithm. The result is that people engaging in digital sound design can use Sound of Touch to leverage their existing intuitions about how different objects will sound when these objects are touched, struck, or otherwise physically manipulated—a feature shared by acoustic instruments and objects from our everyday lives. The Sound of Touch is thus a sonic exploration tool that borrows properties from both acoustic and electronic sound creation, bringing them together in a way that leverages advantages of each.

References

1. Aimi, R. (2006). *Extending physical instruments using sampled acoustics*. PhD thesis, Massachusetts Institute of Technology.

2. Gardner, W. G. (1995). Efficient convolution without input-output delay. *Journal of the Audio Engineering Society. Audio Engineering Society*, *43*(3), 127–136.

3. Merrill, D., & Raffle, H. (2007). The sound of touch. In *Extended abstracts of proceedings of the SIGCHI conference on human factors in computing systems (CHI'07)*.

4. Hunt, A., Wanderley, M., & Paradis, M. (2002). The importance of parameter mapping in electronic instrument design. In *Proceedings of the 2004 Conference on new interfaces for musical expression (NIME'02)* (pp. 1–6).

5. Ryokai, K., Marti, S., & Ishii, H. (2005). Designing the world as your palette. In *Conference on human factors in computing systems (CHI'05)* (pp. 1037–1049).

6. O'Modhrain, S., & Essl, G. (2004). Pebblebox and Crumblebag: Tactile interfaces for granular synthesis. In *Proceedings of the 2004 conference on new interfaces for musical expression (NIME'04)* (pp. 74–79).

7. Cage, J., & Tudor, D. (1960). Cartridge music.

8 The Gamelunch: Basic SID Exploration of a Dining Scenario

Stefano Delle Monache, Pietro Polotti, and Davide Rocchesso

Embodiment, enaction, physicality, and directness of manipulation are considered to be important attributes of interactive artifacts. Exploration in interaction design focuses on ways of skillfully translating these qualities into action [1–4]. In current design practice, however, this process is not always successful. This is especially noticeable in sonic interaction design. As a matter of fact, designers are often content with discrete interactions. Moreover, visual thinking and sketching approaches involving forms of visual storytelling [5] are still predominant. In this context, the sonic dimension is mainly approached by intuition and creativity.

In recent years the classic basic design approach used in post-Bauhaus schools has been revitalized as an effective means to tackle the complexity of contemporary interaction design [6–9]. Within the basic design paradigm, analytical thinking, technically oriented solutions, and sensory stimulation integrate in an iterative process of manipulation and refinement of experimental realizations. Such a practice represents a valuable way to develop consistent reasoning (line of thoughts) around particular themes and interactive contexts. Indeed, basic design investigates specific perceptual effects and their experimental combinations. We argue that a basic approach to sonic interaction design could free the discipline from the dominance of music, allowing the definition of an independent and perceptually based vocabulary, similarly to what happens for visual thinking. In this point of view an ecological approach [10] is particularly effective, and a physics-based sound synthesis not only allows us to describe sounds through configurations, materials, and their properties, dynamics of gesture, but also to suggest a certain immediate visualization of the sketched sound, not to mention possible manifestations through the sense of touch [11–14].

This brief contribution aims at (1) exploiting the value of a basic design approach, (2) formulating design problems in terms of exercises, and (3) developing workbenches and experiences around sonic interaction design that may serve as tools for reflection [15] in order to collect a useful repertoire of design ideas and concepts.

8.1   Description of the Setup

Notable and effective tabletop-like interactive surfaces, such as the Reactable [16], the Table Recorder,[1] and the AudioPad [17], recover direct manipulation of objects (tokens) and support collaborative processing of musical sounds. With a complementary approach, the Gamelunch is a sonically augmented dining table. Its name derives from joining the Balinese Gamelan orchestra and lunch practices [18]. The workbench was developed at University of Verona, in the scope of the EU-funded CLOSED (Closing the Loop of Sound Evaluation and Design) project [19], in order to test and improve the Sound Design Toolkit (SDT), a set of perception-founded and physically consistent models for sound synthesis [20]. The ingredients of the Gamelunch are (1) basic design approach, (2) physics-based sound synthesis, and (3) dining practices. Following the approach to kitchen and household environments undertaken in CLOSED, the dining scenario offers a rich set of immediate and natural gestures for the investigation of interactive continuous sound feedback.

Simple everyday actions such as cutting, piercing, sticking, drinking, pouring, grasping, and stirring have been analyzed in terms of source behavior and resulting sounds. By means of various sensor-augmented everyday objects, continuous interactions are captured in order to drive the control parameters of physics-based sound synthesis algorithms. Mock-ups and design experiences have been sketched around well-defined themes and constraints with the aim of highlighting the role of sound in the perception/action loop [21]. As shown in figures 8.1 and 8.2, the Gamelunch

Figure 8.1
Interactive artifacts, cutlery, water jug, and bowls.

Figure 8.2
The bottom of the table, the Gamelunch in action.

environment includes (1) a table with embedded sensors, acquisition boards, and loudspeakers; (2) bowls and dishes; and (3) graspable sensor-augmented objects such as a water jug, a tray, cutlery, and bottles.

8.2 Sonic Interaction at Work

Various configurations of the Gamelunch were produced in an iterative process of problem setting, redesigning of the interactive artifacts, and documentation and analysis of the results. The collected experiences have been systematized as basic design exercises, prototypical for different kinds of interaction primitives, and published on the Web site www.soundobject.org/BasicSID. Such exercises are useful as specifiers of design issues, and their solution can be sought through class assignment, shared observation, or self-reflection.

Around the workbench three major themes were highlighted:

- Contradiction in sonic feedback
- Rhythmic interactions in cutting
- Creation of a consistent hosting soundscape (Antiprimadonna)

8.2.1 Contradiction in Sonic Feedback

The principle of contradiction is exploited to highlight the role played by sounds in our everyday lives and bring attention to the information being conveyed through the auditory channel, often underexploited or even neglected. Experimental continuous sonic counteractions have been sketched so as to affect the coordination of the different sensory channels, for instance, in terms of perceived effort.

As a general principle, the energy consistency between the gesture and the physics-based generated sound is ensured in order to maintain the feeling of a causal continuous control of the action [1, 2]. Also, the sound feedback is meant to be nonsymbolic and preattentional. Two exercises deal with the manipulation of transformational invariants, namely the temporal development of the actions of pouring water (water-jug sonification) and of cutting (knife sonification). The sound feedback, a braking sound in the case of pouring and a metallic squeaking sound in the case of cutting, gives the feeling of a resisting force opposed to the action, which naturally does not involve a particular effort. A second couple of exercises deal with the manipulation of structural invariants, namely the nature/identity of the material being manipulated, with respect to the action of stirring liquids (sangria-bowl sonification) and the action of dressing the salad (salad-bowl sonification) [22]. The goal is to experiment with a contradictory sound feedback by working on the material properties or identity. Inter-estingly, in some engaging tasks some users started challenging the interaction with the artifacts in an expressive manner, thus creatively pushing the boundaries of the designed space. Even if rigorous experimentation has not been conducted yet, most of the people who experienced the Gamelunch agreed on the effectiveness of the distortion in terms of acquiring consciousness of the role of sonic feedback in everyday actions.

8.2.2 Rhythmic Interactions in Cutting

The theme concerns supportive and expressive feedback for cyclic continuous actions such as cutting vegetables on a cutting board. By playing with the synchronization and balance between senses, kitchen tools augmented with sensors may enrich the enactive experience of food preparation. The objective is to support coordination between the two hands in a cyclic task such as cutting carrots or zucchini in rondelles, where longitudinal translation of the carrot is combined with the rhythmic action of cutting. The exercise tackles the design of a continuous and/or rhythmic sound feedback to give a sense of progress. It should emphasize jitter in the coordinated move-ments and the inherent expressiveness of gestures.

The experimental setup (see figure 8.3) encompasses a beat follower of the sounds of the blade hitting the chopping board, a tracker of vertical knife acceleration, and a color tracker of the portion of the carrot being cut. The sensed data are fused in order to provide a continuous sound feedback for the longitudinal progression and a rhythmic sound feedback for the cyclic action in order to support the coordination between the two hands during the action. Three different control mapping strategies are proposed as rhythmic feedback:

A. Beat with adaptive tempo, synchronous with the real stroke of the knife on the board.

B. Beat with fixed reference tempo at 75 beats per minute; the user has to keep it in upbeat with respect to the knife strokes.

C. Beat with adaptive tempo; the user has to keep it in upbeat with respect to the knife strokes.

A wide palette of physics-based impact sounds is provided for the sonification of the impact of the knife on the cutting board. A number of cross combinations of different virtual materials (glass, wood, metal, hybrids) and sizes are tested to make the distinction between impact and feedback sounds clear. In addition, a continuous

Figure 8.3
Cutting a carrot.

friction sound feedback is used to reinforce the rhythmic feedback and provide an auditory indication of progress of the whole action. By raising its pitch and making the sound sharper, the sonic feedback conveys the feeling of the reducing mass of the vegetables.

A number of experiences were freely performed and audiovisually recorded for qualitative evaluation. Maintaining a perfect synchronism in mode A seems to be more unnatural and, thus, less successful when compared to modes B and C. Mode B seems to require the attitude and skills of a player performing music with a metronome. In contrast, mode C affords maintaining a regular pace while not being strictly tied to a fixed reference tempo as in mode B.

8.2.3 Creation of a Consistent Hosting Soundscape (Antiprimadonna)

The proposed assignment exploits by analogy a famous exercise conceived by Tomás Maldonado of the School of Ulm. In figure 8.4 the Antiprimadonna, literally anti-queen bee, envisages the formal organization of seven vertical bands of variable width

Figure 8.4
Teacher Tomás Maldonado, Antiprimadonna, Giovanni Anceschi 1962–1963.

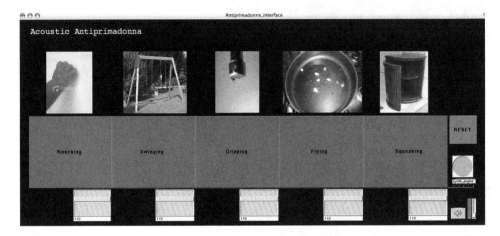

Figure 8.5
Antiprimadonna, graphical user interface with the representation of the five sound panels.

[23]. Among the seven bands, five can be of different colors, but the remaining two should have different black-and-white textures. The arrangement of the bands must be such that none of them plays the part of the queen bee. The objective is to train designers to produce nonhierarchical patterns in order to master the emergence of hierarchies in a controlled way. In its sonic realization, the Antiprimadonna is composed of five sound panels, respectively, two textures and three gestures. The overall composition is a soundscape of sound events and elementary interactions. Figure 8.5 shows the sonic organization of the panels: (1) knocks on wood; (2) metallic, squeaky sound; (3) dripping sound texture; (4) frying sound texture; (5) open/close wooden squeaky sound. The assigned effect is achieved by exploiting the counterpoint of

the temporal patterns, a clear differentiation of the timbre space, and a refined balance of the audio mix.

8.3 Evaluations and Conclusions

Basic design involves a large use of self-reflection, joint observation, and discussion in team: thus, the knowledge advances through iterative acquisition, rationalization, and systematization of experimental findings. Informal evaluation of the Gamelunch environment, originated in various settings, workshops, and public exhibitions, confirmed the validity of the adopted methodology, the effectiveness of the proposed sound feedback in shaping continuous interactions, and the performative potential of the augmented objects. Although the sound identity influences the initial perception of the object being manipulated, it is the articulation of the dynamical properties of the given sound and its fitting with the continuous, exerted action that sensibly embodies the sound feedback in the user/object interaction loop. The collection of a set of basic tutorials and exercises in sonic interaction design will help to broaden the line of investigation and progressively provide guidelines with the aim of improving the awareness of the sound design process.

Note

1. Gmeiner, G. The Table Recorder: Instrument for everyday life's patterns, http://www .fregment.com/.

References

1. Dourish, P. (2001). *Where the action is: The foundations of embodied interaction.* Cambridge, MA: MIT Press.

2. Essl, G., & O'Modhrain, S. (2006). An enactive approach to the design of new tangible musical instruments. *Organised Sound, 11*(3), 285–296.

3. Ishii, H. (2008). Tangible bits: Beyond pixels. In A. Schmidt, H. Gellersen, E. van den Hoven, A. Mazalek, P. Holleis, & N. Villar (Eds.), *TEI 2008—Proceedings of the 2nd international conference on tangible and embedded interaction* (pp. xv–xxv). New York: ACM Press.

4. Varela, F. J., Rosch, E., & Thompson, E. (1991). *The embodied mind: Cognitive science and human experience.* Cambridge, MA: MIT Press.

5. Buxton, B. (2007). *Sketching user experiences: Getting the design right and the right design.* Waltham, MA: Morgan Kaufmann.

6. Anceschi, G. (2006). Basic design, fondamenta del design. In G. Anceschi, M. Botta, & M. A. Garito (Eds.), *L'ambiente dell'apprendimento—Web design e processi cognitivi* (pp. 57–67). Milano, Italia: McGraw-Hill.

7. Findeli, A. (2001). Rethinking design education for the 21st century: Theoretical, methodological, and ethical discussion. *Design Issues, 17*(1), 5–17.

8. Franinović, K. (2008). Basic interaction design for sonic artefacts in everyday contexts. In *Focused—current design research projects and methods* (pp. 95–112). Bern, Switzerland: Swiss Design Network.

9. Lupton, E., & Phillips, J. C. (2008). *Graphic design: The new basics*. New York: Princeton Architectural Press.

10. Gaver, W. W. (1993). What in the world do we hear? An ecological approach to auditory event perception. *Ecological Psychology, 5*, 1–29.

11. Valimaki, V., Pakarinen, J., Erkut, C., & Karjalainen, M. (2006). Discrete-time modelling of musical instruments. *Reports on Progress in Physics, 69*(1), 1–78.

12. Cook, P. R. (2002). *Real sound synthesis for interactive applications*. Natick, MA: A. K. Peters.

13. Rocchesso, D., Bresin, R., & Fernström, M. (2003). Sounding objects. *IEEE MultiMedia, 10*(2), 42–52.

14. van den Doel, K., Kry, P. G., & Pai, D. K. (2001). Foley automatic: Physically-based sound effects for interactive simulation and animation. In *SIGGRAPH '01: Proceedings of the 28th annual conference on computer graphics and interactive techniques* (pp. 537–544). New York: ACM.

15. Schön, D. A. (1983). *The reflective practitioner*. London: Basic Books.

16. Jordà, S., Kaltenbrunner, M., Geiger, G., & Bencina, R. (2005). The reacTable. In *Proceedings of the international computer music conference (ICMC 2005)* (pp. 579–582).

17. Patten, J., Recht, B., & Ishii, H. (2002). Audiopad: A tag-based interface for musical performance. In *NIME '02: Proceedings of the 2002 conference on new interfaces for musical expression* (pp. 1–6). Singapore: National University of Singapore.

18. Polotti, P., Delle Monache, S., Papetti, S., & Rocchesso, D. (2008). Gamelunch: Forging a dining experience through sound. In *CHI '08: CHI '08 extended abstracts on human factors in computing systems* (pp. 2281–2286). New York: ACM.

19. Susini, P., Misdariis, N., Lemaitre, G., Rocchesso, D., Polotti, P., Franinovic, K., et al. (2006). Closing the loop of sound evaluation and design. In *Proceedings of the 2nd ISCA/DEGA tutorial and research workshop on perceptual quality of systems*, Berlin.

20. Delle Monache, S., Polotti, P., & Rocchesso, D. (2010) A toolkit for explorations in sonic interaction design. In *AM '10: Proceedings of the 5th audio mostly conference*, (pp. 1–7). New York: ACM.

21. Rocchesso, D., Polotti, P., & Delle Monache, S. (2009). Designing continuous sonic interaction. *International Journal of Design*, *3*(3), 55–65.

22. Warren, W. H., & Verbrugge, R. R. (1984). Auditory perception of breaking and bouncing events: A case study in ecological acoustics. *Journal of Experimental Psychology*, *10*(5), 704–712.

23. Delle Monache, S., Devallez, D., Polotti, P., & Rocchesso, D. (2008). Sviluppo di un'interfaccia audio-aptica basata sulla profondità spaziale (an audio-haptic interface based on spatial depth). In *Proceedings of 17th CIM Colloquio di Informatica Musicale* (pp. 109–114). Venice, Italy.

9   ZiZi: The Affectionate Couch and the Interactive Affect Design Diagram

Stephen Barrass

The limbic brain that developed in the early mammals is the seat of emotions, ludic playfulness, and maternal behaviors such as empathy and nursing [8]. This evolutionary heritage may provide a shared foundation for affective communications between humans and their pets.

Even a dog that presumably doesn't have any sophisticated intelligence about human-human interaction can tell if it has pleased or displeased its owner. [7]

Affective user interfaces could also allow computers to respond to the emotional state of the user. Reynolds and Picard demonstrated this idea by measuring emotions such as frustration and anger from the force of a computer user's grip on the mouse and the expression in his or her voice when describing what he or she was doing.

Product designers design furniture, kitchen appliances, and cars to provide positive and pleasurable experiences. The Hedonomics movement proposes that interaction designers should also design user interfaces to provide pleasurable experiences [3]. An example of the intersection between product design and interaction design is found in AIBO the robotic dog, marketed by Sony in 1989. AIBO was designed to be playful and to engage empathy by communicating its moods and feelings.

Your AIBO's personality and moods will change depending on the relationship it has with you and its environment.

A gentle stroke on AIBO's "touch" sensors will trigger an adorable response.

AIBO can display various expressions and feelings, like *joy, sorrow,* and even *anger.* Depending on its mood, AIBO will flap its ears or wag its tail. Its face will show you whether it's *happy* or *sad* [11].

Children and elderly people did feel real affection for AIBO [13], and children also felt a sense of ethical responsibility toward it [6]. Researchers into vocal expression have collected 40,000 samples of children speaking to an AIBO [10].

The sound designer Tekemura Nobukazu, who designed the sounds in AIBO, commented that "usually people don't think consciously of what it is like to be *angry* or to *cry*. . . . Humans can obviously use words to express themselves. To create the sounds of emotions was a difficult task."

The International Affective Digitized Sounds (IADS) project has compiled a corpus of 111 nonverbal sounds that convey a wide range of emotions [2]. Participants in IADS studies rate each sound on an *affect grid* with axes of valence and arousal. Valence is a scale from *displeasure* to *pleasure*, whereas arousal is a scale from *sleepy* to *excited*. This model of affect has also been used with visual images, perfumes, facial expressions, and many other sensory stimuli. A rating of the IADS database by the author is shown in figure 9.1. Some examples include a dog growling (106 usages), rooster crowing (120), pig grunting (130), children playing (224), baby crying (260), a girl screaming (276), a thunderstorm (602), and a bar of a Beethoven symphony (810). The results are overlaid with Russell's circumplex of emotions [9] to allow analysis in terms of both the dimensional and categorical theories of affect [12].

9.1 Description of the Work

ZiZi, the affectionate couch created by Stephen Barrass, Linda Davy, Kerry Richens, and Robert Davy, is a couch that provides both physical and emotional support. The couch expresses how it feels by sounds and purring vibrations. ZiZi whines for attention when *bored*, yips with *excitement* when you sit, growls with *pleasure* when you pat her, and purrs with *contentment* when stroked. Making the couch feel contented makes you feel *contented* too.

ZiZi was designed for the Experimenta House of Tomorrow exhibition, which asked installation artists to envisage a domestic situation of the future [4]. The theme led us to imagine how the beeps of the microwave, dishwasher, and the vacuum cleaner might change if more complex sounds were possible. We also explored what would happen if sounds were embedded in objects around the house that currently sit there quietly, and what kinds of sounds furniture would make, and why. Moreover, would this increase the *irritation* we feel in response to meaningless noises, or could it be a *pleasurable* experience? To answer all these questions, we designed a couch that had a character expressed by sounds. The sounds were designed to attract people to sit on the couch and stroke it so that it would purr. ZiZi, the Affectionate Couch was exhibited in the House of Tomorrow at the Black Box Arts Centre in Melbourne in 2004 (see figure 9.2).

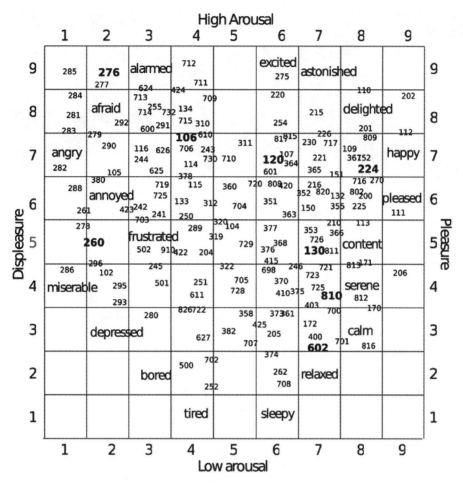

Figure 9.1
Rating of the IADS database on the affect grid overlaid with Russell's circumplex of emotions.

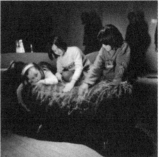

Figure 9.2
ZiZi the Affectionate Couch at Experimenta House of Tomorrow. Photos courtesy of Experimenta
Media Arts.

9.2   Interaction

The fake fur covering Zizi is striped scarlet and brown to attract attention, with an
animal-like form, although it does not have a head or a tail. There are raised arm-
supporting bumps with a mane of fur close to where one's hands fall when seated, to
cue patting. A motion detector inside the couch senses movement up to 3 m away.
The signal from the sensor varies in strength and pattern with the distance and kind
of motion. This signal is analyzed by a microprocessor to identify four states of interaction that we labeled *nothing*, *sitting*, *patting*, and *stroking*. The shape of the ottoman
couch allows people to sit on both sides, and three or four people can interact at the
same time. Those on the couch, and others in the vicinity, can hear her sonic responses
to the interation. This makes ZiZi an example of a *many-to-many* interface where multiple people experience the response to the sum of interaction. When there are fewer
people the responses to individual actions are clearer.

9.3   Sound Design

The sounds animate the couch with an affectionate pet-like character. The sounds
are designed to attract attention, reward sitting, encourage patting, and convey
contentment.

The sound design began with explorations of dolphins clicking and whale songs,
recordings of pet dogs, and synthetic beeps and tones from appliances. The palette of
sounds was constrained by the hardware to sixteen short mono audio files. The final
palette of sixteen was selected to encourage and reward different kinds of interaction
with the couch, with four interaction states containing four sounds in each:

1. *Nothing* Sounds attract attention. Short (½ second) and high-pitched (whimper, whistle, kiss-kiss, tweet-tweet).

2. *Sitting* Sounds reward sitting. Slightly longer (1 second) with sharp attack to produce a haptic quiver as well (smooch, giggle, whizzer, grunt).

3. *Patting* Sounds encourage patting. More sustained (2 seconds) with a more regular pattern of vibration (happy, joyjoy, growl, whizwhiz).

4. *Stroking* Sounds convey contentment. Longer more sustained sounds to produce haptic purring effects from the vibrators (raspberry, yowl, purr, squeal).

Four sounds can be played at once through a four-channel mixer to produce a changing combination that reduces any sense of repetition. The sounds are also routed to nine subwoofer rumble-packs inside the couch that produce vibrations and purring effects that can be felt by people sitting on the couch.

The sound palette can be heard at Barrass [1].

9.4 Observations

Although ZiZi looks like a piece of furniture, her sounds convey the impression that she *likes* attention and *enjoys* being stroked. The audience at the House of Tomorrow exhibition included people of all ages, from toddlers to the elderly. There were no written instructions about what to do. People copied the behavior of others in a "monkey-see, monkey-do" style. Many explained how ZiZi "worked" to each other and to strangers. The effectiveness of the sounds was highlighted at a noisy gallery opening when people sitting on ZiZi behaved as though "she" was a normal couch instead of stroking or playing with her.

Gallery attendants who watched ZiZi for the month of the exhibition reported that the usual period of interaction was several minutes, and that some people went back more than once. When packing up at the end of the exhibit, the attendants referred to ZIZi almost as a person and waved goodbye to her. A teacher wrote to say ZiZi had a marked positive effect on social behavior in her class of autistic children. The popularity of ZiZi in the House of Tomorrow led to curatorial selections for the Experimenta National Tour 2005, Seoul Media City Biennale 2005, Experimenta Under the Radar in the UK in 2006, the International Symposium on Electronic Arts in Singapore in 2008, and Creative Industries week in Shanghai in 2009. ZiZi has been stroked by more than 100,000 people in these exhibitions, and her faux fur has been replaced three times. She is now in the permanent collection of the Museum of New and Old Art in Hobart, Tasmania.

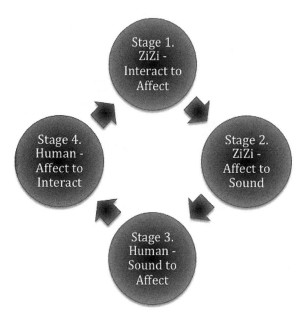

Figure 9.3
Affect interaction loop.

9.5   Discussion

The interaction with ZiZi appears very simple compared to many other installations. Nevertheless, the fondness expressed by attendants and audiences suggests that the interface has been successful in producing empathy and feelings for an inanimate object. The interface also produced playful behaviors and social interactions around it. The effect on visiting autistic children may be explained by their heightened sensitivity to nonverbal sounds [5]. A further development of this sonic interaction design could focus on sounds with positive valences to encourage some behaviors and negative valences to discourage other behaviors.

These ideas led us to consider how to design a couch with a different character and sounds. A more detailed analysis of the interaction with ZiZi is shown in figure 9.3.

In stage 1, ZiZi has an emotional response to the current interaction (nothing, sit, pat, stroke) such as (*bored, excited, happy, content*). In stage 2, ZiZi expresses her current emotion with a sound such as tweettweet, giggle, happy, or purr. In stage 3, the person who is interacting feels an emotional response to the sound that ZiZi makes. And finally, in stage 4, the person interacts with ZiZi depending on his or her own emotional state. This conversational loop has a mapping to or from affect at each stage that can be specifically designed.

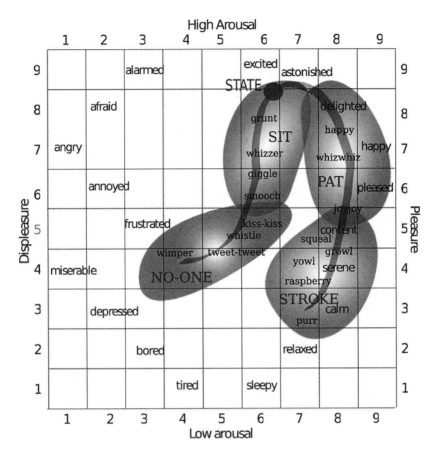

Figure 9.4

Interactive affect design diagram (IADD) for the sounds in ZiZi the Affectionate Couch.

9.6 The Interactive Affect Design Diagram

The relationship between sound and affect can be described by an affect grid. The rating of sounds in the affect grid can also be used to convey affect by sounds. The idea that the affect grid could be used to understand a sound design was tested by rating the sounds in ZiZi, as shown in figure 9.4. The whines and whistles designed to *attract* attention convey boredom and frustration. The sounds in the state of *sitting* convey emotions of excited and astonished with positive valence and high arousal. The sounds of *patting* convey increased pleasure with emotions such as being delighted and happiness. The sounds of *stroking* convey pleasant emotions but with lower levels of arousal, such as contentment and serenity.

The mapping from interaction to affect is described by labeling the interaction states *nothing, sit, pat,* and *stroke* on the affect grid. The sounds that ZiZi makes in each state are shown by shaded regions associated with each label. The transition between interaction states is a continuous path, with a bead showing the current state that moves along it in response to the motion sensor.

The interactive affect design diagram (IADD) transforms the affect grid into a tool for design. The process of designing an interactive affective sonic character begins by drawing a path in the IADD to specify the mapping from interaction to affect. The path for ZiZi is on the pleasure side of the diagram. A grumpy character would have a path in the other half of the diagram, and a more complex character would show both positive and negative valences. The designer then selects a palette of sounds for the character and rates these on the affect grid. It takes less than an hour to rate the IADS corpus, so scaling another palette is practical. The next step is to shade regions that group sounds into interaction states. The final stage is to design the physical affordances of the object to guide the user in understanding how to interact (e.g., attract attention, suggest sitting, cue patting).

In future work we will continue to develop the interactive affect design diagram by applying it in creative projects. In particular, inspired by the limbic communication between humans and other animals, we are interested in exploring the potential for interfaces with positive and negative affect that encourage some forms of interaction and discourage others. The IADD could also generalize to the design of other modalities, such as touch or color, rated in the affect grid.

References

1. Barrass, S. (2010). The sounds of ZiZi, http://stephenbarrass.wordpress.com/2010/03/02/sounds-of-zizi/ [Retrieved March 1, 2010.]

2. Bradley, M. M., & Lang, P. J. (2007). *The international affective digitized sounds* (2nd ed.; IADS-2): *Affective ratings of sounds and instruction manual.* Technical report B-3. University of Florida, Gainesville, FL.

3. Hancock, P. A., Pepe, A. A., & Murphy, L. L. (2005). Hedonomics: The power of positive and pleasurable ergonomics, ergonomics in design. *The Quarterly of Human Factors Applications, 13*(1), 8–14.

4. Hughes, L., Saul, S., & Stuckey, H. (2003). Experimenta House of Tomorrow Exhibition, Black Box Gallery, Melbourne, September 5 to October 3, 2003, http://www.experimenta.org/exhibitions-projects/currentExhibitions/past-2.html.

5. Järvinen-Pasley, A. M., Wallace, G. L., Ramus, F., Happe, F., & Heaton, P. (2008). Enhanced perceptual processing of speech in autism. *Developmental Science, 11*(1), 109–121.

6. Melson, G. F., Kahn, P. H., Beck, A. M., Friedman, B., Roberts, T., & Garrett, E. (2005). Robots as dogs? Children's interactions with the robotic dog AIBO and a live Australian shepherd. In *CHI '05 extended abstracts on human factors in computing systems (Portland, OR, April 2–7). CHI '05* (pp. 1649–1652). New York: ACM.

7. Reynolds, C., & Picard, R. (2001). Designing for affective interactions. In *Proceedings of 9th international conference on human-computer interaction, august 5–10, New Orleans, LA.* Available at http://vismod.media.mit.edu/pub/tech-reports/TR-541.pdf.

8. Panksepp, J. (1998). *Affective neuroscience: The foundations of human and animal emotions.* New York: Oxford University Press.

9. Russell, J. A. (1980). A circumplex model of affect. *Journal of Personality and Social Psychology, 39*, 1161–1178.

10. Schuller, B., Steidl, S., & Batliner, A. (2009). *The interspeech 2009 emotion challenge, interspeech (2009).* Brighton, UK: ISCA.

11. Sony AIBO Brochure. (2007). *AIBO ERS-7 (MIND 3) brochure—2005/2006 edition (latest).* Retrieved from http://support.sony-europe.com/AIBO/1_2_library.asp, March 2009.

12. Stevenson, R. A., & James, T. W. (2008). Affective auditory stimuli: Characterization of the international affective digitized sounds (IADS) by discrete emotional categories. *Behavior Research Methods, 40*, 315–321.

13. Turkle, S., Taggart, W., Kidd, C. D., & Dasté, O. (2006). Relational artifacts with children and elders: The complexities of cybercompanionship. *Connection Science, 18*(4), 347–361.

10 SonicTexting

Michal Rinott

Texting has become a central activity in our digital day and age. In 2008, 75 billion text messages were sent in the United States in one month [1] alone! Texting on mobile phones, using the thumb and the phone pad, has been so influential on a generation of teenagers that researcher Sadie Plant reports that some now use their thumbs for index-finger activities such as turning on lights and pointing [2].

The importance of texting has prompted designers to seek efficient and usable solutions for text entry on mobile devices. One category of approaches, originally created for PDA devices with stylus input, has involved using continuous gestures for writing words, preventing the need to lift the stylus from the surface between letters, as is typical of the number pad interface (e.g., T-cube [3], Quikwriting, [4]). The recent ubiquity of touch screens, most notably the iPhone and table-top systems, has prompted a surge of renewed interest in continuous gesture-based input methods now using the finger (e.g., Swype [5] and ShapeWriter [6]).

Although both show good results in usability and "word-per-minute" efficiency, these gestural solutions are based on focal vision. This is so despite the fact that texting is often performed in situations in which vision is compromised: in the dark, by people with visual disabilities, and while on the move (a recent U.S. insurance survey of 1,503 drivers found that almost 40 percent of those respondents from 16 to 30 years old have said they text while driving [1]).

SonicTexting is a gesture-based text entry system that uses tactile input and auditory output. The goal in making SonicTexting was to create a texting interface in which audio, not vision, would be the central feedback modality. Making texting an auditory—and tactile—interaction is a way to decrease the visual load in mobile situations [7].

Not less importantly, SonicTexting was an attempt to create an engaging interaction that would be challenging and rewarding to master. Rather than maximizing word-per-minute texting efficiency, the emphasis in designing SonicTexting was placed on creating an engaging audiotactile experience. Inspired by the expertise achieved by teens with the number pad interface, SonicTexting was an attempt to tap into the types of audiotactile expertise people gain in playing musical instruments and using gaming controllers. Creating a desirable interaction method that does not rely on vision can include visually disabled users and can show the potential of rich auditory experience in digital artifacts.

SonicTexting was created in 2004 as a Masters Degree project in interaction design. It has been implemented to the level of a working prototype and presented as an interactive installation in various exhibitions.[1] This case study describes the process of designing this particular sonic interaction and discusses the insights gained from observing people using it within different contexts (e.g., a museum, at a conference, at a fair) to designing sonic interactions.

10.1   The SonicTexting System

SonicTexting is an audiotactile system for inputting text using continuous thumb gestures. Sound provides the sole feedback for the gestures, aiding orientation and navigation.

The gestural vocabulary of SonicTexting is based on the Quikwriting model [4]. Quikwriting is a text entry system in which the stylus is never lifted from the surface during writing. The writing area is divided into zones arranged around a central resting zone. To form a character, the user drags the stylus from the central resting zone out to one of eight outer zones, then optionally to a second outer zone, and finally back to the resting zone.

Based on this general gesture model, SonicTexting introduced two main innovations:

• A specialized input device that naturally supports this center-periphery-center motion.

• Feedback for the gesture via continuous, synchronous sound.

The following sections describe the SonicTexting system in detail.

10.1.1   Input Device: The Keybong

The SonicTexting input device, nicknamed the *Keybong,* is a one-handed device that fits in the palm of the hand (figure 10.1). The Keybong consists of a small joystick enclosed in a plexiglass shelling. The joystick movement is limited by a circular boundary. The joystick naturally supports the common gesture pattern of SonicTexting:

Figure 10.1
The Keybong.

moving from a central location, through a specific path, back to the center. The springy return of the joystick to the center requires the user to actively perform only the first part of the gesture.

The Keybong contains a small eccentric motor that provides gentle vibration feedback in the writing process. This tactile layer accompanies and augments the sound layer. The Keybong joystick is also a button: pressing it down clears the entered text. The Keybong form is designed to fit comfortably in the hand, ensure that it is held in a fixed orientation, and be small enough in size to be used inside a handbag or coat pocket.

10.1.2   The Gestural Alphabet

Writing in SonicTexting is performed by moving the Keybong joystick from the center to one of eight "axis" positions around the circle periphery (N, NE, E, SE, S, SW, W, NW) and either returning to the center, or moving around to another position around the circle (a "nested" position) before returning to the center.

The gestural alphabet is presented to the user, for initial learning, via a static visual representation of the letter locations (figure 10.2). It is read as follows: to write an *a*, the controller is moved in the *a* direction (NW), then back to the center. To write a *b*, the controller is first moved to the *a* direction, then moved along the circular periphery toward the *b* (N), then back to the center.

Figure 10.2
The gestural alphabet.

The nested nature of the gesture model—whereby reaching nonaxis letters requires first moving to the axis letter, then moving left or right in a zoomed-in periphery—is communicated through the fractal-inspired design of the map, created by duplicating and rotating a basic graphic element to signify the axis letters and the nested letters.

In the Quikwriting model the letter arrangement is by frequency: frequent characters require shorter gestures. In SonicTexting an alphabetical arrangement was chosen in which the letters ascend in alphabetical order clockwise. This order was selected because of a prioritization of memorability over gesture length, given the relative ease of reaching all letters using the Keybong.

10.2 Sound in SonicTexting

In SonicTexting sound provides continuous feedback during movement—an interactive sonification of the gesture path. Sound is also used after the gesture for a letter-by-letter readback of completed words.

The functions of sound in SonicTexting are:

- To guide the first, outward-bound movement to the axis letter
- To guide the next movement (if needed) around the periphery
- To provide feedback for the entry of the letter

- To provide feedback for the writing of whole words
- To aid the memorization of the gesture paths

These functions are achieved through the following sonic features:

Phonemes   Looped letter phonemes, in a female voice, are played in synchrony to the user's movement. In the current implementation, the phonemes are Italian (e.g., the sounds /ah/, /bhe/, /ch/). As the controller moves outward in one of the eight axis directions, the relevant phoneme is sounded in a loop. When the controller is located between these axis directions (e.g., NNW), the phoneme sounds overlap. Navigating to a letter thus requires a process similar to that of tuning to a station on an analog radio—finding the location of the cleanest sound.

Loudness   Loudness is a function of the distance between the controller and the location on the periphery of the controller range. As the controller moves from center to periphery, the volume of the looped phonemes grows louder. Thus, the user needs to find the loud sound of the desired letter.

Pitch   The letter phonemes are sung in different pitches according to their position around the circle periphery. The pitch ascends clockwise, starting at the N, note by note through one octave (/ah/ sung in Do, /bhe/ in Re, /ch/ in Mi, and so on). In this way the gesture path for every letter has a unique tune, according to its path around the circle.

Tactile "acquisition" feedback   A slight vibration is felt when the user reaches the area of a position and "acquires" it. On feeling this the user can let the controller return to the center. This nonsound element was selected to increase the tactile aspect of this audiotactile experience.

Learnability and expertise   As users gain experience with the system, they memorize the letter locations and gesture paths. An "expert mode" was created for users who already know the gesture paths. In this mode, discrete percussion sounds are played when the controller acquires—moves into a close distance to—a position on the periphery. The sounds for these positions are pitched as in the main sound mode to preserve the gesture tunes. The velocity of the movement determines loudness of the initial part of the sound (the attack), so that faster movements create stronger sounds. Expert mode creates a very compact sound pattern, as opposed to the longer looping phoneme sounds.

Readback   Following completion of a word (after a space character), the letter phonemes are read back to the user in sequence. This serves as a confirmation to the user working "eyes free" that the word has been written correctly. Moving the Keybong in any direction stops the readback and returns to live sound feedback.

10.3   Sonic Design Considerations

In the following paragraphs some insights from the design process are described:

Sonic content   In this writing task, it may be more appropriate to use letter names rather than phoneme sounds (e.g., the letter name "Bee" as opposed to the phoneme /bh/). However, phoneme sounds were selected in order to create a "sonic texture" of speech sounds rather than letters. For this reason also, the readback function uses the phoneme sounds for letter-by-letter readback rather than speech-engine-generated whole words.

Voice selection   Throughout the design iterations, a number of different people gave their voices to SonicTexting. One of the most notable was a very low male voice, which gave the experience of a special quality of darkness. However, the preferred voice was a female voice of a (nonprofessional) singer who spoke/sang the phonemes with a clear, resonant voice.

Spatial aspects   In the initial design stages, the sonic design task was conceived of as a direct sonification of the gesture map. A number of attempts were made to sonify spatial aspects of the space. One attempt was an inhale sound when leaving the center area and an exhale when returning to it. Another was a "bump" sound when the controller moved to another "area." In the final design these were abandoned in the search for the most minimal sonic representation. The current spatial mapping, in which the volume and pitch change through an octave, seems natural to the round space with its eight peripheral positions. The rising pitch with ascending letter order and falling pitch with descending letter order correspond to alphabet songs in different languages, which tend to contain this attribute.

Feedback and feed-forward   In the basic sound mode of SonicTexting, sound provides a means of learning the gesture scheme. In this situation the sound provides feed-forward—guidance as to where to go. In the expert sound mode, sound provides feedback—an indication that the periphery point has been reached. It is assumed that in basic mode the user moves in search of the next letter, whereas in expert mode the user knows the position and needs a minimal form of confirmation. The slight vibration on "acquisition" of the position creates the possibility for silent operation.

10.4   Observations

SonicTexting has been implemented as a working prototype for an installation setting. The Keybong controller was connected to an (unseen) computer, the sound played back through a speaker above the user. A screen showed the static gesture map and a text input line where entered letters could be seen.

No formal usability testing was performed on the prototype. However, the Sonic-Texting installation was presented and experienced in a number of contexts: a design museum, a design exhibition, and an HCI conference. In total over 2,000 people tried the prototype in over 60 days of display. This section describes the main insights gained from observing visitors to the installation.

• The majority of visitors, both adults and children, reacted enthusiastically to the experience and were motivated to learn to SonicText.

• In the exhibition setting, visitors' interpretation of SonicTexting varied: some saw it as a game, others as a kind of musical instrument, and still others as a desirable mobile device feature.

• Most visitors could use SonicTexting to successfully write a word after 1–2 minutes of practice, a much shorter time than had been anticipated. Thus, the first part of the SonicTexting learning curve proved steeper than expected.

• Visitors tended to expect visual feedback to appear on the gesture map. Instructing them to "move in the direction of the letters using sound," and to "seek the pure sound," helped increase their dependence on the audio feedback and thus improved their performance.

• There were large differences among people in the degree to which they could make use of the sound output. Some "caught on" immediately and started using the sound to navigate, but for others "tuning in" to the auditory channel was more difficult to do. Children tended to be very good at this task!

10.5 Conclusions for Designing Sonic Interactions

The SonicTexting experience is in three spaces: the visual space of the gesture map, the tactile space of the Keybong movement, and the auditory space of the sound. In SonicTexting, users need to depend on their hand-ear coordination to find the letters, rather than the hand-eye coordination of the visual map. One user tried to express this by comparing the Keybong with a keyboard: "With a keyboard, the space is laid out in front of you; with the Keybong it is more abstract: the space is in the head, not in the Keybong."

In a visually dominated digital world, people are not accustomed to focusing on sound as a main feedback channel, especially for a traditionally visual activity such as text input. Methods that direct attention to the auditory channel help people change this initial tendency. Careful design of the experience—through good sound quality, lo-fi visuals, presentation of instructions via audio—dispose people to open their ears.

The interpretation of the SonicTexting prototype as a game by some exhibition visitors, as well as the tendency of visitors to return to the installation for additional practice, indicate that the interaction was enjoyable to many. Although it cannot be proven in this study, it is this author's impression that the strong correspondence between movement and sound, and the audiotactile quality of the interaction, are central causes for this enjoyment.

SonicTexting was an academic project and has not been developed into a commercial product. Despite its relative simplicity in computional implementation—recorded voice with real-time volume modifications—it created an engaging experience that people succeeded in using "on the spot." The installation generated interest in different communities—the CHI community, the industrial design community, and the SID community. This is encouraging for SID projects and for students venturing into the field interested in using prototyping to communicate SID ideas.

Acknowledgments

The MaxMsp implementation of SonicTexting was created with the help of Yon Visel. The author thanks Michael Keislinger, Jan Christoph Zoels, and Edoardo Brambilla for different forms of help and support.

Note

1. *Touch Me,* Victoria & Albert Museum, London, UK, June–August 2005;

Manual Labor, Facelift festival, Genk, Belgium, April 2005;

Interactivity Chamber, CHI2005 Conference, Portland, OR, April 2005;

BITE—a Taste of Interactive Installations, Fondazione Sandretto, Turin, Italy, July 2004;

Salone Del Mobile, Triennale Museum, Milan, Italy, April 2004.

References

1. As Text Messages Fly. Danger Lurks. (2008). *New York Times,* September 28. http://www
.nytimes.com/glogin?URI=http://www.nytimes.com/2008/09/20/us/20messaging.html&OQ
=_rQ3D3Q26adxnnlQ3D1Q26orefQ3DsloginQ26adxnnlxQ3D1222720462-3yyQ511eiApAel87t
pPI2vKA&OP=3d613473Q2FhD0Q51hQ5B_Q27Q7DQ3F__xzhzQ5DQ5DShQ5D-hzQ5DhtQ7Dh
zQ5Dg0Q7DQ7DQ23dQ3EedQ609xgJ.

2. Plant, S. (2000). On the mobile: The effect of mobile telephones on social and individual life. http://web.archive.org/web/20080625013404/http://www.motorola.com/mot/doc/0/234 _MotDoc.pdf.

3. Venolia, G., & Neiberg, F. (1994). T-cube: A fast, self-disclosing pen-based alphabet. In *CHI '94: Proceedings of the SIGCHI conference* (pp. 265–270). New York: ACM.

4. Perlin, K. Quikwriting: Continuous stylus-based text entry. In *Proceedings of the 11th annual ACM symposium on user interface software and technology* (pp. 215–216). New York: ACM.

5. Swype—Text Input for Screens. (n.d.). http://www.swypeinc.com.

6. ShapeWriter. (n.d.). http://www.youtube.com/watch?v=sBOyGp25sSg.

7. Pirhonen, A., Brewster, S., & Holguin, C. (2002). Gestural and audio metaphors as a means of control for mobile devices. In *Proceedings of the SIGCHI conference* (pp. 291–298). New York: ACM.

11   The A20: Interactive Instrument Techniques for Sonic Design Exploration

Atau Tanaka, Olivier Bau, and Wendy Mackay

Advances in digital audio technology have had an impact on both musical performance practice and on music listening and consumption. Interaction plays a central role in both domains. In the former, interaction is centered on instrumental gesture, often requiring training and technique. In the latter, interaction tends to take place between a listener and collections of music, where the end user searches, retrieves, and organizes sequences of music titles but ultimately does not directly manipulate sound. We present a sequence of studies that draws on interactive musical instrument design developed in the field of new instruments for musical expression (NIME) coupled with user-centered design (UCD) techniques to investigate the potential of sonic interaction in the context of future personal music usage scenarios.

Whereas UCD implicates the end user in a design exploration process, NIME research is typically focused on specialist applications of interactive music technologies. We were interested to see, through an iterative process that linked UCD and NIME, whether we could identify emergent themes from users' descriptions of interacting with musical content in everyday life and then propose advance forms of sonic interaction as ways to address these themes. In a three-stage process, we (1) study preexisting music listening habits and conduct UCD sessions on the sharing of musical experience, (2) draw on NIME techniques to build, then introduce as design probe, the A20, a multifaceted interactive sonic object, and (3) conduct a user study evaluation of the A20's effectiveness in facilitating the scenarios imagined by users and its potential for generating new design ideas.

11.1   User-Centered Design Sessions

We conducted a series of user-centered design studies to elucidate contemporary personal music player usage. These studies were comprised of ethnographic interviews to establish existing use patterns and participatory design workshops to imagine and generate hypothetical future usage scenarios.

11.1.1   Study 1: Interviews

In the first study, we conducted twenty semistructured interviews focused on existing practice with personal music players. Interviews lasted approximately 25 minutes each and were held in four different university campus locations in Paris. We selected French and American students aged 20–25, young adults in a time of cultural discovery and active sociality. They represented a rich cross-cultural, cross-linguistic, and cross-expertise mix whose fields of study ranged from humanities and liberal arts to design and engineering.

We asked participants to identify specific examples of personal music player use, as well broader themes such as mobile communication and the emotional contexts of music. We used *critical incident technique* [5] to elicit specific recent memories of personal music player use in context and examples of interruptions, or specific moments at which music listening use was interrupted. We also asked participants to describe how and when they share music with friends and family and how music listening was integrated with other activities in their daily lives.

With a focus on personal music player use, our interviews brought forth specific examples of how portable music technology is used socially. Interviewees focused on musical exchange as a way to stay in touch:

If I'm listening and think that this friend will like it, I'll start up MSN to tell them [sic].
    We'll keep the Skype connection open, and I can hear in the background what she's listening to.

Users also speak of how the evocative power of music reminds them of people or events:

. . . the song reminds me of that time with my boyfriend. . . . and when it comes up, I've even texted him the lyrics as a way to say "hi."
    If I was listening to music when reading a book—when I reread the book at another time, even in silence, I hear that same music in my head.

Finally, users indicated an interesting set of personalized strategies for sorting out and searching through collections of music or preparing compilations of music for friends:

I'll download a bunch of stuff without thinking too much and later go through it and start making playlists for friends.

These results indicate that users exhibit a tendency to create personal and social associations through music. They demonstrate ways in which the participants use existing technologies to turn the act of music consumption into acts of socialization.

However, the interviews also revealed that the participants were aware of the isolating and potentially antisocial nature of headphone-based music listening. When asked to speculate about desirable new technology, they repeatedly invented scenarios in which they exchanged musical messages with friends or lovers, or fantasized about eavesdropping on music listened to by strangers. The socializing power of music was overwhelmingly evident in their ideas despite the isolating effects of present-day music players.

11.1.2 Emerging Themes

From these interviews we identified three emerging themes: *communication, associa-tion*, and *navigation*.

Communication can be of varying social or geographical proximity, one-on-one or in a group. Association can be with a person, an event, a memory, or between two pieces of music. Navigation can mean navigating a playlist, navigating in a city, or in a social network. Many people talked about how music reminded them of specific events and people or served as a trigger for contacting someone. Although strangers and close ones were addressed in user scenarios, they were not treated equally. Passer-by exchange was based more on the potential of social exchange, whereas exchange among friends was based on the personalities involved and memories they shared. Users confirmed the evocative power of music to recall and recreate important per-sonal moments.

11.1.3 Study 2: Participatory Design Workshops

In the second study we conducted two participatory design workshops. We utilized the emerging themes identified in the interviews to help guide a structured brain-storming process. Each workshop consisted of three sessions: *scenario building, brain-storming*, and *video prototyping*.

Scenario building was based on individual scenario notation in storyboard form followed by group discussion. Participants were asked to describe recent personal music player use and to reflect back on critical incidents explored during the interview process. Pictorial storyboarding aided the workshop participants to turn an anecdotal account from daily life and represent important steps in abstracted graphical form. This was a crucial step in using real-life events to ground the brainstorming that followed.

Brainstorming sessions took place in breakout groups of three. Participants com-bined aspects from their individual scenarios to form an imaginary meta-scenario and then started imagining how the music-listening activity in those settings could be improved, augmented, or expanded, possibly by new technologies. We used *idea cards* (figure 11.1) in the form of flash cards to help inspire the participants. These cards presented compelling and evocative antonyms as idea-dyads such as these:

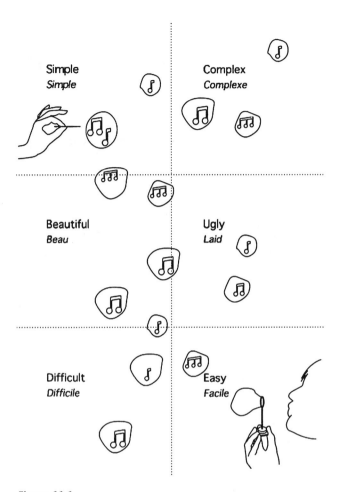

Figure 11.1
Idea cards ready to be cut along horizontal lines for structured brainstorming.

- simple-complex
- difficult-easy
- personal-anonymous
- flexible-rigid

The construction of a meta-scenario allowed the real-world grounding from the previous step to be abstracted beyond the personal and anecdotal and allowed group work. This permitted identifying aspects of the scenario that could be enhanced by new technologies. The idea cards provided conceptual grounding to keep the focus from becoming overly technology oriented. We drew on the emerging themes identified in the interviews to inject ideas about metaphors and new form factors. For example, a card evoking a compass evoked a *navigation* metaphor, and a photo album card became a metaphor for *association*.

Each group then acted out its scenario and filmed it with a simple camcorder to create a *video prototype*. They were given a range of materials—foam, paper, clay, colored markers—to create mock-ups that would help in visualizing an imaginary device, accessory, or concept. The process invited participants to project their storyboard into the physical world, imagining form factors and an actual use dynamic. Through this process of *enaction*, participants were able to test their scenario as an embodied experience [4]. The videos, lasting 2–5 minutes, were then presented to the group for discussion.

The participants' individual written scenarios in part proposed imagined solutions to the stories from the earlier interviews. The scenarios included the following:

Shared Jukebox A three-dimensional virtual interface (mocked up with foam blocks hanging from string) where people can select songs together.

Laura likes Punk Laura is bored with her music. She sees a punk in the Metro, approaches him, docks her MP3 player to his. They exchange music and socialize (figure 11.2).

Marbles Users place marbles in a palette. Each marble represents a piece of music as well as a task in a to-do list, and their placement represents a geographic location. Meeting at work, users share their commuting experience (figure 11.3).

11.2 A20

Parallel to the user-centered design sessions, we used NIME techniques to develop a novel interface for sonic interaction. We sought to draw on qualities of interactive musical instruments in conceiving a platform to test scenarios of future personal music

Figure 11.2
Video brainstorming skit from edited video.

Figure 11.3
Geographic musical marble tasks.

player usage. Qualities of a technologically enhanced musical instrument include that it:

- Is an autonomous, self-contained system
- Allows real-time interaction
- Permits direct object manipulation
- Has the ability to change contexts

In this definition of an instrument, we distinguish the expressive and at times idiosyncratic nature of musical instruments from the utilitarian nature of tools [11]. The act of instrument building thus has different objectives than the task optimization of engineering. The idea here was to draw on the NIME instrument-building tradition to create a generic interface that could nonetheless afford rich, embodied interaction with sound. The hope was that, by focusing on modes of sonic interaction emerging from the UCD studies, we might arrive at a prototype of a kind that would not otherwise arise out of classical product development or task-based interaction design.

As a point of departure for the A20, we were inspired by the multiple visual display modes afforded by the novel form factor of the D20 [9]. The D20 is a visual interface embodied in an icosahedron—a near spherical polyhedron with twenty discrete triangular facets. The geometry of the form affords several display modes: *equator, pie, and binary* (figure 11.4). These display modes afford several distinct types of interaction to be executed by rotating the device (figure 11.5). This range of visual representations and corresponding forms of interaction make the D20 a generic visual interface. For the A20, we sought to adopt the form and build a functional audio device whereby each facet is an independent loudspeaker. The hypothesis underlying the design of

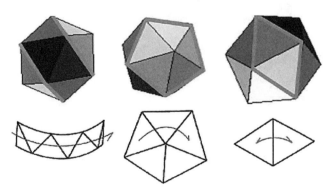

Figure 11.4
Three display modes of the icosahedron: *equator, pie,* and *binary.*

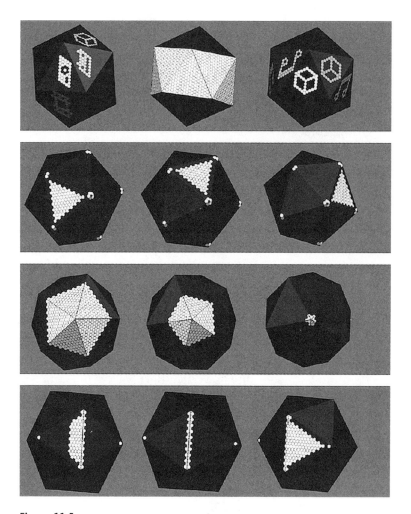

Figure 11.5
Application cases of the D20's display modes.

the A20 is that topological features of the icosahedron form factor that gave the D20 its multiple display modes would translate to the audio domain and yield similar modes of sonic interaction.

The multifaceted loudspeaker draws on prior work in the construction of spherical loudspeakers [1; 13]. This previous work has for the most part been for research on generating and controlling sophisticated diffusion patterns for single sound sources, in part to emulate complex radiation characteristics of acoustical musical instruments. Here we were interested in the multidirectional properties of the icosahedron for the audio display of multiple independent sound sources. Sensors have been added to spherical speakers [12] to add a performative element with the incorporation of sensor-based capture of gesture that is at the heart of NIME. With the A20 we were interested in capturing gestures associated with rotation of the object, to couple with the programmable, multifaceted sound output. In this way, the A20 is both an input device and an output (audio display) device.

We built the twenty-sided frame using rapid prototyping techniques of three-dimensional printing with stereolithography and attached lightweight flat-panel loudspeakers on each facet, with each panel cut to a triangular shape. The object was 22 cm in diameter with each side 14 cm (figure 11.6). Inside the structure were a multichannel computer audio interface allowing each facet to be addressed independently, a triaxial gyroscope sensor to sense rotation, and pressure sensors detecting pressing on the speaker surfaces. Details on the hardware and software of the A20 have been previously reported [2].

Figure 11.6
The A20.

11.3   A20 as Design Probe

We introduced the A20 as a design probe [6, 7] to the subjects from the first study and conducted a user study of the A20. We sought to use the A20 to validate its ability to successfully address the emerging themes from the workshops and interviews. The device was a functioning research prototype that would enable participants to enact their scenarios from the structured brainstorming. We hoped that by injecting a novel interface for sonic interaction, subjects could engage with new forms of interaction with sound and music. The A20, although it had specific functional capabilities, was a generic interface that we hoped would liberate the workshop participants from the strong cultural associations identified with commercial mobile music devices such as mobile telephones and MP3 players.

We conducted two half-day sessions with six and seven participants, respectively. The makeup of the participants was similar to that in the previous workshops. The workshop had two phases: (1) to test the perceptual qualities of the A20 as a multi-faceted audio display device and (2) to test the effectiveness of the A20 in acting out scenarios similar to those from workshop 2. The results from the perceptual studies localizing source facet and identifying dynamic sound diffusion trajectories across facets have been reported [2].

The use of sensor input allowed us to illustrate and test elements of interaction representing each of our three global themes (table 11.1).

The most straightforward example was navigating a playlist by flicking the A20 left or right. Here we provided users a direct comparison with a classic existing personal music player task. After a brief moment of feeling out the sensitivity of the system, users quickly mastered executing this task. Reactions were overwhelmingly positive, users finding it ludic and fun. Some users even got into a performative rhythm of flicking successively or back and forth. This task was an example of *navigation* and exploited the whole-object interaction and audio-only nature of the device to allow

Table 11.1
A20 elements of interaction

Navigation	Flick A20 left/right to move forward and backward in a music playlist
Communication	Clicking on a facet "sends" sound by rotating around equator and fading away
Association	Stimulus stays fixed relative to surrounding space through compensatory panning as user turns

a form of direct manipulation of the musical content. The playlist was no longer a textual representation of the music but the music itself.

The second A20 vignette presented to workshop participants was the possibility of making a musical selection by pressing on a face and having the selected piece rotate around the facets of the object. This was presented as an abstract element of interaction but was immediately seized on by several participants to enact *communications* scenarios. One such scenario involved the user turning in space with the A20 to orient himself with respect to a correspondent—an act of *association* of an entity in real space with the topology of the A20. He then selected a piece of music from a particular face to share with the other user. The circular rotation of the music for him represented a sort of "whoosh" effect of sending the music to his friend.

These individual elements of interaction directly tied to technical functionality of the A20 allowed us to present illustrative vignettes that permitted users to imagine their application in each of the global scenarios. The storyboard scenarios that they created as an outcome of this showed how each participant assimilated (or not) the A20's qualities to extend on these themes and to imagine entirely new ones.

Participants were inspired by the polyhedron form factor to imagine combining cardboard mockups of multiple A20s. Communication and music sharing were then facilitated by the physical affordances of the device to allow linking up, Lego-style. These were applied to situations that extended on ideas that emerged in the earlier workshops.

Party DJ: "Each person contributed to the mix at a party with his A20. They [sic] stack up, and the designated DJ determines the priority of the music by putting it higher or lower in the stack."

The form factor also facilitated users to imagine interesting variations on the *association* theme:

Geo-sounds: ". . . the A20 can be a recorder while traveling to capture impressions from different places. Each face saves a sonic snapshot from a place I visit."

The idea of the A20 as a recording device was not part of the A20 technical specification but recurred in participants' imagined scenarios:

Life recorder: ". . . accompanies you through life from birth, recording sounds from special moments, accumulating a history inside."

Some imagined completely new uses for the A20:

A20 dice: ". . . throw the A20 like a die, and it goes into shuffle mode. Toss it on the floor, and it fills the living room with sound."

Modular gaming: "link together multiple A20s where each block and its multidirectional functionality represent a module in a game. The structure made determines the configuration of the game."

The A20 succeeded as a technology probe on several levels. At the very basic level, it gave workshop participants a physical artifact that facilitated acting out scenarios related to the social user of personal music players. The initial scenarios people presented were surprisingly consistent from the interviews at the beginning of the project through to the final workshop.

11.4   Discussion

By building the A20 based on NIME instrument-building methods, we were able to produce a generic interface for sonic interaction that would not have arisen from a pure UCD approach. Conversely, by applying the A20 to scenarios and themes emerging from the UCD sessions, we put interactive music technologies into contexts not typically encountered in NIME. Participants in the UCD study would not have imagined a technology of the sort proposed by the A20 in the initial workshop sessions. At the same time we did not wish to drive the direction of the UCD sessions by introducing the A20 right away. Instead, by identifying emerging themes that were focused on music use needs, we were able to see in the subsequent user study whether the A20 satisfied the needs as articulated in the earlier workshop.

11.4.1   Communication, Association, Navigation

Revisiting the three emerging themes from the UCD workshops, we found that the A20 was able to address each of these themes with different degrees of success. The most successful cases were in tasks related to *navigation*. The topological nature of the A20's sound output coupled with interaction enabled by rotation sensors gave compelling results. Users were easily able to flick the device to navigate through a playlist. Having the interaction input and display output coincident on the same device created the effect of *direct object manipulation,* where the user did not dissociate audio display from control input. Rather, there was the effect that they were directly interacting with the content of sound.

Connected to the theme of navigation was that of *association*, where different facets of the A20 played distinct musical streams representing what could be different people or different places of the sort articulated in the UCD workshop. Rotating the device allowed the user to navigate toward, or focus on and zoom into, the facet, and therefore person or place, of interest. The perceptual part of the user study demonstrated that users were able to distinguish, if not the exact face that was playing, whether the active facet was near or far, and its general direction. By distributing multiple sound sources over distinct faces and coupling them to the movement of the device users were aided to call on instinctive perceptual mechanisms of *streaming* in the process of sound source separation [3].

Finally, tasks related to the theme of *communication* were the most difficult to validate using the A20. The A20 in its prototype form was networked and communications-ready. However, there was little in the multifaceted form factor that directly afforded communicative use. The multiple facets were successful in conveying a sense of multiple associations. In an attempt to convey a sense of message passing and the dynamic information flow inherent in communication, we sought to use dynamic panning techniques to create motion effects of sound across the facets, to swirl around and then be "sent out" away from the object. This effect was difficult to convey. Where was the sound being sent off to? This also demonstrated a limitation of audio display on the A20. Although it was very effective at displaying topologically distributed sound along the surface of the object, it had no capability to project sound into the space around it.

This was compounded by the fact that we were able to fabricate only one prototype of the A20, making it difficult to enact scenarios of multiple A20s interacting with one another. Interestingly, subjects in the user study did use multiple cardboard mock-ups of the A20 to imagine different communicative scenarios, stacking up A20s to assemble playlists, as well as the game-like scenarios reported in the previous section. This shows that although the multifaceted qualities of the A20 may not directly afford communicative activities, the shape and form factor do inspire a sense of modularity and interconnectedness that could be useful in communications scenarios.

11.4.2 Unexpected Results: The Haptic Channel

One characteristic of the A20 that yielded an unanticipated outcome was the tactile sensation afforded by direct contact of the user's hands with the vibrating loudspeakers. Haptic feedback had not been considered in the original design of the A20.

Tactile interfaces are a rich field of research unto itself, and the A20 was not designed to directly address this area. We nonetheless responded to the unexpected response of subjects in the user study to the haptic channel by creating two tests that focused on the coupling of auditory and haptic perception. In the first, we explored the multimodal potential across sound and touch and to see if the haptic channel, by carrying the same information as sound, could augment auditory perception. In the second, we tested whether the haptic channel and auditory channels could simultaneously display distinct information to be differentiated by the user. We were interested to see in what ways the haptic and auditory channels might enter into *synergy* or into *interference*.

Presented with a haptic stimulus under each hand, users displayed a good capability to distinguish two different haptic signatures. Likewise, when presented with two sound sources each playing on a separate facet of the object, users had absolutely no problem indicating whether they were the same or not. However, when presented with one auditory stimulus and one haptic stimulus, users reported greater difficulty in determining whether the two stimuli were from the same source or not.

These findings run counter to our typical understanding of multimodal interaction [8]. Typically we would expect that a task would be better accomplished when presented across two modes of perception. However, in our case monomodal performance (haptic only or audio only) yielded lower error. This points out the importance of *correlation* across modes. The task we presented to the subject was to differentiate two sources—that is, to detect a situation when there was low correlation between two stimuli. Low correlation was more accurately sensed when the stimuli shared modality and was difficult to perceive across independent modalities.

11.5   Conclusions

Music is a powerful, evocative medium. People readily express their musical tastes and speak of the important place music has in their lives. Even if the average layperson today does not play a musical instrument, he or she nonetheless seeks to be expressive through music. This view of musical engagement that does not separate audience from performer is called *musicking* [10]. If musicking levels the playing field across musician and nonmusician, it seemed that techniques heretofore reserved for specialist music applications might be useful to gain insight into music listening and acts of music appreciation. This was the basis of our idea to introduce NIME techniques in a sonic interaction design process for studying personal music experiences. This project sought to couple specialized design approaches from interactive music with participatory design methods to create a working system that fulfilled the expressive and communicative urge of the everyday music listener. With the A20, we have built a design tool to explore digitally facilitated musicking.

The ways in which users from our final study incorporated the A20 in their scenarios confirmed its generic ability as a vehicle with which to execute basic music engagement scenarios. The new ideas that were generated, as well as the ways in which some basic characteristics of the device were *not* assimilated by users, meanwhile pointed out the expansive qualities tied to its unique form factor. Finally, the A20 afforded certain forms of interaction and usage that inspired users to imagine entirely new applications. In the final scenario storyboards written after participants had experienced the A20's audio display and interaction modes, application ideas emerged that had not been seen in the earlier workshops. These included thinking of a personal music player as a device that could interact with the surrounding environment and fill a whole room with a certain ambiance, or of docking modules to author an instance of a game.

We were interested to see whether rich associative and communicative interactions with sound and music could be deployed in a hand-held digital context on an audio-only interface. Could people enter into modes of direct manipulation interaction with sonic content on the object? Was it compelling that the device itself embodied the music? With this we were interested in exploring whether the medium being "consumed" could also become the interaction medium and potentially an information carrier.

By coupling the disciplines of UCD and NIME, we sought to answer the question of how rich expressive interfaces coming from a top-down development process such as NIME could be used alongside ideas and uses emerging from bottom-up processes such as UCD to define an expansive design space that would facilitate sonic interaction and be an inspiring generator of ideas. This coupling enabled us to make use of advanced interactive music techniques within a participatory context, allowing novel forms of interaction to be studied that otherwise would not have arisen from a pure UCD approach.

The resulting prototype is not an endpoint but instead represents a design probe that opens up a generative design space to further imagine and explore new scenarios and create ideas for a potential real-world system. We feel that the A20 successfully met this challenge, acting as an expansive platform for generating and exploring new ideas. The results from this study can be applied to and used in other contexts, with other media, and with other devices types. This kind of approach in which a research prototype is driven by multiple disciplines offers avenues by which interaction designers can extend design practice.

References

1. Avizienis, R., Freed, A., Kassakian, P., & Wewsel, D. (2006). A compact 120 independent element spherical loudspeaker array with programmable radiation patterns. In *Proceedings of AES'06.*

2. Bau, O., Tanaka, A., & Mackay, W. (2008). The A20: Musical metaphors for interface design. In *Proceedings of new interfaces for musical expression (NIME).*

3. Broadbent, D. E. (1954). The role of auditory localization in attention and memory span. *Journal of Experimental Psychology, 47*(3), 191–196.

4. Dourish, P. (2004). *Where the action is: The foundations of embodied interaction.* Cambridge, MA: MIT Press.

5. Flanagan, J. C. (1954). The critical incident technique. *Psychological Bulletin, 51*(4), 327–358.

6. Gaver, B., Dunne, T., & Pacenti, E. (1999). Cultural probes. *Interactions* (Jan–Feb), 21–29.

7. Hutchinson, H., Mackay, W. E., Westerlund, B., et al. (2003). Technology probes: inspiring design for and with families. In *Proceedings of CHI'03* (pp. 17–24). New York: ACM.

8. Oviatt, S. (1999). Ten myths of multimodal interaction. *Communications of the ACM, 42*(11), 74–81.

9. Poupyrev, I., Dunn, H. N., & Bau, O. (2006). D20: Interaction with multifaceted display devices. In CHI'06 Extended Abstracts (pp. 1241–1246).

10. Small, C. (1998). *Musicking: The meanings of performing and listening.* Hanover, NH: Wesleyan University Press.

11. Tanaka, A. (2009). Sensor-based musical instruments and interactive music. In R. Dean (Ed.), *The Oxford handbook of computer music* (pp. 233–257). New York: Oxford University Press.

12. Trueman, D., Bahn, C., & Cook, P. (2000). Alternative voices for electronic sound, spherical speakers and sensor-speaker arrays (SenSAs). In *Proceedings of the international computer music conference (ICMC).*

13. Warusfel, O., & Misdariis, N. (2001). Directivity synthesis with a 3D array of loudspeakers: Application for stage performance. In *Proceedings of COST'2001.*

Sonification of Human Activities

The following three case studies investigate the role of interactive sonification for monitoring human activities, such as EEG signals, presented in the second case study. Applications of sonification to rehabilitation are described in the first case study. The authors explain how multimodal feedback can be used to improve performance in complex motor tasks. A third case study suggests a novel technique called high-density sonification to monitor complex datasets using sound.

12 Designing Interactive Sound for Motor Rehabilitation Tasks

Federico Avanzini, Simone Spagnol, Antonio Rodà, and Amalia De Götzen

Technology-assisted motor rehabilitation is today one of the most potentially interesting application areas for research in SID. The strong social implications, the novelty of such a rapidly advancing field, as well as its inherently interdisciplinary nature (contents combine topics in robotics, virtual reality, and haptics as well as neuroscience and rehabilitation) are some of the aspects that consolidate its challenging and captivating character. Such prospects justify the considerable amount of attention it has received in the last decade from researchers in the fields of both medicine and engineering, the purpose of their joint effort being the development of innovative methods to treat motor disabilities occurring as a consequence of several possible traumatic (physical or neurological) injuries. The final goal of the designed rehabilitation process is to facilitate reintegration of patients into social and domestic life by helping them regain the ability to autonomously perform activities of daily living (ADLs, e.g., eating or walking). However, such activities embody complex motor tasks for which current rehabilitation systems lack the sophistication needed in order to assist patients during their performance. Much work is needed to address challenges related to hardware, software, control system design, as well as effective approaches for delivering treatment [13]. In particular, although it is understood that multimodal feedback can be used to improve the performance in complex motor tasks [9], a thorough analysis of the literature in this field shows that the potential of auditory feedback is largely underestimated.

12.1 Motivations

Strong motivations for integrating interactive sound into motor rehabilitation systems can be found by examining in some detail the most prominent current research challenges in the field of technology-assisted rehabilitation, as described by Harwin, Patton, and Edgerton [13] in a recent study.

As already mentioned above, the most important challenges are related to recovery of ADLs. The functional movements associated with ADLs typically involve very complex motor tasks and a large number of degrees of freedom of the involved limbs (e.g., arm, hand, fingers). On the one hand, this requires the use of sophisticated sensors and actuators (in particular, multiple-degrees-of-freedom robots have to be used in the case of robotic-assisted therapy). On the other hand, representing such complex motor tasks to the patient is a particularly challenging goal. The simple schematic exercises implemented in current rehabilitation systems help recovery of ADLs only to a limited extent.

ADLs rely on an essentially continuous and multimodal interaction with the world, which involves visual, kinesthetic, haptic, and auditory cues. Such cues integrate and complement each other in providing information about the environment and the interaction itself, both in complex tasks (e.g., walking) and in relatively simpler ones (e.g., a reach and grasp movement). In this regard, auditory feedback has to be used in conjunction with other modalities to continuously sonify the environment and/or the user's movements.

The engagement of the patient in the rehabilitation task is another fundamental aspect to consider. It is common sense that a bored patient may not be as motivated as an engaged patient. In the literature it is widely accepted that highly repetitive movement training in which the participant is actively engaged can result in a quicker motor recovery and in better reorganization [5]. Therefore, an open research challenge is how to increase engagement and motivation in motor rehabilitation.

Several studies have shown that auditory feedback intentionally designed to be related to physical movement can result in attainment of optimal arousal during physical activity, reduce the perceived physical effort, and improve mood during training [19]. Moreover, engagement is strictly related to the concept of presence, that is, the perception of realism and immersion in a virtual environment, commonly used in virtual reality research. In this respect it is known that faithful spatial sound rendering increases the realism of a virtual environment, even in a task-oriented context [14, 33]. Nonetheless, it must also be emphasized that auditory feedback can also be detrimental to patient engagement if not properly designed. This is a general issue in sound design: users will typically turn audio off in their PC interface if the auditory feedback is monotonous or uninteresting/uninformative.

The use of interactive sound in rehabilitation systems is also motivated by technological challenges. The qualities of virtual reality and robotic systems in motor rehabilitation are counterbalanced by their disadvantages in terms of customizability and high costs, and designing low-cost devices and hardware-independent virtual environments for home rehabilitation systems is indeed one of the current challenges for technology-assisted rehabilitation.

In this context the auditory modality can be advantageous over the visual and haptic ones in terms of hardware requirements and computational burdens. High-quality sound rendering is comparatively less computationally demanding than three-dimensional video rendering or haptic rendering and can be conveyed to the patients through headphones or through a commercial home theater system with no need for dedicated, expensive, and cumbersome equipment. In the context of home rehabilitation, auditory feedback may even be used as a sensory substitute for the visual and haptic modalities [1, 27].

Finally, auditory feedback may be in certain cases the only modality accessible to the patient, whereas other modalities (and especially the visual one) are not. A notable example is found in neurorehabilitation treatment following a traumatic brain injury such a stroke: in this case, it has been demonstrated by many studies [10, 22, 35] that it is essential to start the rehabilitation process within the acute phase (typically less than 3 months from the trauma) in order to improve the recovery of ADLs. However, this is not always possible because the acute-phase patient has extremely limited motor and attentional capabilities and, in some cases, may have a limited state of consciousness. Auditory feedback may be successfully used in such situations because it can still be perceived without requiring patients to keep their attention focused on a screen and can be processed with relatively little cognitive effort.

12.2 Auditory Feedback for Continuous Interaction

We now examine a few scenarios, methods, and technologies from recent research on sound modeling and on sonic interaction design, which can be employed and applied in the context of motor rehabilitation tasks.

In order to realize a fully interactive auditory feedback, suitable synthesis models that allow continuous control of audio rendering in relation to user gestures need to be used. One interesting scenario is provided by the PhysioSonic [40] system, which presents a generic model for movement sonification as auditory feedback, in which target movement patterns produce motivating sounds and negatively defined sounds are triggered by evasive movement patterns. Sonification is applied to intuitive attributes of bodily movements and comes in the form of metaphorical sounds (e.g., the sound of a spinning wheel is associated to velocity). Furthermore, sounds can be chosen by each subject and change over time, thus reducing fatigue or annoyance. In an example the authors implement a system for the treatment of shoulder injuries, providing two different training scenarios for the abduction movement in which the arm elevation and velocity are sonified into environmental sounds and reproduction rate of a sound file, respectively. In both cases, all evasive movements add noise or creaking to the audio feedback proportionally to their displacement.

One second relevant example of continuous and interactive sonic feedback related to user gesture is the Ballancer [32], a simple interface composed by a track (approximately 1 m long) and an accelerometer that measures acceleration in the direction of the track's length and thus makes it possible to estimate the track's tilt angle. The movement of a virtual ball, which rolls on the track and stops or bounces back when it reaches the extremities, is rendered in real time both visually on a monitor and sonically through a physically based sound synthesis model that uses the state of the ball and the tilt angle as input controls. The task of the user of this interface is to balance and tilt the track in order to move the virtual ball to a target position on the track and to stop it there. The experimental tests presented [32] demonstrate that the presence of continuous auditory feedback (the rolling sound of the virtual ball) shortens the completion time for this task with respect to the case where no sound is provided. Therefore, the auditory feedback effectively conveys information about such a complex gesture as tilting and balancing. Although the Ballancer is not conceived as an interface for motor rehabilitation, it highlights the potential of continuous auditory feedback in supporting motor learning in complex tasks.

Despite the abundance of literature on the use of human-computer interaction (HCI) methods in the design and evaluation of input devices and interfaces [24], sound started to play a significant role in HCI research only in recent years, and yet few studies [36] were devoted to the application of HCI methods to the design and the evaluation of "new interfaces for musical expression." Orio, Schnell, and Wanderley [30] started the investigation in this direction in 2001, focusing on the evaluation of controllers for interactive systems. The authors mention a target acquisition task that could be compared with the acquisition of a given pitch as well as a given loudness or timbre [39], proposing interesting analogies with HCI studies.

These works inspired a thread of research in the field of sound and music computing, focused on the analysis of simple HCI tasks (e.g., target acquisition) in the auditory domain [31]. Here the aim is not the comparison of different input devices but rather the evaluation of the influence of different kinds of feedback on the user's performance. As an example, de Götzen and Rocchesso [8] performed various tests to evaluate pointing/tuning tasks with multimodal feedback. Their results suggest that when the interaction involves any sort of kinesthetic feedback, the performance is distinctly better with respect to free gesture interfaces, and that these improvements in performance are especially significant with high speeds of the target, that is, when the target should be more difficult to hit. Furthermore, redundant feedback is needed when the task is difficult. These results support the idea of applying predictive HCI laws, along with multimodal feedback, in the field of technology-assisted motor rehabilitation with the purpose of improving patients' performance during the rehabilitation task.

Recent research on novel musical interfaces provides a number of systems and approaches that could be directly applied to rehabilitative applications [28, 34]. Work in mobile sensor performance technology is particularly interesting in this respect. Small sensors (including microphones, accelerometers, and so on) are already being used to detect various kinds of movements and gestures that can affect the produced auditory feedback such as by changing the tempo of a musical accompaniment or by controlling some expressive effects added according to input gesture [3, 4].

The application of all these results to the design of audio feedback for motor rehabilitation systems must take into account the specificities of people involved, which can often be affected by various perceptual deficits. In particular, extensive experimental work is needed in order to assess the influence of audio feedback on motor learning processes; to understand the effect of the combination of auditory feedback with other modalities, such as the visual and haptic ones; and to define criteria and guidelines for the design of the feedback, depending on the required motor task.

12.3 Current Uses of Auditory Feedback in Technology-Assisted Rehabilitation Systems

In recent years auditory feedback has been used in various systems for technology-assisted rehabilitation. The simplest possible use, which can be found in many systems discussed in the literature, consists in employing nonprocessed, prerecorded samples of vocal or environmental sounds in order to improve the involvement of the patient in the task. As an example, Cameirao et al. [2] developed a neurorehabilitation system, composed by a vision-based motion capture device augmented with gaming technologies. In this case audio has a rewarding function: in particular, a "positive sound" is triggered whenever the patient accomplishes the goal of a specific game. In a similar way, speech and nonverbal sound are used by Louriero et al. [23] as a feedback modality, with the role of providing encouraging words and sounds during task execution and congratulatory or consolatory words at the end of the exercise. Despite its simplicity, such use of sound has positive effects on the patient's emotions and involvement.

One second approach to the use of auditory feedback is to actively guide the execution of a motor task, rather than simply triggering it as a response to the patient performance. As an example, Masiero et al. [25] present a robotic device that includes simple auditory feedback: a sound signal is delivered to the patient, and its intensity is increased at the start and the end phase of the rehabilitation exercise in order to signal to the patient the occurrence of these phases. According to the authors, this kind of feedback retains the power of maintaining a high level of attention in the patient. On the other hand, the feedback has no correlation with the quality of a patient's performance. Colombo et al. [6] used a similar type of feedback to guide the user's movement in wrist and elbow-shoulder manipulators.

A more interactive use of sound can be found in the GenVirtual application [7]. This augmented reality musical game is designed as an aiding tool for patients with learning disabilities. Users of this system are instructed to imitate sound or color sequences in the GenVirtual environment, and auditory feedback is provided to help a user memorizing the sequences. Similar approaches can be found elsewhere [12, 20, 21]. However, it has to be noted that no realistic interaction is provided between user and environment, even though sounds are more closely correlated to user movements with respect to the former examples. Moreover, auditory feedback is still realized in the form of triggered prerecorded sounds.

In many systems, auditory feedback is intended to provide generation of soundscapes that can reinforce the verisimilitude and realism of a virtual environment, thus addressing aspects of sound design that are closer to SID research topics and in particular to aesthetic quality issues. To date, a plethora of environments have been developed, ranging from relatively simple driving scenarios (such as car, boat, or airplane [9, 17]) to more complex ADLs [15, 29]. The latter work describes a system that allows patients to practice various everyday activities, such as preparing a hot drink: here the role of audio feedback is to render as realistically as possible the sounds of the virtual objects involved in the activity and manipulated by the patient (e.g., the kettle, the cup). However, a fully realistic sonic interaction is not achieved because of the unidirectional and noncontinuous nature of the relation between user movements and sound generation.

Despite the great variety of uses assigned to auditory feedback, the studies discussed above do not generally provide a quantitative assessment of the effectiveness of sound regarding a patient's performance in the rehabilitation task. Schaufelberger, Zitzewitz, and Riener [37] are among the few authors who have provided such an assessment, although they used healthy subjects. In their work the use of short tonal sequences is experimentally evaluated in the context of an obstacle scenario. In particular, different distances from the obstacle and different obstacle heights are sonically rendered using different repetition rates and different pitches of a tonal sequence. Experimental results provide quantitative indications that, when acoustic feedback is added to the visual one, subjects perform better both in terms of completion time and in terms of fewer obstacles hit.

It has also to be noted that, despite the substantial amount of research, there are very few cases in which technology-assisted rehabilitation systems have made the step from research prototype to a real-world application in a medical context. A relevant example is vibroacoustic sound therapy (VAST), initially conceived for children with profound and multiple learning difficulties and recently developed with frail and mentally infirm elderly people in the context of an interactive multisensory environment (iMUSE) [11].

To conclude this section, we provide a quantitative analysis of current uses of auditory feedback in technology-assisted rehabilitation systems. A detailed review of a large number of systems has been carried out by the authors. Specifically, the systems taken into account for this analysis have been collected based on the works referenced in two recent review articles [16, 38], on a related journal special issue [18], and on the 2006–2008 proceedings of two relevant international conferences: the ICORR (International Conference on Rehabilitation Robotics) and the ICVR (the International Conference on Virtual Rehabilitation). A total of thirty-six systems, described in forty-seven reports have been selected. These systems have been grouped into four different clusters, representing four different macrocategories of auditory feedback: *auditory icons*, *earcons*, *sonification*, and *synthetic speech* (figure 12.1). These categories correspond to those identified by McGookin and Brewster [26].

Such analysis pitilessly reveals that most of the systems do not make any use of auditory feedback. In addition, speech and sonification, despite being the two most attractive alternatives for SID, are used only in a small number of cases. On the other hand, the vast majority of the systems that employ sound use it in the simplest possibly way; that is, they use prerecorded samples triggered by a single event. As a result, it emerges clearly that, although several systems exist that make use of multimodal virtual environments, the consistent use of auditory feedback is very little investigated, and thus its potential is largely underestimated.

12.4 Conclusions

Although current technology-assisted rehabilitation systems exploit only a limited set of possibilities from SID research, several studies show that properly designed auditory

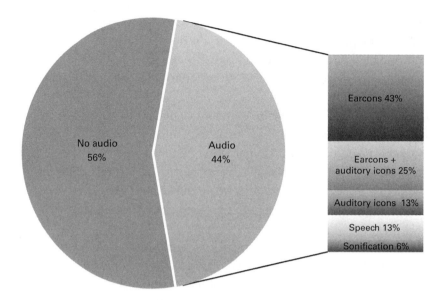

Figure 12.1
Pie chart representing the distribution of auditory feedback techniques for all of the thirty-six reviewed systems.

feedback, able to provide temporal and spatial information, can improve engagement and performance of patients in the execution of motor tasks, can improve the motor learning process, and can possibly substitute for other modalities (as with visually impaired users). Moreover, the relatively limited computational requirements of audio rendering and the low costs of related hardware make it attractive to use auditory feedback in the context of home rehabilitation systems. In light of this, the authors strongly believe that research in technology-assisted rehabilitation may only take advantage from a wary use of the know-how in sonic interaction design.

References

1. Avanzini, F., Rocchesso, D., & Serafin, S. (2004). Friction sounds for sensory substitution. In *Proceedings of ICAD 2004*, Sydney.

2. Cameirao, M. S., Bermudez i Badia, S., Zimmerli, L., Oller, E. D., & Verschure, P. F. M. J. (2007). The rehabilitation gaming system: A virtual reality based system for the evaluation and rehabilitation of motor deficits. In *Proceedings of the IEEE virtual rehabilitation conference*, Venice, Italy.

3. Camurri, A., Volpe, G., Vinet, H., Bresin, R., Maestre, E., Llop, J., et al. (2009). User-centric context-aware mobile applications for embodied music listening. In *1st international ICST conference on user centric media*, Venice, Italy.

4. Castellano, G., Bresin, R., Camurri, A., & Volpe, G. (2007). Expressive control of music and visual media by full-body movement. In *NIME '07: Proceedings of the 7th international conference on new interfaces for musical expression* (pp. 390–391). New York.

5. Colombo, R., Pisano, F., Micera, S., Mazzone, A., Delconte, C., Carrozza, M., et al. (2008). Assessing mechanisms of recovery during robot-aided neurorehabilitation of the upper limb. *Neurorehabilitation and Neural Repair, 22,* 50–63.

6. Colombo, R., Pisano, F., Micera, S., Mazzone, A., Delconte, C., Carrozza, M. C., et al. (2005). Robotic techniques for upper limb evaluation and rehabilitation of stroke patients. *IEEE Transactions on Neural Systems and Rehabilitation Engineering, 13*(3), 311–324.

7. Correa, A. G. D., de Assis, G. A., do Nascimento, M., Ficheman, I., & de Deus Lopes, R. (2007). Genvirtual: An augmented reality musical game for cognitive and motor rehabilitation. In *Proceedings of virtual rehabilitation conference* (pp. 1–6). IEEE, September 27–29.

8. de Götzen, A., & Rocchesso, D. (2007). The speed accuracy trade-off through tuning tasks. In *Proceedings of 4th international conference on enactive interfaces (Enactive 07)* (pp. 81–84). November.

9. Deutsch, J. A., Lewis, J. A., Whitworth, E., Boian, R., Burdea, G., & Tremaine, M. (2005). Formative evaluation and preliminary findings of a virtual reality telerehabilitation system for the lower extremity. *Presence (Cambridge, MA), 14*(2), 198–213.

10. Dromerick, A. W., Edwards, D. F., & Hahn, M. (2000). Does the application of constraint-induced movement therapy during acute rehabilitation reduce arm impairment after ischemic stroke? *Stroke, 31,* 2984–2988.

11. Ellis, P., & van Leeuwen, L. (2009). Confronting the transition: Improving quality of life for the elderly with an interactive multisensory environment-a case study. In C. Stephanidis (Ed.), *Lecture notes in computer science, HCI (5)* (Vol. 5614, pp. 210–219). Berlin: Springer.

12. Gil, J. A., Alcafiiz, M., Montesa, J., Ferrer, M., Chirivella, J., Noe, E., et al. (2007). Low-cost virtual motor rehabilitation system for standing exercises. In *Proceedings of virtual rehabilitation conference* (pp. 34–38). IEEE, September 27–29.

13. Harwin, W. S., Patton, J. L., & Edgerton, V. R. (2006). Challenges and opportunities for robot-mediated neurorehabilitation. *Proceedings of the IEEE, 94*(9), 1717–1726.

14. Hendrix, C., & Barfield, W. (1995). Presence in virtual environments as a function of visual and auditory cues. In *Proceedings of the virtual reality annual international symposium* (p. 74). Washington, DC: IEEE Computer Society.

15. Hilton, D., Cobb, S., Pridmore, T., & Gladman, J. (2002). Virtual reality and stroke rehabilitation: A tangible interface to an every day task. In *Proceedings of the 4th international conference on disability, virtual reality & associated technology* (pp. 63–70). Veszprm, Hungary.

16. Holden, M. K. (2005). Virtual environments for motor rehabilitation [Review]. *Cyberpsychology & Behavior, 8*(3), 187–219.

17. Johnson, M., der Loos, H. V., Burgar, C., Shor, P., & Leifer, L. (2003). Design and evaluation of driver's seat: A car steering simulation environment for upper limb stroke therapy. *Robotica*, *21*(1), 13–23.

18. Kanade, T., Davies, B., & Riviere, C. N. (2006). Special issue on medical robotics. *Proceedings of the IEEE*, *94*(9), 1649–1651.

19. Karageorghis, C., & Terry, P. (1997). The psycho-physical effects of music in sport and exercise: A review. *Journal of Sport Behavior*, *20*, 54–68.

20. Krebs, H., & Hogan, N. (2006). Therapeutic robotics: A technology push. *Proceedings of the IEEE*, *94*(9), 1727–1738.

21. Krebs, H., Hogan, N., Aisen, M., & Volpe, B. (1998). Robot-aided neurorehabilitation. *IEEE Transactions on Rehabilitation Engineering*, *6*(1), 75–87.

22. Lauro, A. D., Pellegrino, L., Savastano, G., Ferraro, C., Fusco, M., Balzarano, F., et al. (2003). A randomized trial on the efficacy of intensive rehabilitation in the acute phase of ischemic stroke. *Journal of Neurology*, *250*(10), 1206–1208.

23. Louriero, R., Amirabdollahian, F., Topping, M., Driessen, B., & Harwin, W. (2003). Upper limb robot mediated stroke therapy—Gentle's approach. *Autonomous Robots*, *15*, 35–51.

24. MacKenzie, I. S., Sellen, A., & Buxton, W. (1991). A comparison of input devices in elemental pointing and dragging tasks. In *CHI91* (pp. 161–166). New York.

25. Masiero, S., Celia, A., Rosati, G., & Armani, M. (2007). Robotic-assisted rehabilitation of the upper limb after acute stroke. *Archives of Physical Medicine and Rehabilitation*, *88*, 142–149.

26. McGookin, D. K., & Brewster, S. A. (2004). Understanding concurrent earcons: Applying auditory scene analysis principles to concurrent earcon recognition. *ACM Transactions on Applied Perception*, *1*(2), 130–155.

27. Meijer, P. B. (1992). An experimental system for auditory image representations. *IEEE Transactions on Bio-Medical Engineering*, *39*(2), 112–121.

28. Misra, A., Essl, G., & Rohs, M. (2008). Microphone as sensor in mobile phone performance. In *NIME '08: Proceedings of the 7th international conference on new interfaces for musical expression*, Genova, Italy.

29. Nef, T., Mihelj, M., Colombo, G., & Riener, R. (2006). Armin—robot for rehabilitation of the upper extremities. In *IEEE international conference on robotics and automation (ICRA 2006)* (pp. 3152–3157). Orlando, FL.

30. Orio, N., Schnell, N., & Wanderley, M. M. (2001). Input devices for musical expression: Borrowing tools from HCI. In *NIME'01* (pp. 1–4). Seattle, WA.

31. Pirhonen, A., Brewster, S., & Holguin, C. (2002). Gestural and audio metaphors as a means of control for mobile devices. In *CHI2002* (pp. 291–298). Minneapolis, MN.

32. Rath, M., & Rocchesso, D. (2005). Continuous sonic feedback from a rolling ball. *Multimedia, IEEE, 12*(2), 60–69.

33. Rauterberg, M., & Styger, E. (1994). Positive effects of sound feedback during the operation of a plant simulator. *Lecture Notes in Computer Science, 876,* 35–44.

34. Reben, A., Laibowitz, M., & Paradiso, J. (2009). Responsive music interfaces for performance. In *NIME '09: Proceedings of the 7th international conference on new interfaces for musical expression* (pp. 37–38), Pittsburgh, PA.

35. Rosati, G., Masiero, S., Carraro, E., Gallina, P., Ortolani, M., & Rossi, A. (2007). Robot-aided upper limb rehabilitation in the acute phase. In *Proceedings of the IEEE virtual rehabilitation conference* (p. 82). Venice, Italy.

36. Rubine, D. (1991). *The automatic recognition of gesture.* Doctoral dissertation, Carnegie-Mellon University, School of Computer Science, Pittsburgh, PA.

37. Schaufelberger, A., Zitzewitz, J., & Riener, R. (2008). Evaluation of visual and auditory feedback in virtual obstacle walking. *Presence (Cambridge, MA), 17*(5), 512–524.

38. Sveistrup, H. (2004). Motor rehabilitation using virtual reality. *Journal of Neuroengineering and Rehabilitation, 1,* 10.

39. Vertegaal, R. (1994). *An evaluation of input devices for timbre space navigation.* Doctoral thesis, University of Bradford, Bradford, UK.

40. Vogt, K., Pirró, D., Kobenz, I., Höldrich, R., & Eckel, G. (2009). Physiosonic—movement sonification as auditory feedback. In *Proceedings of the 15th international conference on auditory display (ICAD2009)*, Copenhagen, Denmark, May 18–21.

13 Sonification of the Human EEG

Thomas Hermann and Gerold Baier

This application overview summarizes the ideas, techniques, and potential application domains of the sonification of the human electroencephalogram (EEG). We start by explaining why sonification is a particularly promising tool to understand EEG data; then we review early uses of sound for the investigation of the electric brain activity, followed by a design-oriented explanation of different techniques for sonifying EEG data, many of which have been introduced by the authors. We supply sonification examples on an accompanying Web site so that the information content and aesthetics of the sonifications can be directly perceived.[1] Finally, we give an outlook on how EEG sonification can change clinical procedures in the near future, focusing particularly on the potential of interactive sonification.

13.1 Electroencephalography

Electroencephalography (EEG), the noninvasive measurement of the time-variant electric potential on the scalp, is a well-established clinical technique to record and interpret electrophysiological processes in the human brain. Applications range from aiding the diagnosis of neurological disorders (e.g., epilepsy) to the monitoring of sleep disorders and to enabling brain-computer interfaces for locked-in patients. To this date, the primary data inspection technique is to look at visually displayed EEG signals as shown, for example, in figure 13.1, although this allows only a limited access to the multivariate data. In fact, although with some experience global signal behavior and univariate features can be extracted fairly well, it is difficult to extract more subtle and in particular multivariate information such as phase differences between, for instance, the top and the bottom channels in figure 13.1. An alternative widely used visual display is to show the instant spatial distribution of electric potentials, usually by color-mapped plots (topographic mappings), yet these fail to display the temporal evolution of the signal.

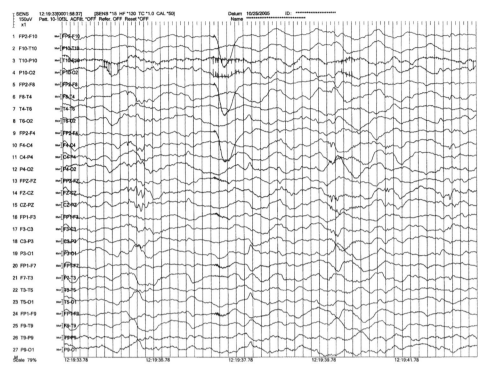

Figure 13.1
Typical EEG data visualization, 27 channels.

In contrast, the *sonification* of EEG addresses our listening skills and enables us to interpret multivariate signal patterns by listening. Sonification is particularly useful for such biomedical data because (1) the information is encoded in rhythmical patterns, a structure that is ideally detected by listening; (2) EEG data are spectrally organized, which corresponds to the excellent spectral resolution of our auditory system; and (3) the patterns are spatially organized on the scalp, connecting well to our ability to perceive spatial sound. For all these reasons, sonification offers the potential to better bridge the gap between the complex data space and our perceptual spaces than visualizations. Indeed, picking up mean field potentials created from distributed sources is, in principle, not so different from a set of microphones that pick up sound signals from a performing orchestra. Although a multitrack visual display of a recorded sound signal fails to properly understand the music, listening enables us to fuse the different streams into one coherent audible percept.

Certainly, the analogy is not so direct, and we cannot expect a direct playback of EEG waves as sound to be sufficiently useful for interpretation by our listening system, which is why a rational data-driven sound synthesis is required. In addition, there are several other motivations for using sound for the display of EEG data: because sound frees listeners from addressing a certain location, as necessary with a visual display, they can move freely and even use the visual sense otherwise, for instance, to observe a patient or to attend to other tasks simultaneously. Sound also enables visually impaired users to inspect the data and can be used even in dark environments. If properly done, the sonifications may contribute to a pleasant and unobtrusive soundscape, for example, in clinical monitoring, which can even be ignored as long as no problem arises, enabled by the human skill of habituation. Yet, as soon as something abnormal (pathologic, clinically relevant) happens, attention is drawn immediately to the signal. Finally, using sonification in combination with visualization means a multimodal representation of data with the potential for the analyst to gain a deeper understanding and a higher felt immersion and familiarity with the data. A further aspect is that sonification offers an attractive closure of an interaction loop for the patients themselves: they might be able to learn to self-regulate activity by attending their own brain wave sonification, offering the potential for new therapeutic approaches for the treatment of epilepsy and other disorders.

13.2   The Origins of the Use of Sound in Medicine

Listening became a popular technique in medicine during the nineteenth century [3], initiated by the efforts of Leopold Auenbrugger, Nicolas Corvisart, and Rene Theophile Hyacinth Laennec, the latter inventing the stethoscope and developing the technique of auscultation to diagnose disease states of the human body, specifically diseases of the chest and heart, by means of listening to audible signs of pathology [20]. Laennec's stethoscope, ancestor of the modern binaural instrument, developed into an iconic symbol for the practicing physician. It demonstrates that listening can reveal relevant details of abnormal anatomy and physiology that are not accessible to the sense of vision. It is interesting to note how these developments in the early nineteenth century also inspired the creation of a language to describe and characterize the sound and to teach novices to understand signs of perceived sound correctly [22].

In the field of neurophysiology, the conversion of data to sound started with the observation that action potentials of nerve cells produced click sounds when the recording electrodes were connected to a speaker membrane [23]. This soon led to the now common practice of auditorily monitoring neural activity during recordings in animals.[2]

13.3 EEG Sonification Techniques—An Overview

Sonification, the systematic and reproducible synthesis of informative sound using data [13], offers a wide set of techniques to achieve specific bindings between data and sound. Here, we focus on those approaches that find application to the sonification of EEG data.

13.3.1 The EEG

The neurophysiological origin of the human EEG was first demonstrated by Hans Berger in 1929. He discovered and described rhythmic activity during normal and altered states of the brain. Only a few years later, Edgar Adrian and Matthews mention that they worked "one of us listening to the rhythm from his head in a loud speaker" [1]. Anecdotally, in the era of chart recorders, listening to the rhythm of the pen scratching on paper was employed by technicians to discover rhythmic abnormalities such as spike-wave activity in long-term monitoring. In analogy with the neurophysiological experiments mentioned above, it might appear obvious to consider the direct playback of amplified EEG measurements, a technique called *audification* in sonification research. However, the direct real-time playback often does not prove to be useful because the relevant rhythms in EEG are mainly between 1 and 30 Hz and thus below the audible frequency range of the human ear. Unfortunately, hardly any sophisticated techniques to create more complex sounds from EEG were available during most of the twentieth century, and it was only with the development of computers with programmable digital audio that the basis was laid for new approaches to rediscover the potential of sound to understand the EEG.

13.3.2 Audification

As mentioned above, the most direct technique is a data-to-sample interpretation of EEG time series data as a sound signal. Certainly, some amplification and also some temporal compression are common practice to adjust the volume and to shift the spectral range to a suitable audible range. In addition, audifications mainly sound like noise, and if compressed in time to shift the signals to the auditory domain, the time compression makes it difficult to inspect the clinically relevant time windows of 10 seconds commonly used for standard display. EEG audification is illustrated with sound example S1,[3] which presents a transition from normal to epileptic EEG compressed to a 3-second sample. Basically, muscle artifacts can be heard as events in the first half, and the rough rhythm in the second half corresponds to rhythmic epileptic spiking. Being nonmusical and more noise-like, the sound may fit well into working environments, where musical sounds would quickly disturb. However, the degree of information is limited because important details of the data, such as shifts in synchronization between channels, remain inaudible as long as the technique does not preserve their audibility. Although audification is a strong method for many data types such as in seismography [12], its benefits for clinical applications of EEG analysis have not been shown convincingly so far. Therefore, more powerful techniques are needed.

13.3.3   Parameter-Mapping Sonification

Arguably the most widely used technique in sonification is to map values of a measured variable to values of a sound synthesis parameter such as frequency, brightness, or amplitude. In fact, this specific sonification technique is so widespread that it often is referred to as *sonification* as such. For EEG time series, typically recording time is maintained as the organizing parameter of the sound so that the sonification is a temporally valid data representation, although stretching or compression is often applied for data inspection [17]. Sound example S2 illustrates a parameter-mapping sonification in which several EEG channels are mapped to frequency deviations of several continuous pitched tone streams, leading to pitch contours that represent variations in the data. Because a decoupling of real-time recording and sonification is possible, the mapping can be adjusted to shift any rhythm of interest to the range suitable for rhythm detection by the human ear. Incidentally, an epileptic rhythm with a frequency of approximately 3/second (as audible in S2), is just right for a real-time sonification. Therefore, the sound makes it possible to distinguish epileptic from normal EEGs quite well. Although this continuous parameter-mapping sonification seems intuitive in light of the continuous data, so-called discrete mapping sonifications are promising because they allow using the time axis in more complex ways: for instance, data variables can be mapped to attack times, decay times, and so forth, and they can deliver acoustically less dense sonifications. As a result, effects of masking that can suppress some information streams are reduced.

13.3.4   Event-Based Sonification

Event-based sonification is a special case of discrete parameter mapping sonification where, from the underlying data, some relevant events of interest are extracted from the underlying data and then represented by sound events. Event-based sonification of a human EEG has been thoroughly investigated by the authors in past years [4–10]. For EEG sonification we used short, pitched percussive sound events to label local maxima in the time series. To display several channels we use pitch to identify the y-coordinate on the scalp with increasing values from the neck to the forehead and stereo panning to indicate the x-coordinate (left to right hemisphere). Thereby we were able to create sonification even for 27-channel data without risk of the sound masking relevant detail by its event density. Certainly, some signal conditioning of the time series is a useful first step to remove unwanted rhythms such as low-pass filtering to remove high-frequency events.

Sound example S3 demonstrates a transition from normal EEG to epileptic pattern for two selected data channels, audible as different tones. Some muscle artifacts can be heard before the rhythmic pattern of the epilepsy sets in. Sound example S4 gives an example for a denser 27-channel event-based sonification where some additional elements are brought in to tune the sonification for task-specific use: on the one hand, events are differentiated between interevent times, interpreted as estimated event frequency: events above 8 Hz are displayed by waterdrop-like sound events; below that interevent frequency they are associated with more solid sound events. The loudest event stream signals events that correspond to significant activity at a certain frequency. To detect events at high frequencies, a hierarchic detection of maxima and minima is done, searching for maxima of maxima and other compound events. The approach is described in more detail by Baier, Hermann, and Stephani [9]. We have recently demonstrated a multichannel online sonification in a concert hall,[4] which showed that the spatiality of sound provides an additional dimension, allowing the perception of activity on the scalp as sound events coming from spatially distinct directions.

Other approaches for event-based sonification demonstrate that the sounds can have very different aesthetics. The "Listening to the Mind Listening" concert [11] at the Sydney Opera House in 2004 [14] brought together several pieces that sonified a 5-minute EEG dataset in real time, showing that the sound ranged from noise-like patterns to data-driven jazz-music-like sonifications. For serious use, however, the capability to use the sound for an interpretation of the underlying data is crucial, and this will in most cases dominate aesthetic choices. However, because sound is also largely interpreted on an emotional level, and particularly when used continuously in monitoring situations, careful design for compatibility with the environment is crucial.

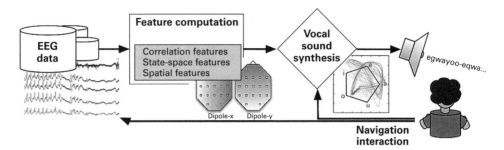

Figure 13.2
Illustration of vocal EEG sonification.

13.3.5 Vocal EEG Sonification

A recent approach for EEG sonification is the vocal sonification introduced in 2006 by the authors (figure 13.2) [15, 16]. It is motivated by three different guiding rationales: (1) A sonification should possess the feature of statistical convergence, which means that with increasing density and availability of spatial data, the sound should converge to a limit that represents the full spatiotemporal continuum. This ensures that with ongoing development of the EEG such as an increased number of electrodes, the sonification will still be usable. (2) The sonification should use a form of sound where the anticipated human listeners are highly skilled in interpreting differences. Speech-like sounds are sounds where humans gather expertise from early years of their life. Finally (3), for the sake of communication, the sonification should bring sounds in a form that can be easily imitated and communicated—the chosen vowel sounds can even be written down by using text to some extent. This is a critical aspect that might help with the acceptance of the sonification.

To account for the generality claimed in (1), we selected data features that can be computed for all sorts of EEG measurements, such as spatial features, correlation features, and other multivariate features. The mathematical complexity may at first glance obscure a direct interpretation of the sound with respect to the variation in the data, but above all we aim at the creation of holistic auditory gestalts. It is known that human listening is organized in a way to automatically pick up gestalt patterns from complex stimuli so that an appropriate correspondence between data features and auditory percepts will help the listener to intuitively learn interpretations that provide correct information.

In the sonification, a transition from unvoiced speech signal to voiced speech signal occurs when the multivariate state vector leaves the subspace of normal background EEG. Thus, a voiced sound implies an overall deviation from normality, indicating a possibly pathologic dynamics. The pitch of the voiced signal changes with the average velocity of the state trajectory. Thereby, for example, epileptic spikes with their abnormally steep slope are easily detected. Further details on the mapping can be found in Hermann et al. [15, 16]. Sound example S5 represents a short fragment in which epileptic spike-and-wave complexes become audible as vowel sequences, similar to "wa-yoooo-eq-wayoooo." Sound examples S6 and S7 demonstrate that a similar sequence can be found in signals from other patients with the same form of epilepsy, and artifacts (examples S8 and S9) and sleep sounds (examples S10 and S11) are characteristically different. It takes some time to become familiar with this type of EEG representation, but humans are very capable of learning, and we expect that the sounds become intuitive when they are worked with for extended periods.

13.3.6 Hybrid Model-Based Approaches

Recent experiences indicate that some intelligence is necessary in the way data are processed before they are brought to sound synthesis. The clearer the task and context, the easier it becomes for the preprocessing of data to have relevant information acoustically stand out. Data mining and machine learning approaches are strong ingredients for successful sonifications. One example where complex preprocessing occurs in EEG sonification is the use of coupled nonlinear oscillator chains as adaptable filters to extract suprathreshold activity in certain frequency bands of EEG signals [2]. This approach was applied to render the sonification in sound example S12 for epileptic EEG. In this approach, twelve different oscillator chains are tuned to twelve frequencies within an octave. Activity leads to sound events from which a decrease in frequency over time can be heard.

13.4 Applications of EEG Sonification

The sonifications presented above make it clear how different in sound and character the sonifications are: some sound musical, others more like noise or speech; some compress the data drastically, whereas other techniques allow real-time use or the detailed investigation of split-second fragments of the data over extended time. Equally heterogeneous is the scope of applications: even in the restricted field of epileptology, sonification finds diverse applications. The first is data analysis, in which we hope to support physicians in better differentiating and diagnosing different types of epilepsy. Reducing the required time while increasing the level of certainty is a good goal to achieve here. For that, sonification must integrate with the workflow of physicians and augment the traditionally used diagnosis tools.

For nurses and other personnel sonification may help to better control EEG equipment, to detect more quickly whether electrodes lose contact to the skin, or to detect when some other error arises. Changing sound characteristics immediately draw the attention, which could help to spot causes of errors. For clinical monitoring, EEG sonification may help to reduce the stress of observing an array of visual monitors for several hours—we hope that with sonification, observers can even close their eyes and relax, being drawn to relevant phenomena by the real-time sonifications. When quick reaction is required (e.g., at seizure onset), sound may help to accelerate the appropriate response.

Although most examples throughout this overview demonstrate sonifications of epileptic seizures, EEG sonification has wider application potential. Apart from epilepsy, EEG is a standard tool for many other examinations, such as to analyze photosensitivity, sleep disorders, and depression, among others. Here, basically the same applications (monitoring, operational support, analysis) are possible. Another field is that of brain-computer interfaces and research in diverse fields (psychology, neurocognition, linguistics) where an EEG is recorded in response to specific stimuli to investigate the neural processing of information in the brain. Here sonification can help to monitor data acquisition and to generate hypotheses about patterns otherwise overlooked.

There are further application areas of EEG sonification, such as in interactive arts and music. Composers and artists have been attracted by the idea to make brain waves an element of their works. Often, however, the translation from data to sound is neither transparent nor systematic.

13.5 EEG Sonification and Interaction

A direct closed interaction loop in the context of the human EEG is to provide auditory feedback in real time to the person whose brain waves feed the sonification. Interaction then occurs at the level where the user changes his/her mental processes in reaction to the sound. In work with EEG feedback on amyotrophic lateral sclerosis (ALS) patients with insufficient vision, simple auditory clues were provided to represent slow cortical potentials [19]. Improving on this, the POSER (Parametric Orchestral Sonification of EEG in Real-Time for the Self-Regulation of Brain States) [18], a complex event-based sonification approach, was developed in the context of brain-computer interfaces. In experiments with voluntary subjects the authors found that listeners exposed to POSER feedback were capable to some extent of controlling their own brain activity significantly by using the auditory feedback given by the sonification. Other more complex auditory data representations have been suggested but have not been tested quantitatively [21]. These results suggest that sonification may be an effective therapeutic means in future applications, for example, for locked-in patients who find it easier to listen than to focus on a visual display or for children with attention deficit disorder. The POSER uses MIDI (musical instrument digital interface) to trigger keyboard notes on conditions, leading to more musical sonifications. These sonifications attract the listener's attention by suggesting the idea of being able to compose music by mental activity. An important role in such settings will be played by filter settings. As mental activity is distracted or disrupted by active listening to exterior sounds, EEG sonifications must allow for sufficient time to switch from a state of perception to a state of attention to mental activity. For example, when an EEG is used for the suppression of epileptic seizures, auditory feedback need only be given when indications of a preseizure state are found in the EEG. Individual tuning of such auditory feedback systems to avoid extreme mental states, to support deep meditation states, or to develop specific mental qualities is perhaps only speculative. Nevertheless, it may be a therapeutic avenue to take in the future.

Another dimension of interaction occurs when data analysts inspect EEG data: interaction with a computer program is common, including the interactive adjustment of scale or filter frequencies, leading in real time to adapted visualizations. We think that interactive sonic scanning through EEG data can increase the awareness of qualitative states and help in quicker detection of relevant segments in the data. Particularly, in epilepsy monitoring, the process of visually scanning a continuous recording of a couple of days is time consuming and exhausting. A specific application here would be to use compressed sound events to display characteristics of manually selected segments in order to identify seizure states. Epileptic seizures with their specific rhythmicity and frequency evolution might be quickly recognizable when presented auditorily. Another form of interaction sometimes required is the interactive control of sonification parameters or data selection parameters, such as the frequency range of interest to be pronounced acoustically. These controls are mostly implemented as graphical user interfaces, and real-time interaction supports a better understanding of parameters and their auditory effects.

Sonic interaction design thus touches many different facets, from interactive control via navigation or scanning, reaction to signals in monitoring, to interaction in the sense of a closed-loop sonification system. Sonic interaction design needs to be aware of the sound aesthetics, the compatibility of interactive sound with other environmental sounds, and with the primary task the sonification is aiming at. An interesting question is how EEG sonification can be successfully integrated into an already existing workflow. How will users learn, accept, and appreciate sonification? How will the sound interfere with their secondary goals? Ideally, sonification will become so intuitive that people will not be so aware of its presence but start to miss it when the sound is suddenly off. Designing slow transitions of workflow that maintain operation while making users familiar with a new modality may also be regarded as a dynamic aspect of sonic interaction design.

13.6 Conclusion

This application overview summarized different methodological and application-specific aspects of human EEG sonification, from audification to possible clinical use scenarios. Sound examples were given to illustrate the different approaches, and aspects of interaction were addressed. We hope that it becomes apparent how influential sonification could be both for clinical practice and biomedical research. We are still at a young stage of research, and substantial efforts are needed to demonstrate and quantify the utility of sonification in medical procedures. Once sonification has been established in one domain, we hope that other domains with related data types such as fMRT, NIRS will begin to open up for consideration of sonification as an additional, auditory information source to understand the complex processes in the human brain.

Acknowledgment

We thank Prof. U. Stephani, Dr. H. Muhle, and Dr. G. Wiegand (Universitätsklinikum Schleswig-Holstein, Kiel, Germany) for providing clinical EEG recordings for research purposes, for their support with the interpretation of the data, and for their continued encouragement with this project.

Notes

1. See http://techfak.uni-bielefeld.de/ags/ami/publications/HB2012-TOT.

2. For a reference to the impact of listening see, for example, the Nobel prize lecture of David Hubel, available at http://www.nobelprize.org/nobel_prizes/medicine/laureates/1981/hubel-lecture.html.

3. Sound examples are available at http://techfak.uni-bielefeld.de/ags/ami/publications/HB2012
-TOT.

4. See http://sonification.de/publications/wienmodern2008.

References

1. Adrian, A., & Matthews, H. (1934). The Berger rhythm: Potential changes form the occipital lobes of man. *Brain*, *57*, 354–385.

2. Baier, G., & Hermann, T. (2004). The sonification of rhythms in human electroencephalogram. In S. Barrass and P. Vickers (Eds.), *Proceedings of the international conference on auditory display (ICAD 2004)*, Sydney, Australia: International Community for Auditory Display (ICAD).

3. Baier, G., & Hermann, T. (2008). Temporal perspective from auditory perception. In S. Vrobel, O. E. Rössler, & T. Marks-Tarlow (Eds.), *Simultaneity: Temporal structures and observer perspectives* (Vol. 3, pp. 348–363). Singapore: World Scientific.

4. Baier, G., & Hermann, T. (2009). Sonification: Listen to brain activity. In R. Haas & V. Brandes (Eds.), *Music that works—Contributions of biology, neurophysiology, psychology, sociology, medicine and musicology* (Vol. 3, pp. 11–24.). Vienna/New York: Springer.

5. Baier, G., Hermann, T., Lara, O. M., & Müller, M. (2005). Using sonification to detect weak cross-correlations in coupled excitable systems. In E. Brazil (Ed.), *Proceedings of the international conference on auditory display (ICAD 2005)* (pp. 312–315). Limerick, Ireland: International Community for Auditory Display (ICAD).

6. Baier, G., Hermann, T., & Müller, M. (2005). Polyrhythmic organization of coupled nonlinear oscillators. In *IV' 05: Proceedings of the ninth international conference on information visualisation (IV'05)* (pp. 5–10). Los Alamitos, CA: IEEE Computer Society.

7. Baier, G., Hermann, T., Sahle, S., & Stephani, U. (2006). Sonified epilectic rhythms. In T. Stockman (Ed.), *Proceedings of the international conference on auditory display (ICAD 2006)* (pp. 148–151). London: International Community for Auditory Display (ICAD), Department of Computer Science, Queen Mary, University of London.

8. Baier, G., Hermann, T., & Stephani, U. (2006). Multivariate sonification of epileptic rhythms for real-time applications. In *Abstracts of the American Epilepsy Society* (no. 1.002). West Hartford, CT: American Epilepsy Society.

9. Baier, G., Hermann, T., & Stephani, U. (2007). Event-based sonification of EEG rhythms in real time. *Clinical Neurophysiology*, *118*(6), 1377–1386.

10. Baier, G., Hermann, T., & Stephani, U. (2007). Multi-channel sonification of human EEG. In B. Martens (Ed.), *Proceedings of the 13th international conference on auditory display* (pp. 491–496). Montreal: International Community for Auditory Display (ICAD).

11. Barrass, S., Whitelaw, M., & Bailes, F. (2006). Listening to the mind listening: An analysis of sonification reviews, designs and correspondences. *Leonardo Music Journal, 16*, 13–19.

12. Dombois, F. (2001). Using audification in planetary seismology. In N. Z. Jarmo Hiipakka & T. Takala (Eds.), *Proceedings of the 7th international conference on auditory display* (pp. 227–230). Helsinki University of Technology: ICAD, Laboratory of Acoustics and Audio Signal Processing.

13. Hermann, T. (2008). Taxonomy and definitions for sonification and auditory display. In P. Susini and O. Warusfel (Eds.), *Proceedings of the 14th international conference on auditory display (ICAD 2008)*. Paris: ICAD.

14. Hermann, T., Baier, G., & Müller, M. (2004). Polyrhythm in the human brain. In S. Barrass (Ed.), *Listening to the mind listening—Concert of sonifications at the Sydney Opera House*. Sydney, Australia: International Community for Auditory Display (ICAD).

15. Hermann, T., Baier, G., Stephani, U., & Ritter, H. (2006). Vocal sonification of pathologic EEG features. In T. Stockman (Ed.), *Proceedings of the international conference on auditory display (ICAD 2006)* (pp. 158–163). London: International Community for Auditory Display (ICAD), Department of Computer Science, Queen Mary, University of London.

16. Hermann, T., Baier, G., Stephani, U., & Ritter, H. (2008). Kernel regression mapping for vocal EEG sonification. In P. Susini and O. Warusfel (Eds.), *Proceedings of the international conference on auditory display (ICAD 2008)*. Paris: ICAD.

17. Hermann, T., Meinicke, P., Bekel, H., Ritter, H., Müller, H., & Weiss, S. (2002). Sonification for EEG data analysis. In R. Nakatsu and H. Kawahara (Eds.), *Proceedings of the international conference on auditory display* (pp. 37–41). Kyoto, Japan: International Community for Auditory Display (ICAD).

18. Hinterberger, T., & Baier, G. (2005). POSER: Parametric orchestral sonification of EEG in real-time for the self-regulation of brain states. *IEEE Transactions on Multimedia, 12*, 70.

19. Kübler, A., Kotchoubey, B., Kaiser, J., Wolpaw, J., & Birbaumer, N. (2001). Brain-computer communication: Unlocking the locked-in. *Psychological Bulletin, 127*, 358–375.

20. Laennec, R. (1821). *A treatise on the diseases of the chest.* (First English translation by J. Forbes). London: T&G Underwood.

21. Rutkowski, T., Vialatte, F., Cichocki, A., Mandic, D. P., & Barros, A. K. (2006). Auditory feedback for brain computer interface management—An EEG data sonification approach (pp. 1232–1239). *KES 2006, Part III, LNAI*. Berlin: Springer.

22. Sterne, J. (2003). *The audible past.* Durham/London: Duke University Press.

23. Wedenskii, N. (1883). Die telephonischen Wirkungen des erregten Nerven. *Centralblatt für die medizinischen Wissenschaften (in German), 26*, 465–468.

14 High-Density Sonification: Overview Information in Auditory Data Explorations

Johan Kildal and Stephen A. Brewster

From the discipline of information visualization [1] we know that to explore and analyze any set of information first it is necessary to obtain an overview of it. From that vantage point a user is able to focus the exploration on areas of interest in the information space, retrieving detail information as required. In instances when information needs to be explored nonvisually, overview information must also be extracted as a first step, but there are few methods to support this. Support for nonvisual overviews is primarily needed by visually impaired users to be able to use and understand the visualizations commonly found in everyday life, for example, graphs and charts in school textbooks or the financial pages of a newspaper. Such support is also needed in the increasingly common mobile interaction contexts so as to overcome the limitations of small displays and a busy visual channel [2]. In addition, some large collections of scientific data can benefit from nonvisual overviews that are presented alone or in combination with data visualization techniques [3]. Thus, developing techniques for obtaining overview information nonvisually from a variety of information sources has become a question that is receiving increasing attention in research [4–6].

We have studied the problem of nonvisual numerical data mining, starting from the process of obtaining overview information [7]. A user-centered iterative design process was followed in which interface design and interactive techniques were evaluated both with visually impaired and with sighted-blindfolded participants. As a result of this research, we developed *high-density sonification* (HDS), a parameter-mapping data sonification (PMDS) technique to present overview information interactively during the exploration of numerical data sets and to support the transition from overview to more detailed analysis of data where required.

PMDS techniques provide means for accessing numerical information nonvisually. Data relations are made perceivable via the mapping of features of the data onto parameters of sound [8]. These techniques have been used to convey detailed views of data sets in the form of auditory graphs [9], and they have shown that the optimal strategy is to map relative numerical values to pitch of sound and to lay them out temporally [10]. With HDS, instead of considering single data points from the whole data set sequentially, data values are grouped together into meaningful subsets (*data slices*). For example, in a data set that contains 10 years' worth of monthly rainfall data in the biggest cities in the world, meaningful data slices could contain all the data for each city, all the data for each separate month, all the data by ranges of latitude, or any other sorting of the data that can reveal interesting results for the purpose of the analysis. Then, each data slice is sonified simultaneously, rendered as a single auditory event in the form of a complex musical chord. When generated interactively, such chords can be compared and successfully interpreted by the user to extract overview information regarding the selected arrangement of data into slices. After that, the user can locally and gradually unfold the grouped data points of a slice into an auditory graph form and focus on parts of the data that require closer exploration, including fully detailed information retrieved in speech form. For instance, if in the example above and mapping values to pitch the data had been arranged by city, once the driest city was identified (the chord with the lowest perceived overall pitch) the user might want to examine if there was a trend toward less rain in that city over the years (a lowering trend in that auditory graph), or if rainy seasons happened at regular intervals (peaks follow a regular rhythmic pattern in the graph). Research with pitch-and-time auditory graphs showed that the speed of rendering the points in the graph functioned as a perceptual filter, where increasing the speed resulted in detail becoming blurred while revealing overview information [11]. In this sense HDS is the particular case of a collection of auditory graphs in which each graph is rendered at the maximum speed. Thus, HDS opens the possibility for exploring collections of auditory graphs, each one compressed into a single auditory event, in the form of a "graph of graphs," where overview information (trends and patterns) from the whole data collection is conveyed.

Before outlining its main properties, HDS is defined here in its most general form:

High-density sonification (HDS) is the simultaneous sonification of some of the data points in a data set. The resulting single auditory event is a representation of that data sub-set, which can be compared against the representations of other data sub-sets that are generated using the same data-to-sound mapping strategy. [7]

This definition of HDS does not limit its use to any particular numerical data structure, data-to-sound mapping strategy, or method for grouping data into subsets for sonification. The authors chose to focus their research on the particular case of tabular numerical data (a very general case, given the fact that every set of raw data is transformed into a table before data can be perceptualized [12]) and to make use of a well-tested value-to-pitch mapping strategy. The method used to group values together for simultaneous rendering should depend on the meaning of the data and the aims of the exploration. For tabular data, meaningful explorations can be easily obtained by grouping data that share the same value in all but one coordinate axis. Following with the example about rainfall data, they could be presented on a two-dimensional (2D) table in which cities were arranged along one axis (maybe sorted by latitude) and months chronologically arranged along the other axis. The same data could also be arranged along the three axes of a three-dimensional (3D) table, for instance, cities, months of the year, and years. Thus, as illustrated in figure 14.1, in the case of the 2D table arrangement, cells could be grouped by rows (all the data for a particular month) or columns (all the data about a city). Similarly, for the 3D table arrangement in the example, this method provides three possible arrangements of the cells into slices (into 2D tables): all the data corresponding to a city, to a year, or to one of the 12 months over the years. In a general case, similar data arrangements can be conceived for tabular structures with higher dimensions. The interactive sonification of a data table that results from using this method of grouping cells is called the *canonical HDS* of a data table. All the information in a table can be conveyed with each one of the different groupings of cells that the canonical HDS produces. However, each arrangement will convey a different aspect of the overview

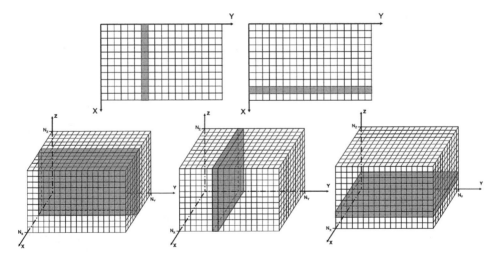

Figure 14.1
Data subset patterns for canonical HDS of 2D tables (above) and 3D tables (below). The canonical sonification of 2D tables generates two complementary views of the data set, whereas for 3D tables it generates three complementary views.

information, revealing trends and patterns that appear along one of the axes (the influence of the latitude, or the global evolution of rainfall in the populated areas of the world, for the 2D arrangement of data in our example). Thus, the different views of the data that the canonical HDS of a table produces are called *complementary views* (somewhat like looking at the same object from different angles: each view reveals certain characteristics of the object, and putting all views together can offer an overview description of the whole). The canonical HDS of a table is formally defined as follows:

Given a data table with *n* orthogonal dimensions, its canonical HDS consists of *n* complementary views. Each view is generated along one of the axes, rendering in a single auditory event all the cells that share the same value on that axis. As a result, the set of auditory events generated within a view is orthogonal, meaning that each cell from the whole data set belongs to one and only one auditory event within a complementary view. [7]

Although the research reviewed here concentrated mainly on the use of the canonical form of HDS, it should be noted that other methods for grouping data are also possible within the definition of HDS, and specific types of data sets and exploration goals might benefit from them (for example, concentric rings of data to analyze the density of population around a big city).

For explorations to produce information that can be interpreted effectively, they should be interactive [13–15] and aided by a system that supports focus+context [12]. To explore 2D tables using HDS, we developed a form of nonvisual focus+context interaction in a system called *TableVis* [7]. This system consisted of a tangible rectangular surface (a graphics tablet augmented with tangible borders) on which information was accessed using a stylus. The information was scaled to fill the area inside the tangible borders, as shown in figure 14.2. In this way, by making use of proprioception (the sense of the position of the limbs) and kinesthesis (the sense of the movement of the limbs), users could tell how far apart one hand was from the other and judge approximately where on the tablet the stylus had been placed. Similarly, they could also change the position of the pen by jumping directly to specific areas, such as the center of the rectangle or any of the corners [16].

An important property of HDS was evaluated in a study [7, 17] that investigated the effect that the size of the table had on the performance of users extracting overview information from 2D tables. Three sizes of tables were considered in the study: small (4 rows × 7 columns = 28 cells), medium (24 rows × 7 columns = 168 cells), and large (24 rows × 31 columns = 744 cells). Most explorations at any table size took between 20 and 40 seconds, with at least 80 percent of the overview information

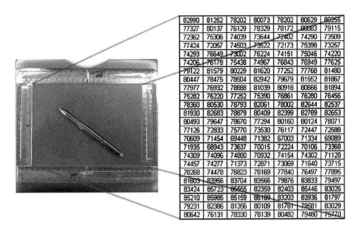

Figure 14.2
Tabular numerical information scaled to fill the area inside the tangible borders.

retrieved being correct and many explorations reaching well above 90 percent correctness. The scores in efficiency (time taken for an exploration), effectiveness (correctness of the exploration), and subjective workload (how hard it was to do an exploration) suffered statistically significant variations when the table size was increased. However, variations in efficiency and subjective workload were significant only between the smallest and the largest table sizes. The ratios of decrease in efficiency and increase in subjective mental workload with respect to the increase in the number of cells in the tables explored were on average as low as –5.46 percent and 5.08 percent, respectively. In turn, scores in effectiveness were significantly decreased only between the medium and large sizes, with an average ratio of variation of –25.7 percent with respect to the increase in the number of cells. These results suggest that when using HDS as part of a focus+context system, there is no need for zooming or computer-aided information filtering in order to complete explorations for overview information, at least in the broad range of sizes considered. Thus, HDS can provide some of the synoptic capacity typical of vision when a data visualization is being examined.

Another important property of HDS relates to the common technique of calculating the arithmetic means of rows and columns to perform initial data analyses, which are a good source of overview information. With HDS and making use of a value-to-pitch mapping strategy, users can estimate relative arithmetic values in the data by solely using auditory perception. Judging the relative perceived overall pitch of the chords generated with HDS (i.e., by deciding with a first impression which chords sound with a higher or lower overall pitch), users can sort data subsets by their relative arithmetic mean value. The reliability of this technique was assessed through various studies with different types of real or realistic data sets and exploration tasks [11, 17, 18], finding that reliability was above 70 percent, often reaching levels close to 100 percent. In a large study (chapter 8 in Kildal [7]) designed specifically to challenge this technique by presenting very similar pairs of chords (generated from fifteen data points per set, with constant highest and lowest values), it was found that 62 percent of the pairs were sorted by pitch consistently (with 95 percent significance) by the participants in the study. That same sorting was also found to be a correct estimation of the relative arithmetic mean values in 85 percent of those cases. This property is important for a data exploration technique because it permits users to always work with the raw data and effortlessly estimate arithmetic mean values, without having to perform any mental calculations or create additional data representations that increase the complexity of the data mining task, particularly when this happens in strictly nonvisual conditions.

In conclusion, HDS is an important addition to the literature in PMDS techniques for providing, as part of a nonvisual focus+context system, a successful solution to the problem of obtaining overview information from numerical data sets nonvisually. HDS offers high performance at acceptably low levels of mental workload, all of which vary slowly with the increase of the amount of information explored, thus reducing the need for explicit zooming and filtering. Additionally, users can perform effortless basic statistical analysis of the data perceptually (estimation of arithmetic means), permitting them to work with only the raw data view. HDS also bridges seamlessly with auditory graphs when closer views of the data are needed through the variation of the time between sonified single points, which acts as a perceptual filter of detail in the data. A general definition of HDS has been formulated, and the particular canonical implementation of this technique has been defined and described. However, the general definition of HDS provides a frame for other implementations to be designed and evaluated, which can reveal further strengths of this technique in specific data-mining contexts.

References

1. Shneiderman, B. (1996). The eyes have it: A task by data type taxonomy for information visualizations. In *IEEE symposium on visual languages* (pp. 336–343). Boulder, CO: IEEE Computer Society Press.

2. Roto, V., & Oulasvirta, A. (2005). Need for non-visual feedback with long response times in mobile HCI. In *Special interest tracks and posters of the 14th international conference on World Wide Web* (pp. 775–781). Chiba, Japan: ACM.

3. Kramer, G., Walker, B. N., Bonebright, T. L., Cook, P. R., Flowers, J., Miner, N., et al. (1999). *Sonification report: Status of the field and research agenda. Prepared for the National Science Foundation by members of the International Community for Auditory Display.* Santa Fe, NM: ICAD.

4. Harrar, L., & Stockman, T. (2007). Designing auditory graph overviews: An examination of discrete vs. continuous sound and the influence of presentation speed. In *International conference on auditory display (ICAD)* (pp. 299–305). Montreal: McGill University.

5. Brown, A., Stevens, R. D., & Pettifer, S. (2006). *Audio representation of graphs: A quick look.* In *International conference on auditory display (ICAD)* (pp. 83–90). London: ICAD.

6. Hussein, K., Tilevich, E., Bukvic, I. I., & SooBeen, K. (2009). Sonification design guidelines to enhance program comprehension. In *IEEE 17th international conference on program comprehension, 2009. (ICPC '09)* (pp. 120–129).

7. Kildal, J. (2009). *Developing an interactive overview for non-visual exploration of tabular numerical information.* Ph. D. thesis, Department of Computing Science, The University of Glasgow.

8. Hermann, T. (2008). *Taxonomy and definitions for sonification and auditory display.* In *International conference on auditory display (ICAD).* Paris: ICAD.

9. Walker, B. N., & Cothran, J. T. (2003). Sonification sandbox: A graphical toolkit for auditory graphs. In *International conference on auditory display (ICAD)* (pp. 161–163). Boston: ICAD.

10. Flowers, J. H. (2005). Thirteen years of reflection on auditory graphing: Promises, pitfalls, and potential new directions. In *First international symposium on auditory graphs (AGS2005) at international conference on auditory display (ICAD)* (pp. 406–409). Limerick, Ireland: ICAD.

11. Kildal, J., & Brewster, S. A. (2005). Explore the matrix: Browsing numerical data tables using sound. In *International Conference on Auditory Display (ICAD)* (pp. 300–303). Limerick, Ireland: ICAD.

12. Card, S. K., Mackinlay, J. D., & Shneiderman, B. (1999). Information visualization. In S. K. Card, J. D. Mackinlay, & B. Shneiderman (Eds.), *Readings in information visualization. Using vision to think* (pp. 1–34). Waltham, MA: Morgan Kaufmann Publishers.

13. Stevens, R. D. (1996). *Principles for the design of auditory interfaces to present complex information to blind people.* PhD thesis, The Department of Computer Science, University of York.

14. Hunt, A., & Hermann, T. (2004). The importance of interaction in sonification. In *International conference on auditory display (ICAD).* Sydney, Australia: ICAD.

15. Hunt, A., Hermann, T., & Pauletto, S. (2004). Interacting with sonification systems: Closing the loop. In *International conference on information visualisation (IV)* (pp. 879–884).

16. Kildal, J., & Brewster, S. A. (2006). Exploratory strategies and procedures to obtain non-visual overviews using TableVis. *International Journal on Disability and Human Development*, 5(3), 285–294.

17. Kildal, J., & Brewster, S. A. (2006). Providing a size-independent overview of non-visual tables. In *International conference on auditory display (ICAD*, 8–15*)*. London, UK: Queen Mary, University of London.

18. Kildal, J., & Brewster, S. A. (2007). EMA-Tactons: Vibrotactile external memory aids in an auditory display. In *IFIP interact* (pp. 71–84). Rio de Janeiro, Brazil: EMA-Tactons.

Sound in Virtual Reality

The following two case studies examine the role of sound in virtual and augmented reality. The fields of virtual and augmented reality can benefit from SID research, especially in the auditory simulation of complex events such as impacts. This is the topic of the first case study. Moreover, spatial sound plays an important role in VR simulation, as described in the second case study.

15 Simulating Contacts between Objects in Virtual Reality with Auditory, Visual, and Haptic Feedback

Jean Sreng and Anatole Lécuyer

Nowadays virtual reality (VR) techniques are increasingly used in industrial processes, in particular for virtual prototyping, which consists of replacing physical prototypes by digital mock-ups. VR techniques were found [13] to improve productivity and shorten design, development, engineering, and training times on industrial (such as automotive or aeronautics) products. For instance, in virtual assembly/maintenance simulations, designers and maintenance engineers can virtually validate if an industrial part can be assembled and/or disassembled (such as in figure 15.1) and thus adapt the shape of the part or the maintenance procedure without the need of a real mock-up.

Such virtual processes rely on a realistic physical simulation of virtual objects interacting together. In particular, the considered objects cannot interpenetrate each other. This essential characteristic highlights one of the most important features of the virtual interaction: the simulation of *contact* between objects. The contact can be defined as the fact that two bodies are touching each other. It tightly governs the movement of objects, constraining their trajectory with respect to their direct surrounding. This notion includes the dynamic contact or *collision* between two moving objects and the static contact of one object touching another. Particularly, this contact information is central in industrial assembly/disassembly simulations by providing not only a simple assemblability validation but a more global understanding of the interaction between geometries, notably through *interactive* manipulation.

Providing the information of contact interactively in virtual environments raises many challenges. Because of the growing complexity of simulated scenes and limited computational resources, interactive simulations cannot always provide realistic auditory, visual, or haptic[1] contact rendering in real time. The algorithmic complexity associated with collision detection and physical simulation generally allows a simplified, often idealized simulation of the interaction. For instance, the material properties of the simulated objects or their deformation are not usually considered.

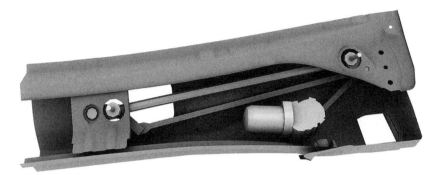

Figure 15.1
Complex-shaped objects usually involved in a virtual prototyping simulation. The manipulation of the light-colored object generates multiple simultaneous contacts with the outer part, blocking the movement.

Consequently, realistic contact-rendering techniques have only been demonstrated in simple contexts involving, for instance, simple geometries or only point-based interaction. Furthermore, there are only few platforms that seamlessly integrate auditory, visual, and haptic feedback of contact. Such platforms are, moreover, often highly dependent on a specific physical engine or collision detection method.

In this chapter we propose an integrated approach for multimodal rendering of the information of contact in VR simulations. A review of related work is first presented on the various techniques and improvements that can be used to enhance the rendering of contact in VR. Then, we describe a formulation of the information of contact independently from the underlying simulation that is adopted. In particular, we stress the importance of considering not only the *state* of contact but also its temporal evolution. Then, we propose examples of audio, visual, tactile and force-feedback rendering techniques relying on our formulation. Finally, we demonstrate our multimodal architecture in a complex virtual prototyping environment.

15.1 Related work

Rigid-body simulation is a wide area of research in virtual reality and computer graphics. Among the various collision detection algorithms used, we can differentiate two main categories: spatial detection algorithms providing information about object intersection on a time-step basis, and spatiotemporal detection algorithms providing information about trajectories' intersection. The spatiotemporal algorithms can detect the exact time [28] at which a collision event occurs. This time can be used to provide a more physical output (for instance, conservation of momentum). This approach, also referred to as an *event-based* detection collision, is, however, generally more computationally expensive. Even if multifrequential approaches can be used conjointly with event-based methods [23], spatial detection algorithms are more commonly employed [17, 20, 21] in complex real-time haptic simulations proper to virtual prototyping, for instance.

Various forms of feedback can be added to the virtual scene to improve the perception of simulated contact between solid objects, mainly auditory, visual, and haptic/tactile feedback.

Auditory feedback of contact has been widely studied in many different contexts from auditory perception psychology to real-time sound synthesis. Ecological psychology studies have shown that subjects often tend to perceive sound as causing events [38]. This attitude is referred as "everyday listening" in the work of Gaver [12], which discusses the use and synthesis of contact sounds. Numerous studies have been conducted on the synthesis of impact and friction sounds using modal synthesis [37]. Simulations using finite element methods have also been used to generate more realistic sound from the analysis of an object's shape [22] and used in real-time rendering techniques adapted to large-scale environments [27]. When used in rigid-body simulations, the impact sounds can be generated using events, as proposed by Takala and Hahn [35] in their description of a methodology for computer animation based on triggered events. Besides the impact event, the sound synthesis engine can also rely on more physical information such as force profiles or interaction velocities, which can be given by ad hoc physical simulation or by haptic interfaces [1, 5, 9, 39, 41] to generate more accurate impact sound or continuous friction sound.

Visual feedback is generally the main, if not the only, feedback in many virtual simulations. The contact between two solid objects in such simulations can often be directly perceived through the sole graphical rendering of the virtual scene. However, in many situations, the perception of the collisions is not easy, especially for virtual scenes involving objects with a complex geometry. To improve the perception of contact, several rendering effects have been used such as drop shadows or interreflexion effects [15]. Several studies were focused on the use of visual aids to improve the perception of contact between virtual objects using graphic effects to convey information about the proximity, the contact, or the contact force [34, 40].

Haptic interfaces have also been used to improve the perception of contact between solid objects, especially for virtual prototyping purposes [17, 20]. The realism of the haptic interaction is based on the ability of the haptic device and the underlying physical simulation to render forces at a high frequency, up to several kilohertz [4].

However, it is not possible to render stiff features such as a contact between rigid objects at such a frame rate using the classical closed-loop feedback used to regulate force output of haptic interfaces [10]. Even if efforts have been expended to overcome these limitations [6], closed-loop haptic feedback is inherently restricted to smooth, low-frequency forces that are unsuitable for a realistic feedback. To improve the realism of impact between rigid objects, rendering techniques based on open-loop haptic feedback have been developed [16, 18, 33] by superimposing an event-based high-frequency transient force pattern over the traditional force feedback. The contact interaction can also be improved by using haptic texturing [7, 24, 30] to simulate the irregularities at the surface of a rigid object.

Many multimodal platforms using haptic, audio, and visual feedback have been developed. For instance, DiFilipo and Pai developed an audio and haptic interface that simulates a contact sound closely related to the haptic interaction [9]. Their approach was, however, limited to a two-degrees-of-freedom (2-dof) interaction. Lécuyer et al. proposed a multimodal platform integrating visual, haptic, and audio feedback of contacts [19] to evaluate the effect of the various types of feedback on users' performance. The interaction was, however, limited to 3 dof, and the audio feedback used was an auditory alarm beep triggered on collisions. Likewise, some multimodal platforms dedicated to virtual prototyping have been experimented with [8, 14].

15.2   The Contact State and Temporal Evolution of the Contact State

In most VR applications involving physical simulations, the contact information can be provided by the simulation algorithm at each instant. In most cases, this information is determined for each time t only by considering the state of the simulation (position of objects, etc.) at that moment. A snapshot of the simulation at an instant t is sufficient to determine this information of contact, regardless of the past or future state of the simulation. This information describes the contact essentially by simply differentiating two states: *the free motion* and *the contact* (figure 15.2).

This contact information, being determined at each time step of the simulation, constitutes a continuous time-independent information flow describing the contact state.

However, the contact is also characterized by the discontinuity between the free motion and the constrained movement. This behavior is an important aspect that is also found in the formulation of the contact equations and in perceptive processes. This phenomenon can only be apprehended by considering the movements of the

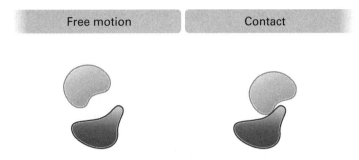

Figure 15.2
The different states described by the information on contact.

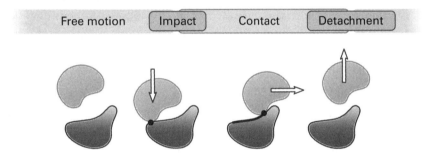

Figure 15.3
The different events associated with the information on contact.

simulated objects. Thus, we can highlight this behavior using the description of the *temporal evolution* of the contact state information.

From this evolution we can particularly identify some important *events* corresponding to the transitions between the two contact states (figure 15.3) described earlier:

• The transition between the free motion and the contact states is described by the *impact event* corresponding to the collision.

• Likewise, the transition from the contact state to the free motion corresponds to the *detachment event*.

It is also possible not only to describe the transitions but to derive more general information from this temporal evolution. For instance, the evolution of the proximity information can also be considered.

The temporal evolution of the contact state enables us to provide a higher level of description of the contact behavior. This higher level of contact description has two main benefits:

• It provides richer information than the contact states alone, which can be employed to elaborate specific contact rendering techniques.

• It highlights the discontinuous nature of the contact represented by the transitions between free motion and contact by explicitly identifying impact and detachment events.

This discrete event-based information, and more particularly the impact event, can be used conjointly with specific rendering algorithms. Indeed, in most cases, the physical simulation is unable to accurately render the fine transient physical phenomenon happening at the instant of impact in real time and thus only provides a coarse rendering. Specific algorithms focused on the rendering of impact can thus be used with the discrete information of the impact event to improve the global rendering of contact. For instance, open-loop haptic rendering [18] or auditory synthesis methods [37] are directly based on this high-level information.

15.2.1 Formulation of the Information on Contact

We propose to rely on the following formulation (figure 15.4) based on proximity to describe the information on contact:

• Position p, direction u, and distance λ information between two proximity points between two object boundaries

• Contact force f information such that $\lambda > \varepsilon \Rightarrow f = 0$

From this formulation we can, in particular, identify two states:

• *Free motion* for $\lambda > \varepsilon$ (in this case, $f = 0$)

• *Contact* for $\lambda \leq \varepsilon$ (in this case, $f \neq 0$)

From the temporal evolution of the contact formulation we can, in particular, identify two events:

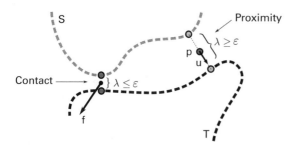

Figure 15.4
Formulation of the information on contact.

- *Impact* at the transition from *Free motion* to *Contact*
- *Detachment* as the transition from *Contact* to *Free motion*

15.2.2 Rendering of the Information on Contact

This formulation of the information on contact, enclosing contact *state* information as well as its *temporal evolution*, constitutes a first general basis on which we can develop multimodal rendering techniques.

From a structural point of view, the multimodal rendering is achieved through several independent loops corresponding to the different auditory, visual, tactile, and kinesthetic sensory channels. These loops use the information from the simulation to achieve the rendering. In this study, we propose to adopt the same contact information description presented in the previous sections to elaborate all the multimodal rendering techniques of the contact information (figure 15.5). The raw rendering information provided by the virtual simulation is then first processed to provide the information on contact using the formulation we adopted because the raw rendering information delivered by the simulation may not, in most cases, directly correspond to our formulation.

This formulation of the information on contact is not necessarily the only information used for all rendering techniques in the setup. For instance, the haptic control loop, usually tightly related to the physical simulation, generally requires some specific control information directly provided by the physical loop.

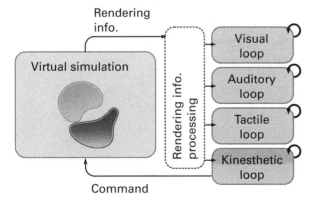

Figure 15.5
Multimodal rendering of the information on contact.

15.3   Computation of Contact States and Events

In this section we introduce a method to generate the different contact events using only generic information about the simulation such as the position of each object. Indeed, although most physical simulations provide rendering information about the position of contact points or objects (used in visual rendering, for instance), they do not generally directly provide more higher-level information such as the instant and location of impact or detachment (this is particularly true for discrete collision detection methods; continuous methods can directly provide the impact information but not the detachment).

15.3.1   Computation of Contact States and Events Based on Object Velocities

The different contact events can be determined using the movement of the bodies during contact, namely, the linear relative local velocity v of the twist between two bodies a and b at the contact point p and the normal n at the surfaces of the bodies (figure 15.6). The contact condition on the object movement is the orthogonality between v and n.

The beginning and the end of a contact can be expressed as the temporal limits of this contact condition, respectively, for the impact (the angle between v and n is obtuse) and the detachment (the angle between v and n is acute).

We can generate the impact and detachment events considering the magnitude and the direction of the normal velocity. The friction from the contact state is determined during the contact using the angular velocity and the tangential velocity (figure 15.7):

Further details on the method used to detect and generate the different contact states/events and forces can be found in Sreng et al. [32].

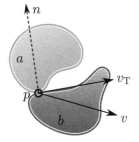

Figure 15.6
Contact normal and velocities of objects in contact.

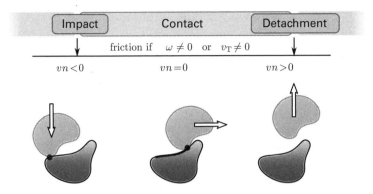

Figure 15.7
Generation of events.

Table 15.1
Examples of rendering techniques of contact states and contact events

| Modality | Contact state and contact events | | |
	Impact	Friction	Detachment
Visual	Particle	Particle/pencil	Particle
Audio	Impact sound	Friction sound	
Tactile	Impact vibrations	Friction vibrations	
Haptic	Impact force transient		

15.4 Multimodal Rendering Based on Contact States and Events

In order to illustrate our approach, we can give examples of sensory feedback based on impact and detachment events and friction state. The generated events can be used with several rendering techniques to enhance the perception of contact. Among the numerous possibilities of association between an event and a rendering technique, we chose to implement an intuitive rendering of contact for all the modalities (table 15.1).

15.4.1 Software Architecture of Event-Based Rendering of Contact

The event-based techniques can be easily integrated over a classic multi-refresh-rate architecture using a physical 6-dof simulation and various devices (audio at 48 kilohertz, visual at 60 hertz, tactile and kinesthetic at 1 kilohertz).

Such rendering architecture is described in figure 15.8. The states and events generator is integrated in addition to a classical rendering architecture. The generated states and events are then used by the different devices.

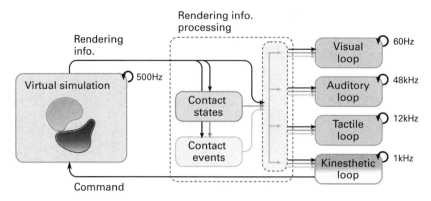

Figure 15.8
Rendering architecture using contact states and events.

15.4.2 Examples of Audio Display of the Information on Contact

We implemented a realistic spatialized audio feedback based on modal sound synthesis of impact and friction [27, 37] to illustrate the impact event and the friction state.

We used an off-line modal analysis of the different objects to produce a sound related to the material, the geometry, and the location of the interaction. To each object, we associate a modal model (sum of exponentially damped sinusoids). Further details about the parameters chosen for the model can be found in Sreng [31].

The input profile given to the filter is generated with the different impact events and friction states.

Audio Impact Event

For each impact, we generate a Gaussian-like input profile of width proportional to the normal velocity to convey the hardness of the impact. This profile can be adapted depending on the impacting materials.

Audio Friction State

During the friction state, we generate a fractal noise input profile that represents the surface roughness [36]. We used a simple multifrequency formulation similar to the procedural texture synthesis techniques used in computer graphics [25]. The spatial noise function representing the roughness of the surface can be expressed as the sum of elementary noise functions. From this spatial description of the surface, we can generate an input profile using the tangential velocity [31].

After first subjective testings, we achieved more realistic results with this technique than with the classic filtered white noise passed through a bandpass filter of frequency proportional to the tangential velocity, which is commonly used to provide the perception of changing pitch at different velocities [37].

Next, the signal from the modal filter is high-pass filtered to attenuate the friction input profile over the modal response to achieve a more realistic rendering. The cutoff frequency is adjusted by considering the lowest mode frequency of the modal model.

The resulting sound output is then spatialized in the virtual soundscape at the position m of the contact point using the VBAP algorithm [26] (figure 15.9).

With audio latencies as low as 8 milliseconds, the achieved sound synthesis was revealed to be quite realistic. During the first informal evaluation, users reported that

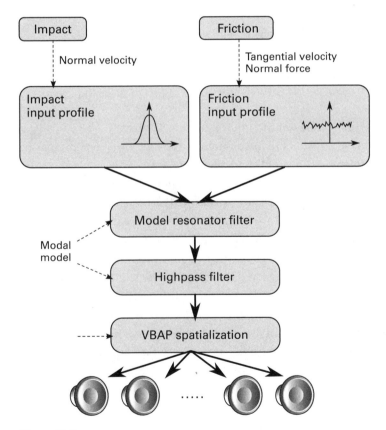

Figure 15.9
Sound synthesis diagram.

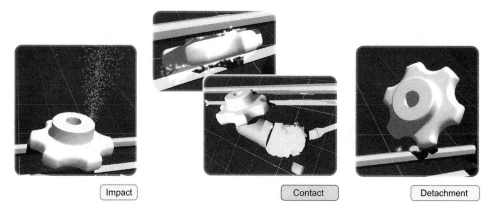

Figure 15.10
Visual display of the contact states and events. Particle effects on impact; pencil and particle effect on contact (friction), particle effect (bubble blow) on detachment.

the sound generated corresponded perfectly with the interactive manipulation, thus providing a better immersion in the virtual scene.

15.4.3   Examples of Visual Display of the Information on Contact

As an illustration, we have implemented two different types of visual feedback using the impact and detachment events and the friction state.

We chose to implement an intuitive visual representation of impact based on particle effects (figure 15.10). On each impact, visual particles are emitted from the contact point with the same direction. On a detachment event we use the same technique of particles to produce a kind of "bubble blow" effect (figure 15.10). We chose to implement two types of visual feedback on the friction state: a first feedback using the same *particle effect* as used for impact and detachment events (figure 15.10) and a second feedback using a *pencil effect* at the contact point. The pencil effect (figure 15.10) reproduces the behavior of a pencil at the contact point. The thinness of the drawn line depends on tangential velocity to make the line thinner if the object moves quickly. The color of the line is given by the contact force f according to the classic blue-green-yellow-red color gradient.

15.4.4   Example of Tactile Display of the Information on Contact

The signal simulated using modal synthesis described in section 15.4.2 to produce audio rendering can be similarly used to provide tactile rendering. Indeed, this signal resulting from the modal resonator represents the vibration of the object's surface.

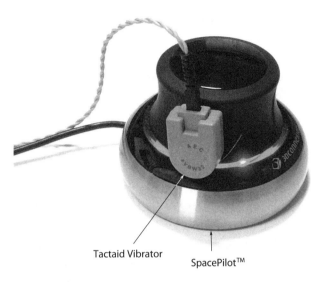

Tactaid Vibrator   SpacePilot™

Figure 15.11
Tactile feedback using a Tactaid vibratory device attached to a passive SpacePilot™ manipulator.

Instead of using this signal to generate sound, we can use a tactile vibration device to directly reproduce the simulated vibratory signal from the modal synthesis directly on the skin.

In order to produce complex vibrations, we used a Tactaid VBW32 device. This device delivers vibrations by using a magnet/coil inertial transducer, thus potentially enabling the synthesis of arbitrarily shaped vibrations (in contrast to devices based on a motor with an eccentric rotating mass).

We tested this vibratory system with a passive isometric device (SpacePilot™): the Tactaid was tightly connected (see figure 15.11) to the manipulated part of the Space-Pilot™ (the moving upper part). This way, the vibrations can be sensed through the fingers while one is manipulating the device.

Because of its simple magnet/coil system, this device behaves as a second-order band-pass filter with a central frequency of 250 hertz. We consequently adapted the modal model and the input force profiles to achieve a convincing rendering:

• The modal model was chosen with resonant modes around 250 hertz.

• The impact force profile chosen was a Dirac impulse. This force profile provided a crisper rendering of impact than the Gaussian-shaped force profile used for sound rendering.

First trials suggested that this vibratory feedback enhances the illusion of presence and distal attribution, providing an intuitive way to feel the virtual contact.

15.4.5 Example of 6-dof Force-Feedback Display of the Information on Contact

Last, we have implemented a high-frequency event-based 6-dof haptic force feedback rendering using the *impact event* on a Virtuose6D™ interface.

A classic haptic feedback was first implemented with a two-channel architecture consisting of traditional closed-loop coupling and event-triggered impulses. The closed loop we used is a classic proportional-derivative control. However, this control scheme cannot allow a realistic contact sensation because the virtual stiffness is bounded by stability issues [10]. To overcome this limitation, event-based haptic approaches have been proposed by Hwang, Williams, and Niemeyer [16] and Kuchenbecker, Fiene, and Niemeyer [18] to improve the rendering of contact by superimposing high-frequency force patterns. However, these approaches were only demonstrated in a 1-dof context.

Here we propose an extension of existing methods: an implementation of 6-dof event-based force-feedback rendering. The main idea behind this technique is to devise appropriate high-frequency 6-dof force patterns by considering each contact as an elementary three-dimensional impulse generating a high-frequency vibration. Futher details can be found in Sreng [31] and Sreng, Lécuyer, Andriot, and Arnaldi [33].

Eventually the final torques applied to the haptic device are the sum of this high-frequency 6-dof force pattern and the classic closed-loop torques.

15.4.6 Technological Demonstrator

These implementations have been successfully integrated in a multimodal VR platform (figure 15.12). The hardware setup used was a 3.4-gigahertz Pentium®4 PC used with a TechvizXL immersive visualization platform, an RME Fireface 800 soundcard with twelve speakers, and a Virtuose6D™ haptic device from Haption. The simulation was performed by the XDE physics engine of the CEA LIST, and the three-dimensional visualization was achieved using the OGRE v1.4 library.

The first tests are promising, suggesting that these implementations can successfully improve the user's perception of contact. Future work is now necessary to formally evaluate the influence on users' manipulation performances and preferences.

15.5 Conclusion

We have presented an integrated approach to conveying the information of contact in VR simulations with auditory, visual, and haptic feedback. We have presented a

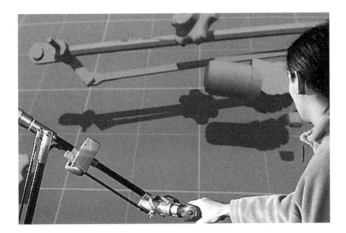

Figure 15.12
Hardware setup.

formulation of the information on contact independent of the underlying simulation. This way, we have stressed the importance of considering not only the *state* of the contact but also its temporal evolution, particularly through *impact event* and *detachment event*.

We have proposed a method to generate the different events using only the position information of each virtual object. This reduces the dependence on a specific collision detection engine. Similarly, we have proposed a technique to generate individual contact forces by simply using global force information. This reduces the dependence on a specific physical engine.

Then, we have presented several examples of sensory feedback. First, we have implemented a realistic audio feedback of impact and friction sounds. Second, we have described a visual feedback implementation based on particles and pencil effect. Third, we have presented a tactile rendering of impact and friction. Last, we have introduced a 6-dof force-feedback rendering algorithm based on the impact event.

These implementations have been integrated in a multimodal architecture dedicated to virtual prototyping and assembly/disassembly simulation. They can be easily extended to other event-based effects on various rigid-body simulations thanks to our generic modular approach.

Future work could concern formal evaluations of the proposed rendering techniques of contact. This evaluation could quantify the effect of the different techniques on the user manipulation in complex virtual scenes, such as virtual assembly contexts.

Similarly, although we have provided good perceptively realistic feedback well suited for virtual prototyping applications, the audio techniques we presented can be improved to match more closely the physical realism. For instance, a nonlinear impact model [2, 29] and more complex friction models [3] or friction textures (similar to the graphic ones) can be used to improve the realism of the force input profile. A more complex vibration model [39] can also be adopted to simulate a wider range of materials than modal synthesis. Finally, a sound wave radiation model can also be added to improve the spatial feeling [11].

Note

1. Haptics refers to the sense of touch, such as tactile stimulation or force-feedback, for instance.

References

1. Avanzini, F., & Crosato, P. (2006). Integrating physically based sound models in a multimodal rendering architecture: Research articles. *Computer. Animation and Virtual Worlds,* *17*(3.4), 411–419.

2. Avanzini, F., & Rocchesso, D. (2001). Modeling collision sounds: Nonlinear contact force. In *Proceedings of Conference on Digital Audio Effects.*

3. Avanzini, F., Serafin, S., & Rocchesso, D. (2005). Interactive simulation of rigid body interaction with friction-induced sound generation. *IEEE Transactions on Speech and Audio Processing,* *13*(5), 1073–1081.

4. Boff, K. R., & Lincoln, J. E. (1988). *Engineering data compendium: Human perception and performance.* Technical Report 3, Armstrong Aerospace Research Laboratory, Wright-Patterson AFB.

5. Cadoz, C., Luciani, A., & Florens, J.-L. (1984). Responsive input devices and sound synthesis by simulation of instrumental mechanisms: The cordis system. *Computer Music Journal,* *8*(3), 60–73.

6. Colgate, J. E., Grafing, P. E., Stanley, M. C., & Schenkel, G. (1993). Implementation of stiff virtual walls in force-reflecting interfaces. *Virtual Reality Annual International Symposium, 1993* (pp. 202–208). IEEE.

7. Diane, M., & Minsky, R. R. (1995). *Computational haptics: The sandpaper system for synthesizing texture for a force-feedback display.* PhD thesis, Cambridge, MA.

8. Díaz, I., Hernantes, J., Mansa, I., Lozano, A., Borro, D., Gil, J. J., et al. (2006). Influence of multisensory feedback on haptic accessibility tasks. *Virtual Reality (Waltham Cross),* *10*, 31–40.

9. DiFilippo, D., & Pai, D. K. (2000). The AHI: An audio and haptic interface for simulating contact interactions. In *Proceedings of symposium on user interface software and technology* (pp. 149–158).

10. Diolaiti, N., Niemeyer, G., & Barbagli, F. J., Salisbury, K., & Melchiorri, C. (1984). The effect of quantization and coulomb friction on the stability of haptic rendering. In *Proceedings of the symposium on haptic interfaces for virtual environment and teleoperator systems* (pp. 237–246).

11. Funkhouser, T., Jot, J.-M., & Tsingos, N. (2002). *Sounds good to me!* SIGGRAPH Course Notes.

12. Gaver, W. W. (1993). What in the world do we hear? An ecological approach to auditory source perception. *Ecological Psychology, 5*(1), 1–29.

13. Gomes de Sá, A., & Zachmann, G. (1999). Virtual reality as a tool for verification of assembly and maintenance processes. *Computers & Graphics, 23*(3), 389–403.

14. Howard, B. M., & Vance, J. M. (2007). Desktop haptic virtual assembly using physically based modelling. *Virtual Reality (Waltham Cross), 11*, 207–215.

15. Hu, H. H., Gooch, A. A., Thompson, W. B., Smits, B. E., Rieser, J. J., & Shirley, P. (2000). Visual cues for imminent object contact in realistic virtual environments. In *Proceedings of the IEEE Visualization Conference* (pp. 179–185).

16. Hwang, J. D., Williams, M. D., & Niemeyer, G. (2004). Toward event-based haptics: Rendering contact using open-loop force pulses. In *Proceedings of the conference on haptic interfaces for virtual environments and teleoperator systems* (pp. 24–31).

17. Johnson, D. E., & Willemsen, P. (2003). Six degree-of-freedom haptic rendering of complex polygonal models. In *Proceedings of the symposium on haptic interfaces for virtual environment and teleoperator systems* (pp. 229–235).

18. Kuchenbecker, K. J., Fiene, J., & Niemeyer, G. (2006). Improving contact realism through event-based haptic feedback. *IEEE Transactions on Visualization and Computer Graphics, 12*(2), 219–230.

19. Lécuyer, A., Mégard, C., Burkhardt, J.-M., Lim, T., Coquillart, S., Coiffet, P., et al. (2002). The effect of haptic, visual and auditory feedback on an insertion task on a 2-screen workbench. In *Proceedings of the immersive projection technology symposium*.

20. McNeely, W. A., PuterBaugh, K. D., & Troy, J. J. (1999). Voxel-based 6-DOF haptic rendering using voxel sampling. In *Proceedings of the ACM SIGGRAPH 1999* (pp. 401–408).

21. Merlhiot, X. (2007). A robust, efficient and time-stepping compatible collision detection method for non-smooth contact between rigid bodies of arbitrary shape. In *Multibody Dynamics*.

22. O'Brien, J. F., Cook, P. R., & Essl, G. (2001). Synthesizing sounds from physically based motion. In *Proceedings of the ACM SIGGRAPH 2001* (pp. 529–536).

23. Ortega, M., Redon, S., & Coquillart, S. (2007). A six degree-of-freedom god-object method for haptic display of rigid bodies with surface properties. *IEEE Transactions on Visualization and Computer Graphics*, *13*(3), 458–469.

24. Otaduy, M. A., Jain, N., Sud, A., & Lin, M. C. (2004). Haptic display of interaction between textured models. In *Proceedings of IEEE visualization conference* (pp. 297–304).

25. Perlin, K. (1985). An image synthesizer. In *Proceedings of the annual conference on computer graphics and interactive techniques, SIGGRAPH '85* (pp. 287–296).

26. Pulkki, V. (1997). Virtual sound source positioning using vector base amplitude. *Journal of the Audio Engineering Society. Audio Engineering Society*, *45*(6), 456–466.

27. Raghuvanshi, N., & Lin, M. C. (2006). Symphony: Real-time physically-based sound synthesis. In *Proceedings of symposium on interactive 3D graphics and games*.

28. Redon, S., Kheddar, A., & Coquillard, S. (2000). An algebraic solution to the problem of collision detection for rigid polyhedral objects. In *Proceedings of the international conference on robotics and automation* (pp. 3733–3738).

29. Reissel, L.-M., & Dinesh, K. P. (2007). High resolution analysis of impact sounds and forces. In *Proceedings of worldhaptics*.

30. Siira, J., & Pai, D. K. (1996). Haptic texturing—a stochastic approach. In *Proceedings of the international conference on robotics and automation* (pp. 557–562).

31. Sreng, J. (2008). *Contribution to the study of visual, auditory and haptic rendering of information of contact in virtual environments*. PhD thesis. INSA Rennes.

32. Sreng, J., Bergez, F., Legarrec, J., Lécuyer, A., & Andriot, C. (2007). Using an event-based approach to improve the multimodal rendering of 6dof virtual contact. In *Proceedings of the ACM symposium on virtual reality software and technology* (pp. 165–173). ACM.

33. Sreng, J., Lécuyer, A., Andriot, C., & Arnaldi, B. (2009). Spatialized haptic rendering: Providing impact position information in 6dof haptic simulations using vibrations. In *Proceedings of the IEEE virtual reality conference* (pp. 3–9). IEEE Computer Society.

34. Sreng, J., Lécuyer, A., Mégard, C., & Andriot, C. (2006). Using visual cues of contact to improve interactive manipulation of virtual objects in industrial assembly/maintenance simulations. *IEEE Transactions on Visualization and Computer Graphics*, *12*(5), 1013–1020.

35. Takala, T., & Hahn, J. (1992). Sound rendering. *Computer Graphics*, *26*(2), 211–220.

36. Thomas, T. R. (1999). *Rough surfaces* (2nd ed.). London: Imperial College Press.

37. van den Doel, K., Kry, P. G., & Pai, D. K. (2001). Foleyautomatic: Physically-based sound effects for interactive simulation and animation. In *Proceedings of the ACM SIGGRAPH 2001* (pp. 537–544).

38. Warren, W. H., & Verbrugge, R. R. (1984). Auditory perception of breaking and bouncing events: A case study in ecological acoustics. *Journal of Experimental Psychology. Human Perception and Performance, 10*(5), 704–712.

39. Yano, H., Igawa, H., Kameda, T., Muzutani, K., & Iwata, H. (2004). Audiohaptics: Audio and haptic rendering based on a physical model. In *Proceedings of symposium on haptic interfaces for virtual environment and teleoperator systems* (pp. 250–257).

40. Zahariev, M., & MacKenzie, C. (2003). Auditory, graphical and haptic contact cues for a reach, grasp, and place task in an augmented environment. In *Proceedings of International Conference on Multimodal Interfaces* (pp. 273–276).

41. Zhang, Y., Fernando, T., Xiao, H., & Travis, A. R. L. (2006). Evaluation of auditory and visual feedback on task performance in a virtual assembly environment. *Presence (Cambridge, MA), 15*(6), 613–626.

16 Sonic Interaction via Spatial Arrangement in Mixed-Reality Environments

Mike Wozniewski, Zack Settel, and Jeremy R. Cooperstock

The recent growth in wireless and mobile computing technology has extended the potential domain of computer applications to the everyday world, far from the generic desktop. The resulting domain, known as *mixed reality*, which blends real and digital information, offers great potential for sonic interaction. Many new techniques can be explored as we transition from purely virtual simulation to applications that incorporate real-world features. Examples include location-based audio delivery, distributed performance, virtual music venues, subjective multiuser experience, and topographical musical arrangement. By representing users in a three-dimensional model, movement and spatial arrangement can be used to provide natural sonic mappings, capitalizing on familiar movements and gestures. In this manner it is possible to virtualize the components of musical performance, including microphones, loudspeakers, effects boxes, and the interconnections among them. Rearranging these items in space and adjusting perspective provide a technique for generic musical interaction with sonic material, offering potential for the discovery of new artistic forms.

16.1 Background

Virtual worlds are becoming important meeting places for many sorts of activity, and the opportunities to exploit these worlds for creative effect have not been lost on artists. The growth of mobile computing, coupled with improvements in tracking technology, also facilitate the creation of hybrid spaces where virtual elements, or even fully coherent three-dimensional scenes, can be overlaid on top of a real space, thereby extending virtual worlds into the physical. Examples include the *Tactical Sound Garden* (TSG) [6], where users were able to "plant" sounds in urban environments, listen to previous creations, and even "prune" the sounds left by others with the adjustment of a few parameters. Similarly, *Alter Audio* [5] allowed cell phone users to compose musical arrangements in space over time. The system allowed sampling of audio via cell phone microphones and provided modifiers in the form of sound effects, synthesizers, and the ability to rearrange these in space. Tanaka proposed an ad hoc (peer-to-peer) networking strategy to allow multiple musicians to share sound simultaneously using hand-held computers [8], and his later work with Gemeinboeck [9] capitalized on location-based services on 3G networks to acquire coarse locations of mobile devices. They proposed the creation of *locative media instruments*, where geographic localization is used as a component of a musical interface. Other examples such as *GpsTunes* [7] and *Melodious Walkabout* [2] have used heading information to provide audio cues that guide individuals in specific directions, demonstrating the interactive potential of spatial audio in augmented reality scenarios.

16.2   Sonic Interaction via Spatial Arrangement

Although the case studies provided later in this text vary in terms of interaction, there is a unifying paradigm used throughout. Specifically, all sonic elements and interactions among them are represented within a virtual three-dimensional model, and there is a relation between these virtual items and their real-world counterparts.

Although arranging sounds in space is hardly new to the world of audio, where the physical arrangement of live instruments is often adjusted to achieve a desired mix (e.g., trombones in the back, violins up front), we can gain expressive power by "virtualizing" a spatial audio configuration. The flexibility of computer simulation allows for additional control mechanisms to adjust the arrangement and interplay of sonic elements in real time. Because it draws on the life experience of sound in the world around us, this approach tends to be intuitive, with interaction based on a familiar conceptual model.

16.2.1   Spatial Signal Bussing

Our work in this domain began with the artistic motivation to reintroduce physicality and human motion into digital performance works. The objective was to maintain a growing number of digital signal processing (DSP) units without having to interrupt a musician from playing his instrument, that is, keeping hands on the saxophone instead of on computing equipment. Although foot pedals and the like have long been used for such purposes, we wanted something more direct and transparent, allowing the audience to see and recognize the effector responsible for change in the sound. This requirement could be satisfied through the use of large-scale motion and gesture as a more natural control paradigm and, ideally, with a reduction in the number of abstract actions needed to be memorized. As a first instance of this kind of control, we designed a configuration of several microphones on stage, arranged in a circle around the performer, each routed to a particular DSP unit. While playing, a performer could physically approach a particular microphone to apply the corresponding DSP effects. Although this configuration was easy to manage and satisfied the objectives of directness and transparency, it remained quite limited in compositional range.

To address this shortcoming, we subsequently "virtualized" the configuration for signal bussing. Using computer simulation, all microphones and DSP were modeled in a *virtual audio scene*, with input provided by an instrumental sound source (live sax), tracked using an orientation sensor. Because the simulation knew where the saxophone was pointing, the performer could physically aim his instrument to surrounding virtual microphones, represented as DSP targets in the scene, and the sound would be captured and processed accordingly by the desired DSP module. Additionally, we allowed the performer to spread the sound to more than one target as a function of the *roll* of the instrument, that is, rotation around the pointing axis.

16.2.2 Sonic Perspective and Listening

A fundamental aspect of sonic interaction emerges when virtual motion is also extended to listeners and audience members, allowing listeners to obtain radically different perspectives of a single spatial audio work. Just as a sound source can be aimed toward a DSP target, a virtual microphone can be controlled to capture (listen to) a selected region of the audio scene. Such microphone steering turns the user into an *active listener* with a specific sonic perspective and the power to determine the quality of mix to be captured and heard. Moreover, this permits the emergence of virtuosity in the mixing task, using simple actions (e.g., moving, turning, focusing), and the development of new expressive forms involving real-time montage in live performance applications.

16.2.3 Interaction via Motion

Taking this approach a step further, we may incorporate motion and dynamic control over the arrangement and interplay of sonic elements. For example, virtual DSP targets or sound source elements, such as phase-locked audio loops, can be pulled closer or pushed away from the performer to control the amount of signal being processed, thereby replacing the sonic interactions that would be performed traditionally with faders, knobs, and pedals. Physical motion can also be incorporated into the virtual environment to explore and play with elements scattered throughout the space, thus allowing both composers and listeners to create new sonic blends of sound sources in their proximity and to interact with each other during performance.

16.2.4 Virtual Performance Venues

We have virtualized different types of performance situations where multiple networked performers can explore and play together (see *Ménagerie Imaginaire* in section 16.3.1). What distinguishes our approach from other network-based jamming implementations [1, 4] is that our players share not only audio streams but the actual sonic resources of the virtual environment as well. Typically, these virtual venues have zones at various locations designated for different types of musical interaction. At one location, for example, there may be a number of synchronized loops to accompany players who come there to jam. As players move through the area, the loops they hear will fade from one to another yet remain rhythmically synchronized to maintain a coherent groove for all performers.

16.2.5 Mixed Reality

Our interaction paradigm involves the modeling of sound elements in virtual space and the coupling of that space and its elements to the real word. Following the taxonomy of Milgram's Virtuality Continuum, illustrated in figure 16.1, our work tends to be more associated augmented virtuality (a virtual environment augmented with real-world information) in that it involves attaching sensors to a physical instrument and representing its position within a virtual scene.

Although it is perfectly valid to organize sounds and effects in surreal or conceptual spaces, such as those one might find in a video game, we seek to provide direct, physical, and natural real-world interaction with audio content. Thus with each new application we create, we move further toward the left side of the continuum, as opposed to the right side, where interaction is mainly limited to virtual content and users require a screen and controllers (e.g., joysticks, gamepads, or other sensors) to apply motions.

Furthest to the left we find augmented reality applications, which maintain a user's presence among real-world objects and virtual content, allowing users to capitalize

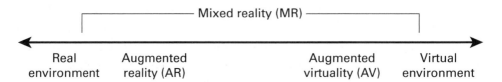

Figure 16.1
Milgram's Virtuality Continuum [3] shows that an application can sit on a scale between virtual reality (total immersion in a synthetic world) and true reality as more physical objects are added to the representation.

on everyday knowledge and movement to drive interaction. The human body can be tracked and modeled so that hand gestures, head movements, and even walking can be mapped to corresponding virtual motion. Other physical objects can also have virtual counterparts and, conversely, virtual objects, such as the DSP targets mentioned earlier, can be associated with physical objects that the user might point toward or even pick up to perform some processing on an audio signal. In such a case, the environment's real features are used as interface elements for interaction. Furthermore, the potentially massive scale of real-world space also lends well to massively multiuser implementations in which hundreds or thousands of users can participate.

16.3   Case Studies

Building on the capabilities described above, we now present examples of the artistic works and proof-of-concept applications we have developed. Each serves to contrast and comment on the different kinds of sonic interaction and implementation strategies we have explored to date. Because our objectives with this work have, to date, been artistic, our primary concern was the creative possibilities afforded to the artist. Nevertheless, we are now undertaking an experiment to investigate the effectiveness of our innovations insofar as they impact on listener perception and understanding of a musically complex scene.

16.3.1 *Ménagerie Imaginaire*: A Performance in a Virtual World

This piece was performed at NIME[1] June 7, 2007 (New York) and at the Pure Data Convention[2] August 23, 2007 (Montreal, Canada).

One of our first performances to capitalize on movement and spatial arrangement of audio in a virtual environment, this piece features a guitarist and a harmonica player who travel throughout a shared virtual landscape and jam in several preestablished zones, each with accompaniment pertaining to a unified spatiotemporal score. Player movement in the environment was purely virtual, using joystick control. Later, an active listener was added to capture the sonic (musical) activity for presentation to the audience in the style of a music video, where the camera's perspective is organized aesthetically to follow the players throughout the performance. With this addition the trajectories of the musicians were predetermined and automated, resulting in a fixed spatiotemporal structure on which the players could improvise by directing their sounds to a number of effects units that were "attached" to each musician as an artist's palette.

16.3.2 *Blairatta Policeme*: A Scored Performance in Mixed Reality

This work was commissioned by Arraymusic[3] and performed live at ARRAYMUSIC Studio, June 3, 2008 (Toronto, Canada).

Blariatta is a chamber piece scored for a jazz trio and three moving musicians, where performers and audience share a common space with an overlaid virtual audio scene containing four signal-processing zones, marked on the floor. The sound of a musician passing through a zone is captured, delayed, and reemitted there. Thus, specific locations in the space are associated with a particular contrapuntal musical function, and cannons are created by the superposition of musical material in the score. Live signal processing, which is usually difficult to manage because of inherent instabilities of feedback delay, was, in this case, limited in time and space: each player's microphone, sent to a delay unit, was open only while the performer was in a zone. Perhaps most interestingly, the mixed-reality interface design involved a spatial organization of musical material, both mobile sound sources and signal-processing zones, coupled to locations in the physical space. The score provided instructions to players describing what, as well as "where," to play (figure 16.2). Similarly, audience members could roam the performance space, emphasizing particular layers of musical material (cannons) according to their proximity to the four zones. The resulting musical experience was globally linear, described in a conventional score, but at the same time locally nonlinear because the final rendering of the work depended on the unique movement (perspective) of each listener (figure 16.3).

16.3.3 *Audio Graffiti*: Multiuser Audio Tagging and Listening Installation

This work was installed at ICMC[4] August 21, 2009 (Montreal, Canada) at the Twelfth Biennial Arts and Technology Symposium[5] March 4–6, 2010 (New London, CT), and at Nuit Blanche October 1, 2011 (Toronto, Canada).

Audio Graffiti operates on a wireless network, allowing multiple participants to contribute to a collaborative, ever-evolving musical texture. Participants connect using a small mobile computer with a connected headset and a position-tracking tag.[6] Users can start and stop the capture of their microphone signal, which is then time-tagged and assigned a playback offset relative to a global rhythm cycle (of a certain number of beats). Playback of each sample is periodically triggered by a scheduler, and its amplitude may diminish over time until it fades out and is removed from the scene, as illustrated by the shrinking spheres seen in figure 16.4.

The "graffiti wall" provides users with a strong conceptual and physical representation of where the tags are placed; no graphical display is needed. The wall also helps to support an active listening paradigm in which each user listens via an audio capture

Figure 16.2
A moment from *Ménagerie Imaginaire* when a performer directs his harmonica into the virtual scene. His screen shows a subjective view, while the screen behind provides an objective rendering for the audience.

cone. Audio zoom is controlled by users' proximity to the wall, as their listening cones respectively encompass a lesser or greater number of tags. Pulling back from the wall yields a "big picture" of all the sonic contributions, whereas coming close excludes nearly all but one or two particular audio tags.

16.4   Implementation Details

16.4.1   Motion Tracking: Binding Real and Virtual

The representation and virtualization of real-world objects rely extensively on motion tracking, which can be accomplished using a wide range of methods, including a global positioning system (GPS), which provides approximate spatial position but is limited to outdoor use, and camera-based or radio frequency systems for indoor use. In order to maintain a corresponding audio perspective, it is also important to capture the orientation of a listener's head. There are a number of commercially available units

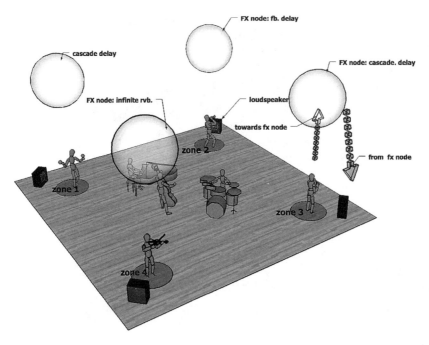

Figure 16.3

A schematic overview of *Blairatta Policeme*. Performers play music and move through the space according to a score. On entering a zone, their instruments' signals are bussed to the corresponding effects node and are heard by the loudspeaker. Throughout the piece, the effects nodes shift positions, carrying their containing sound during their displacement.

to choose among that provide full three-dimensional orientation with an absolute reference.

16.4.2 Audio Streaming versus Playback

Reliable audio streaming between users poses a formidable challenge, especially in mobile wireless implementations. To conserve bandwidth, most wireless audio protocols use compression algorithms, which are often associated with significant latency, thus limiting the possibilities for musically synchronized participants. In response, we developed *nStream*, our own low-latency uncompressed audio streaming protocol [10], which minimizes delay while maintaining an acceptable audio quality for musical applications, even when running on wireless transports. Although limitations of scale are still encountered in both wired and wireless conditions, such protocols are often limited more by the rate at which packets are injected into the network than by bandwidth.

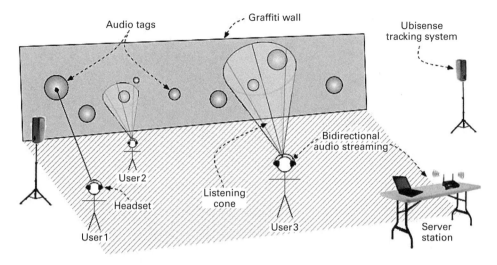

Figure 16.4
Audio Graffiti. One user projects an audio tag onto the wall, represented by a sphere.

16.4.3   Audio Tags

Live audio streaming is not always necessary. In *Audio Graffiti*, for example, user audio is captured discretely and subsequently uploaded as a file for asynchronous transfer to other participants. This permits a larger number of simultaneous users with greater reliability and less reliance on bandwidth and network quality. Each discrete sample or *audio tag* is associated with a three-dimensional location so that it can be rendered as a point source where it was recorded. Alternatively, audio can be represented as *contours*—temporal streams with a particular trajectory that are played as a user travels along its path, much as a playhead moves along a segment of audio recording tape. The latter may be more appropriate for audio segments associated with movement during the recording.

16.4.4   Scalability and Distribution

Effective mixed-reality applications can be deployed in a small area serviced by one wireless router, providing a workable radius of approximately 50 m at best. In such a case, it is feasible to transmit audio to wireless headsets, with all audio content modeled and simulated on a single machine. However, as the spatial extent of an application increases, connectivity issues arise. In such cases, content must be rendered in a distributed manner, local to each user, who must receive updates when the scene changes. To meet this challenge, we developed the *SPIN Framework,*[7] which provides the necessary scalable software layer for maintaining distributed *SPatial INter-action* over a network. A client-server model is used to provide the necessary control-level synchronization between machines so that one shared environment is experienced by all users.

16.5   Discussion: Spatial Organization and Artistic Creation

Spatial arrangement as a technique for composing music is by no means new. The ability to organize music in space has great implications for how it is composed and, in turn, experienced. A virtual scene can be subdivided into a variety of zones where different sounds and sonic behaviors are encountered by performer or listener. Similarly, the sonic perspective of each listener can contribute actively to a unique and subjective experience. Our notions of harmony and counterpoint are extended by considering the "proximity" of musical material in relation to the listener, and per-former/listener movement can create "musical transitions" from one zone to the next, ultimately serving to define the global form of the musical experience.

Working with these constructs, a composer must consider alternative approaches to the creation, performance, representation, and experience of music. For ideas, we look to other disciplines, such as sculpture and architecture, where objects are experienced from multiple points of view, in a nonlinear fashion. In general, spatial music pieces can include a topological description or "map" that identifies the locations of sonic elements, delineates zones, and describes the extents and limits of interaction. Although there is no inherent order in which one visits the content of a work, nor speed at which the listener transitions between locations, it is possible to create linear works in time. Trajectories can be used, for example, to constrain listener movement in the scene, thereby creating timed and sequential "readings" of the work. More-over, because the piece is deployed via computer simulation, it is also possible to define behaviors or events that occur based on user motion. For example, certain sounds might be played only if a user passes through a particular location. The extent to which the creator predefines trajectories or leaves the system free and open for exploration situates the work in a continuum between performance and art installation.

Of particular interest, our approach offers the possibility to work with linear forms, nonlinear forms, or both simultaneously. Over time, we have explored several aspects of artistic design in these environments, ranging from game-like navigable virtual environments such as the work *Ménagerie Imaginaire* to choreographed live (mixed-reality) performances such as *Blairatta Policeme*. We have even deployed multiuser installations such as *Audio Graffiti*, where users interact in the same virtual scene over a period of days, establishing groups of audio tags at different locations in the space. In composition, we have developed techniques and strategies for the arrangement of sonic material so that it remains coherent in nonlinear playback. Because sounds from one zone may be audible in another zone, it is sometimes necessary to create boundaries between sounds that do not mix well or establish transitions between adjacent sounds that work well "coming or going." In a sense, the composer must lay out the work topologically, becoming a sort of sonic cartographer whose score maps out musical activity in both space and time.

16.6 Conclusion

We began this exploration with a simulation technology for placement of virtual effects around a performing musician. However, a formalization of the interaction paradigm has led to works that can be deployed in large-scale mixed-reality scenarios, where potentially large numbers of users can simultaneously interact with the system. This allows for experimentation with these exciting forms and representations, provides new opportunities for artistic creation, and challenges the social institutions for musical performance and diffusion. With new forms come new performance venues, and with these come new modes of audience participation. We look forward to working in this domain, where multiple individuals have the power to create their own interpretations, where the traditional boundaries of musical arrangement fall apart, and where we begin to interact on vast scales of spatial and social extent.

Notes

1. http://www.nime.org/2007.

2. http://convention07.puredata.info/en/index.html.

3. http://www.arraymusic.com.

4. http://www.icmc2009.org.

5. http://www.conncoll.edu/CAT/sym2010.

6. We used the Ubisense tracking system (www.ubisense.net/en), which uses triangulation of UWB radio pulses from active RF tags and operates with a spatial resolution of 1 foot or less because consumer-grade GPS cannot operate at such a fine (pedestrian) scale and loses accuracy when close to walls.

7. http://spinframework.org.

References

1. Bouillot, N. (2007). nJam user experiments: Enabling remote musical interaction from milliseconds to seconds. In *Proceedings of the international conference on new interfaces for musical expression (NIME)* (pp. 142–147). New York: ACM.

2. Etter, R. (2005). Implicit navigation with contextualized personal audio contents. In *Adjunct proceedings of the third international conference on pervasive computing* (pp. 43–49). Munich: Springer.

3. Milgram, P., & Kishino, F. (1994). Taxonomy of mixed reality visual displays. *IEICE Transactions on Information and Systems, E77-D*, 1321–1329.

4. Renaud, A., Carot, A., & Rebelo, P. (2007). Networked music performance: State of the art. In *Proceedings of Audio Engineering Society (AES) 30th conference on intelligent audio environments* (pp. 1–7). Saariselkä, Finland: Audio Engineering Society.

5. Shea, G., & Gardner, P. (2006). Alter audio: Mobile and locative sound experiences. http://www.mobilenation.ca/mdcn/alteraudio.

6. Shepard, M. (2005). Tactical sound garden [tsg] toolkit. *30 60 90 09. Regarding Public Space, 1*(1), 64–71.

7. Strachan, S., Eslambolchilar, P., Murray-Smith, R., Hughes, S., & O'Modhrain, S. (2005). GpsTunes: Controlling navigation via audio feedback. In *International conference on human computer interaction with mobile devices & services (MobileHCI)* (pp. 275–278). New York: ACM.

8. Tanaka, A. (2004). Mobile music making. In *Proceedings of New Interfaces for Musical Interaction* (pp. 154–156). Hamamatsu, Japan: NIME.

9. Tanaka, A., & Gemeinboeck, P. (2006). A framework for spatial interaction in locative media. In *Proceedings of new interfaces for musical expression (NIME)* (pp. 26–30). Paris: IRCAM.

10. Wozniewski, M., Bouillot, N., Settel, Z., & Cooperstock, J. R. (2008). Large-scale mobile audio environments for collaborative musical interaction. In *International conference on new interfaces for musical expression*, Genova, Italy.

17 Heigh Ho: Rhythmicity in Sonic Interaction

Cumhur Erkut, Antti Jylhä, and Davide Rocchesso

The emotional and experiential values of sound suggest an opportunity for establishing a deep relationship between people and interactive devices, products, and services. The human appreciation of rhythm is a good resource for grounding this relationship, alongside the related fundamental capabilities of human nature, such as equilibrium, effort, and pace [1, 2]. In particular, two human sensitivities strongly support the use of auditory displays for providing feedback in repetitive tasks [3]. The first is our ability to aurally construct a continuous support for discrete, repetitive events in the form of tempo or rhythm. The second is our sensitivity to detect, follow, or adapt to the changes in this construct.

Rhythmic interaction, in this respect, provides new perspectives on interactive sound by addressing emergent research topics and application areas [4–6], including mobile technology [4], interactive music [7, 8], and ambient entertainment [5, 9]. In addition, commercial products[1] readily make use of rhythmic interaction in physical exercise. These devices provide spoken feedback on their user's performance or select music from their library based on their pace. Notwithstanding the research and commercial interest, a thematic overview of rhythmic interaction, to our knowledge, has not been worked out. We hope that this contribution and the multimedia examples on the companion Web page[2] help to fill this gap.

Based on interaction paradigms and metaphors that relate to *communication*, be it musical or conversational, the applications mentioned above are sensitive to time and have an internal representation of tempo. They collect the user input by using various sensors (accelerometer, microphone, camera, IR baton, etc.) and perform time or tempo tracking on this input with specific measures on *when* and *how* to match the internal and user tempo. Finally, they render a multimodal output that provides feedback to the user. Nevertheless, the nonspeech auditory feedback is either completely dismissed or entirely based on the music in the application itself.

We sense opportunities for sonic interaction design here. First, these systems can be enhanced by continuous sonic feedback. Second, our field can be extended with specific design guidelines and lessons learned from these systems that embody the communication representation of interaction. This thematic overview, first and foremost, is an attempt to establish this link.

The structure of this chapter is as follows. In section 17.1 we summarize some key terms and concepts that will be useful in our discussion. Then, in section 17.2 we briefly review the selected application domains and summarize the relevant work there. We particularly aim to highlight their interaction paradigms and mechanisms. In section 17.3 we present our experiments with rhythmicity in sonic interaction design. This section should be read in conjunction with the Gamelunch case study (Delle Monache, Polotti, and Rocchesso, chapter 8 in this book). Finally, we conclude our discussion in section 17.4.

17.1 Background

Rhythms can occur at different *scales* and affect a variety of aspects of human life and behavior. On a large scale, our social interactions are largely affected by the rhythms of daylight, by our recurrent physiological needs, and by conventional timetables. On a much shorter scale, rhythms affect human-tool interactions in a wide spectrum of everyday activities and specific jobs. Using a saw, or a hammer, or a sickle involves cyclic gestures that induce a rhythmic *entrainment*. It is there that auditory feedback plays a fundamental role not only to help keep a certain pace and to spare pains but also to induce *synchronization* and *mutual awareness* in cooperative work among humans and with machines.

There is much information in the sonic by-product of rhythmic interactions. Events repeated at a regular pace determine the *tempo* of a cyclic action. When patterns of *pulses* emerge out of local variations of intensity, timbre, or timing, and these patterns are iterated in time, rhythms become characterized by a *meter*, from which listeners may eventually extract a tempo, in ways that are an active area of investigation in music cognition [10, 11].

Deviations and *drift* among repetitions of patterns are also natural indicators of expressive qualities of interaction, such as effort or flow. It is interesting that these qualities are largely invariant across application areas and across gestures. For example, the drifts related to starting or stopping have been shown to be similar in walking and in music phrasing [12]. Shifts in tempo and intensity envelopes are so salient that they can affect the psychological states of subjects participating in an action. This seems to be the secret of drumming in shamanism [13], which indeed needs enactive involvement rather than mere exposition.

Even when they are marked by discrete events producing impulsive auditory manifestations, cyclic gestures are always *continuous*, and these continuous trajectories may reveal important facts about an action. Sometimes, immediate visual or proprioceptive cues fill the space between discrete events, as in drumming [10, 14], thus contributing to the whole *multisensory* experience of a rhythm.

Whatever the application area of rhythmic gestures is, in music, sport, or crafts, exploitation of rhythms goes together with development of *virtuosity*. A skilled person will invariably be capable of matching speed with accuracy, consistency with adaptability, and effectiveness with a minimum of attention and effort [10]. The trade-offs that are usually found in human-machine interaction (e.g., in Fitts's law) are somehow overcome by experts in cyclic gestures. Virtuosity transforms the fatigue of repeating some gestures thousands of times into joy of performing [15]. Even in prosaic contexts, a tool that affords expressive manipulations easily becomes the instrument of a performance. Indeed, many modern musical instruments are evolutions of tools for work and survival.

17.2 Rhythmicity in Interactive Systems

Even musically untrained people have well-established ways for interacting with music, such as humming, tapping, or clapping along. They also enjoy imitating a musical process (conducting or "air" instrument playing). Can we build on these music-appropriate interaction metaphors in interaction design involving new media? This was the driving question that has guided and informed Borchers and Muhlhauser [8] in designing various interactive music systems for exhibitions in years to come. They have identified two dimensions of using music as an interactive medium: *interaction direction* (input, output) and *information type* (program control, application data). They provided the status of musical interaction along these dimensions, pointing out rhythm to be the most accessible concept for computer modeling in musical interaction.

As an example of design outcomes, consider *Personal Orchestra,* which aims to provide a glimpse into the conducting experience to untrained users. The movement and direction of an infrared baton control the tempo and volume of the recorded music and put emphasis on orchestral groups. The internal variable of their system is event time. They aim to match the time of the playback by linearly adjusting the playback time in small temporal intervals based on the detected time and *movie time*, that is, the beats premarked on the video. This dual concept of user and playback time was later formalized as *semantic time* [16]. Other highlights of this research line include the continuous, high-quality audiovisual output, development by interaction design patterns [8], a high-quality audio time-stretching algorithm [16], systematic user evaluation, and other user observations.

The conductor scenario is also an exemplar in the framework of virtual interactive humanoids. However, this line of research builds on the interactions between people in everyday conversation, where mutual coordination and anticipation naturally occur [5]. This interaction paradigm is an extension on the customary mechanism of turn-taking in conversation analysis and modeling. Other examples of conversational virtual agents include a dancer, a conductor, and a trainer humanoid, which all respond to audio and rhythm in different ways. For instance, the conductor application [17] listens to an ensemble of trained musicians and includes a tempo tracker and a score follower. If the ensemble plays too slowly or too fast, the conductor leads the musicians into the correct tempo. In an earlier version of the implementation, the detected and the intended tempo were linearly interpolated by a fixed coefficient. Later this coefficient has been varied according to the number of beats since the start of the tempo correction. The corrections occur at the start of a new bar. The animated conducting gestures of the virtual conductor provide feedback to the musicians. There is no audio output of the system.

The issue of *agency* has been central in the rhythmic interaction between human percussion players and a robot [7]. The interaction schemes used in this research line include a number of approaches for call-and-response and accompaniment-based collaboration. The robot listens to the human performers by using microphones attached to their drums. Both low-level rhythmic features (onset, amplitude, beat, and density detection) and higher-level rhythmic constructs such as the stability of the rhythm are detected by computation. The rhythmic response of the robot is determined by the analyzed features as well as specific interaction modes. Feedback is given in the form of acoustic drum sounds played by the two robotic arms. The mechanical delays of the arms are compensated by anticipation. The interaction mode during the beat detection is of special interest to us. In this mode, human players tend to adjust the robot's tempo. This leads to an unsteady input beat that is hard to track. This problem is solved by interleaving the listening and playing modes and by locking the tempo in the playing mode.

Just as in musical interaction, in everyday conversation we rely on many cues to create rapport with our conversation partners. For instance, when we talk while walking in a group, a common walking pace emerges. Conversation partners align their walking gait with each other and coordinate their conversation accordingly. When people converse on mobile phones, however, important components of this subconscious, nonverbal communication are lost. Can this loss be compensated by augmenting the communication channel? This fundamental question was tackled by Murray-Smith et al. [18], who investigated gait synchronization, measured by accelerometers while users converse via mobile phones. Specific analysis tools have been developed, and techniques from synchronization theory have been used to infer the level of alignment. The primary means of feedback was vibrotactile. One surprising outcome was that the remote mobile participants could synchronize their walking behavior based on the voice channel alone.

17.3 Rhythmicity in Sonic Interaction Design

If the goal is to develop systems and interfaces that support rhythmicity in sonic interaction and that afford the development of virtuosity, what are the ingredients that we may try to use and mix?

With sensors and signal processing, we should try to capture the prominent pulses and the emerging tempo as well as continuous underlying accelerations. Sometimes rhythmic gestures are associated with achieving a goal, and the progress toward that goal should be sensed as well. An example in which all these sensing problems are present is the gesture of chopping vegetables, whose sonic augmentation is described in the Gamelunch case study (Delle Monache, Polotti, and Rocchesso, chapter 8 in this book).

As far as auditory display is concerned, in many cases we want to give a sonic feedback to guide the action in a closed loop. Often user and machine have to synchronize at a certain pace, and the feedback strategy is crucial in determining how effectively this is achieved. In fact, synchronization is performed through anticipation [19], and different rhythmic feedbacks can change how the user internally exploits the anticipatory error that supports the feedback loop. The case of augmented chopping described in the Gamelunch case study illustrates how different feedback strategies can be developed and tested in design practices.

Two authors of this thematic overview have implemented a basic, stripped-down rhythmic interaction framework around the theme of hand clapping. The aim was experimenting with these ingredients plus some of the concepts we indicated in section 16.2 [20, 21]. First, we wanted to operate around the interaction directions presented by Borchers and Muhlhauser [8]. We have used the sounds people are both familiar with and sensitive to, that is, real hand clap sounds as an input and synthetic hand clap sound as an output. We have thus aimed at a consistent soundscape that can integrate program control (real hand claps) and application data (synthetic hand claps and music). Second, we wanted to determine to what degree we could induce the presence of virtual interlocutors as agents primarily with an auditory display, instead of virtual humanoids, robots, or vibrotactile displays. Third, especially the later work we have reported [21] brought about an opportunity to try out different goals, scales, and musical interaction scenarios informed by communication and conversation. The remainder of this section describes this framework in detail.

17.3.1 Hand-Clapping Interface

We have developed and showcased a hand-clapping interface for rhythmic interactions with the computer [20, 21]. The interface captures the hand-clapping sounds of the user, performs a low-level analysis on the occurrence times, tempo, and type of the hand claps, and uses this information in interactive applications. The interface is essentially a piece of software[3] fit for any platform that has a microphone for capturing the sounds.

We have demonstrated rhythmic interaction through the interface with two examples: interacting with the tempo of music and with a virtual crowd of clappers. Both focus on communicating the tempo of the action. In the former the tempo of music is matched to the tempo of the user's clapping, thus creating a closed loop of cyclic sonic interaction. The user perceives the tempo of the music and aims at altering it. This, as a process, captures the need for negotiating the overall tempo with the computer due to the user's anticipation of the timing of the next beat and the current tempo of the piece. An important cue in the feedback is the interplay between the music and that of the user's own clapping and the level of synchrony between them. It is also possible to draw a link to haptics, as the user perceives her clapping in the haptic domain, too.

The process of negotiation by mutual coordination becomes even more evident with a virtual crowd of clappers. The application, built with the hand clap synthesis engine ClaPD [22], aims at modeling the process of an audience applauding to a performer and getting synchronized, for example, to demand an encore. Each clapper in the virtual crowd is modeled as an agent listening to the clapping of the user and adapting its clapping to the user's tempo and phase. At first, the crowd claps asynchronously, but by starting to clap to a steady rhythm, the user can entrain the crowd to the tempo of her clapping. According to our experiments, the user also experiences a degree of entrainment by the synthetic audience.

17.3.2 iPalmas

Extending on the hand-clapping interface, we have developed iPalmas, an interactive Flamenco tutor and rhythm machine [21]. Flamenco is characterized by its pronounced rhythmic content, and hand clapping is instrumental in its creation. Also, clapping is used in Flamenco to communicate tempo changes and upcoming transitions between different parts in the performance. It effectively communicates the meter of the piece.

iPalmas is essentially a tool for Flamenco beginners to get acquainted with the different forms of its rhythmicity by hand clapping. The interactions take place between the user practicing Flamenco hand clapping, and a virtual tutor instructing the user to develop elementary Flamenco skills. These skills include being able to perform a suitable rhythmic pattern accurately and steadily with correct accentuation, reacting correctly to rhythmical cues provided by the tutor, for example, to stop clapping on the correct beat and to react to gradual tempo changes. The system can also model these rhythmic interactions autonomously by letting the virtual tutor lead a crowd of synthetic clappers that aim to follow the lead.

The input to the system is again the sound of the user's hand claps, on which both low-level (onsets, tempo, clap type) and higher-level (temporal variance, matching the accent pattern) analyses are performed. Feedback from the system is multisensory, and it takes place through both sound (the synthetic clapping sounds of the tutor and virtual clappers) and vision (different performance metrics on the user's clapping are visualized). The user can monitor online how well she is able to meet the goals set by the tutor. The system is intended to set new targets as the user improves her skills in order to bring these skills closer to virtuosity. The rhythmic variance is an important factor in communicating the skill level of the user, assuming that better performers are better at maintaining a consistent tempo. As the user gets better in clapping, that is, claps steadily to the correct tempo with the correct accentuation, the tutor starts to give more challenging tasks to the user such as by speeding up the tempo.

iPalmas is a work in progress and aims at providing a variety of contextual sounds for feedback surpassing plain hand clap sounds, which would bring rhythmic interaction closer to a more genuine performative level.

17.4 Conclusions

In this thematic overview, we have provided some fundamental concepts related to the rhythmicity in sonic interaction design. We have reviewed selected interactive systems that are sensitive to user rhythms and rely on interaction paradigms and metaphors that relate to musical or conversational communication. Based on this material, we have seized an opportunity to derive a specific design rationale for rhythmic sonic interaction in section 17.3. We are following these guidelines in the examples presented here and in chapter 8 of this volume. Yet, these exploratory implementations are the first steps in utilizing rhythmicity in sonic interaction design; realizing its full potential in cognition, affect, and motor function [1] stands out as a challenge for our community.

Notes

1. See, for instance, Nike+iPod™ at http://www.apple.com/ipod/nike/ or Phillips Activa™ at http://www.usa.philips.com/c/workout-monitors/176850/cat/ (accessed November 19, 2012).

2. http://blogs.aalto.fi/rhythmicity.

3. The current implementation has been built on the graphical computer music software Pd (http://puredata.info).

References

1. Thaut, M. H. (2005). *Rhythm, music, and the brain: Scientific foundations and clinical applications. Studies on new music research.* New York: Routledge.

2. Rocchesso, D., & Bresin, R. (2007). Emerging sounds for disappearing computers. In N. Streitz, A. Kameas, & I. Movrammati (Eds.), *The disappearing computer* (vol. 4500 LNCS, pp. 233–254). Berlin: Springer.

3. Rocchesso, D., Polotti, P., & Delle Monache, S. (2009). Designing continuous sonic interaction. *International Journal of Design, 3*(3), 13–25.

4. Lantz, V., & Murray-Smith, R. (2004). Rhythmic interaction with a mobile device. In *Proceedings of the 3rd Nordic conference on human-computer interaction (NordiCHI)*, Tampere, Finland (pp. 97–100).

5. Nijholt, A., Reidsma, D., van Welbergen, H., op den Akker, R., & Ruttkay, Z. (2008). Mutually coordinated anticipatory multimodal interaction. In *Proceedings of human-human and human-machine interaction 2007* (vol. LNAI 5042, pp. 70–89). Berlin: Springer.

6. Spillers, F. (2008). "Synch with me": Rhythmic interaction as an emerging principle of experiential design. In P. Desmet, S. Tzvetanova, H. P. Hekkert, and L. Justice (Eds.), *Proceedings of the 6th conference on design & emotion* (Hong Kong SAR.), CD-ROM proceedings.

7. Weinberg, G., & Driscoll, S. (2007). The interactive robotic percussionist: New developments in form, mechanics, perception and interaction design. In C. Breazeal, A. C. Schultz, T. Fong, & S. Kiesler (Eds.), *Proceedings of ACM/IEEE international conference on human-robot interaction (HRI'07)*, La Jolla, CA (pp. 233–234).

8. Borchers, J., & Muhlhauser, M. (1998). Design patterns for interactive musical systems. *Multimedia, IEEE, 5*(3), 36–46.

9. Ruttkay, Z., Zwiers, J., Welbergen, H., & Reidsma, D. (2006). Towards a reactive virtual trainer. In *Proceedings of the international conference on intelligent virtual agents (IVA)*, Marina Del Rey, CA (vol. 4133 of *LNCS*, pp. 292–303). Berlin/Heidelberg: Springer.

10. Dahl, S. (2005). *On the beat: Human movement and timing in the production and perception of music*. Doctoral thesis, KTH School of Computer Science and Communication, Stockholm, Sweden.

11. Ladinig, O., Honing, H., Haden, G., & Winkler, I. (2009). Probing attentive and preattentive emergent meter in adult listeners without extensive music training. *Music Perception, 26*(4), 377–386.

12. Friberg, A., Sundberg, J., & Frydén, L. (2000). Music from motion: Sound level envelopes of tones expressing human locomotion. *Journal of New Music Research, 29*(3), 199–210.

13. Hodgkinson, T. (1996). Siberian shamanism and improvised music. *Contemporary Music Review, 14*(1&2), 59–66.

14. Petrini, K., Dahl, S., Rocchesso, D., Waadeland, C. H., Avanzini, F., Puce, A., et al. (2009). Multisensory integration of drumming actions: Musical expertise affects perceived audiovisual asynchrony. *Experimental Brain Research, 198*, 339–352.

15. Sennett, R. (2008). *The craftsman*. New Haven, CT: Yale University Press.

16. Lee, E., Karrer, T., & Borchers, J. (2006). Toward a framework for interactive systems to conduct digital audio and video streams. *Computer Music Journal, 30*(1), 21–36.

17. Reidsma, D., Nijholt, A., & Bos, P. (2008). Temporal interaction between an artificial orchestra conductor and human musicians. *ACM Computers in Entertainment, 6*(4), 1–22.

18. Murray-Smith, R., Ramsay, A., Garrod, S., Jackson, M., & Musizza, B. (2007). Gait alignment in mobile phone conversations. In *Proceedings of the international conference on human computer interaction with mobile devices and services (MobileHCI)*, Singapore (pp. 214–221).

19. Aschersleben, G. (2002). Temporal control of movements in sensorimotor synchronization. *Brain and Cognition, 48*(1), 66–79.

20. Jylhä, A., & Erkut, C. (2008). Sonic interactions with hand clap sounds. In *Proceedings of Audio Mostly*, Piteå, Sweden (pp. 93–100).

21. Jylhä, A., Ekman, I., Erkut, C., & Tahiroglu, K (2011). Design and evaluation of human–computer rhythmic interaction in a tutoring system. *Computer Music Journal*, *35*(2), 36–48.

22. Peltola, L., Erkut, C., Cook, P., & Valimaki, V. (2007). Synthesis of hand clapping sounds, *IEEE Transactions on Audio, Speech, and Language Processing*, *15*(3), 1021–1029.

18 Barking Wallets and Poetic Flasks: Exploring Sound Design for Interactive Commodities

Daniel Hug

18.1 Background

18.1.1 The Sound of the New Everyday

The obtrusive boxes in our houses and offices are set to disappear into household appliances, entertainment systems, mobile communication devices, clothes, and places we might not even know about. The "new everyday" [1] is permeated by networked technologies and complex, seemingly autonomous, devices. Some of these devices operate entirely without human intervention, and others require some degree of operation or manipulation. The latter have been labeled "interactive commodities" [8]. The term "commodity" stands for the pragmatic, everyday nature of these artifacts.

Sound is of particular interest for such objects, and not only because of small or absent screens, peripheral use, and its relative economical implementation [6]. Interactive artifacts are often "black boxes," objects whose inner workings are unknown. Sound can provide means to give such devices a "display," a means of expression and communication, and a sonic identity. Sound also plays an important role in mediating the presence of artifacts and agency in social contexts.

In designing these sounds it is important to understand the many ways in which artifacts become meaningful. A central aspect is their material, multisensory objecthood through which they become mediators for socio-cultural interactions, expressing our personality, status, emotions, and attitudes, and work as narrative tokens of experiences and thus can be understood as "participants" in the course of action [5, 11, 19]. The latter reached a new dimension considering computerized artifacts that are procedural and interactive in nature, even anthropomorphized or "magical" [12].[1]

18.1.2 Creating Prototypical Cases Using a Workshop-Based Method

Sound design for interactive commodities deals with a design potential associated with currently emerging technologies and applications. These are going to be part of our everyday life, and in order to specify and explore the associated opportunities and challenges, it is necessary to create prototypical instances of such "potential experiences." Such "experience prototypes" [3] then can be evaluated with established methods. The research also aims at answering fundamental aesthetic questions related to sound design for interactive commodities rather than solving technical issues. Therefore, a "Wizard of Oz" prototyping strategy was adopted, where a technical system is simulated by a human controller, the "wizard" [10]. All cases reported here have been presented in a theatrical performance using this method, and often participants could also interact with the simulated artifacts.

As a research framework a system of workshops was devised. In each iteration the workshops followed the same process from conception to the design of prototypes and their evaluation. A detailed description of the method and the workshop structure is given by Hug [9].[2] At the time of the writing of this article, three workshops have been carried out. The initial workshop was held on October 21–22, 2008 at the Helsinki University of Technology, hosted by Cumhur Erkut and Inger Ekman. The second iteration was carried out between December 3 and 19, 2008 at Zurich University of the Arts (ZHdK) in the design department. A third version was held at the Media Lab of the Helsinki University of the Arts between March 9 and 13, 2009, hosted by Antti Ikonen. Altogether forty students participated in the workshops.

In the following, aspects are presented that formed the conceptual basis for the projects developed and thus could be discussed and evaluated.

A Typology of Interactive Commodities

In order to provide a first approach to structure the complex field of interactive commodities, a tentative typology of interactive commodities intended to support to both analysis and design was proposed [8]. This typology was tested during the process of producing the cases described here and is therefore outlined here for reference.

• *Authentic interactive commodities* are simple objects that are self-contained in terms of form and function and have an essential, static identity that is not questioned, thus "authentic." Such objects can be enhanced with electroacoustic sounds in order to support their identity, but the sounds should be very closely related to the given sonic quality of the object, using only moderate "sweetening."[3]

• *Extended interactive commodities* are more or less self-contained objects as well; however, they are endowed with additional functionalities through interactive technologies. In this case the sound of the object itself remains unaltered. The sonic enhancement emphasizes the "additionality" and should not be acoustically integrated into existing object sounds.

• *Interactive placeholders* are physical stand-ins for virtual artifacts. Their meanings and functions can—within the constraints of their appearance and operation modality—be completely redefined through sound. The sound is defined by the type of operation of the simulated object and its functionality, respectively.

• *Omnivalent interactive commodities* comprise complex systems that are not functionally defined through their physical communication and essentially rely on software. The requirements for the sound design can be detached entirely from the device itself, which hardly disposes of any considerable acoustic component or sonic identity.

Fluid Identities and Schizophonia

The identity of an artifact becomes more fluid through interactive software-based technologies. However, the possibility of integrating a practically limitless range of sounds into physical objects by means of miniaturized electroacoustic devices and of controlling them through computer technology make the relationship between a physical object and its sound arbitrary. Not only can the sounds naturally occurring from an interaction with a physical object be extended with signals, but it is also possible to transfer physical qualities, even a virtual presence of other objects (and the associated semantic qualities), into artifacts. This creates inherently schizophonic artifacts. *Schizophonia* is the term coined by R. M. Schafer to denote the separation of sound from its natural sources by electroacoustic means [17], which exactly describes the essential qualities of electroacoustically enhanced interactive artifacts. Because it is unknown how their sounds will be interpreted, this aspect was integrated in the workshop conception and evaluation.

Narrative Sound Design and Its Limitations

An important source of inspiration for designing semantically and aesthetically rich sounds for interactive objects is fictional media, in particular film. Here an immense body of knowledge and know-how about how to convey meaning through the narrative qualities of sounds have developed, and the high quality level satisfies fundamental requirements for sounds that are to become a meaningful, enriching, and appreciated part of our everyday life. Thus, it seems worthwhile to investigate possibilities of leveraging this design knowledge. However, it is unknown whether these narrative strategies from immersive media can be applied in physical artifacts of everyday use. Interactive commodities are used nonimmersively, pragmatically, and in the periphery of attention. They are intertwined with our everyday live experience, becoming intrinsic elements of our identities, intentions, and activities. Different interpretative strategies are thus to be expected.

In order to investigate the potential of narrative design strategies, the development and application of narrative metatopics formed an important element of the workshops. *Metatopics* are high-level narratives that emerged from the analysis of thirty preselected short clips from films and video games through a moderated analysis process in which all participants participated. These clips depicted salient interactions with complex, "magic," or interactive artifacts. An example of a narrative metatopic could be "transformation," which was often expressed sonically. The metatopics then were associated with identified sound design strategies and served as material to build design hypotheses that, at the end, could be evaluated. A summarizing overview over classes of metatopics can be found in section 18.3.1.

18.2   Selected Design Cases from the Workshops

In the following, a selection of the most salient cases that emerged from the workshop series is presented and discussed. I have grouped them according to commonalities, further explained below. The discussion focuses on aspects that contribute to a heuristic framework. Therefore, the specific design strategies employed to create the sounds are usually not explained. More details about this aspect can be found in Hug [9]. All descriptions are based on observations, the concept descriptions provided by the prototype authors, and the recorded discussions with the workshop participants.

18.2.1   Private Spaces

"Smart Home"
Luigi Cassaro, Christoph Böhler, Miriam Kolly, and Jeremy Spillmann

The group presented an interpretation of a sofa as an artifact oscillating between retreat for a single person and a sociable place for several people. Its sonic design focuses on the act of sitting down. The sonic authenticity of the sofa is maintained by using subtle modulations of sounds obtained from recording actual interactions with the sofa and combining these with artificial sounds that share spectral and temporal similarities. When one person sits down, possibly exhausted from work, the sofa produces a subtle, reassuring, full-bodied impact sound combined with a gentle breathing. When two people take a seat, the sofa produces both the impact sound and a crackling sound, which expresses a positive tension associated with friendship or love. When several people are crammed on the sofa, it assumes a more discrete mode and just produces a short impact sound, leaving the sonic space to the conversations of the people sitting down.

Furthermore a fridge was demonstrated that judges the user morally based on the selection of food contained in it. Here the principle of an "extended commodity" is followed, in that the sounds communicate an extension of the fridge as an autonomous system endowed with moral values. The sound design conveys a pressure release in correspondence with the actual process of opening the fridge, using aerodynamic sounds with a gentle, breathing quality. These rich overtone spectra are enhanced with synthesized harmonic layers. The harmonic quality is varied according to the system's judgment of the user's selection of food, for example by a downward pitching sonic gesture of increasing dissonance. This can be considered an enunciation of moral judgment by the system. The design also explores the metatopic of charging and discharging of "magic energy" following the stereotypical enunciation of magic known from film.

"Dangerous Microwave"
Sakari Tervo, Juha-Matti Hirvonen, and Joonas Jaatinen

This microwave is operated as usual, only the sound of switching on is enhanced with an energetic sound taken from an electronic door in a science fiction movie, conveying the powerful quality of the device and the function it offers. The sounds of the microwave's engine are increasingly mixed with aggressive sounds of electric static that were obtained by holding a microphone closely to the back of an actual microwave. These usually inaudible sounds convey an increasing hazardous potential. The sounds are effective on two levels: first, they contribute to the quality of an authentic, powerful, yet malfunctioning electrical device. Second, they contain sonic caricatures that are familiar from film, where they represent dangerous electrical equipment or weaponry, thus conveying a narrative of increasing evil or danger. These two qualities are mixed in a temporal dramaturgy that expresses the metatopic of life cycle: the sounds slowly rise in level and density and finally dissolve in a short outburst.

"Poetic Thermos Flask"
Samy Kramer

A thermos flask is turned into a high-tech tool for both making and storing tea. The user touches the metallic can with the finger, which produces a fine "click." The user then can move the finger vertically along the flask. This produces a synthesized bubbling sound mixed with an aerodynamic hiss and high-pitched chimes with a digital quality, conveying the "magic" of the interaction. As the finger moves up, the chimes pitch up, and the amount of bubbles increases, representing higher energy levels and temperature. Once a desired state has been reached, the finger is taken off the flask, which produces a metallic "clack," reminding one of a solid switch. Here we have a good example of a differential in power: the gentle touch of the flask and the effortlessness of the interaction contrast, but still correspond, with the energetic, impulsive, assertive sound. The user performs a gentle, ancient ritual with a modern high-tech device.

18.2.2   Wearables and "Pocketables"

"Moody Hat"
Kai Jauslin, Monika Bühner, and Simon Broggi

This group demonstrated a hat with an attitude. While lying somewhere it attracts a potential wearer by whispering and making strange and breathing sounds. When somebody picks it up it attaches itself violently to the person's head, which is accompanied by a synthetic hum, oscillating with increased frequency toward a final outburst. Depending on how the person touches the hat, it creates various sounds, which have a close link to the gestural interaction. Touching it in the "appropriate" way results in a tonal scale being played; "inappropriate" touching results in the sound reminiscent of breaking glass. The hat was perceived as animated and emotional. In particular, the initial phase of interaction marks a transition from dead matter to living being, which was a prominent metatopic in the film clips analyzed and is expressed by sounds that convey a rise in energy level due to activation. The manipulative gestures are generic but could control various functions that the hat could offer, which makes it an extended artifact.

"Angry Wallet"
Su-Jin Whang

This participant presented a wallet that monitors the user who is through browsing shopping catalogs to prevent unnecessary expenses. When the user opens the wallet to make a transaction, it may make a bark-like sound in disapproval. If the user insists and folds the wallet open to take money out, the wallet snarls, and if she nevertheless extracts a credit card, the wallet growls very aggressively. The barking and growling noises are derived from the sound of the wallet's zipper, which was slightly defamiliarized by cutting, changing its pitch and playback speed. This design strategy helps to integrate the artificial sounds into the wallet's sonic and gestural identity, rather interpreting than breaking it. The resulting ambiguous sonic identity prevents an overstated indexicality that would result in an undesired cartooning effect. Moreover there is an ambiguity of agency, as the sounds are impersonating the wallet's animated character and personality while at the same time linked directly to the continuous gestures of the user (opening wallet, taking out card, etc.).

18.2.3   Professional Tools for Professionals

"Ancient Tape Measure"
Emil Eirola, Tapani Pihlajamäki, and Jussi Pekonen

The scenario from this group featured a "living" measuring tape. Soon after it is extended, the user needs to struggle in order to use it, and finally it becomes uncontrollable and retracts. The sound design followed the metatopic of controllability and loss of control. Sounds were derived from related filmic examples, associating the forceful extraction with resonant and stable metallic sounds that are transformed to metallic sounds with a wobbling structure. This prototype explores resistentialism[4] by exaggerating the already tricky handling of a measuring tape and characterizes the tape as being rather hostile, using a deep, reverberating metallic impact sound for the retraction of the tape. A magic quality is expressed using related filmic stereotypes, which are perceived as mediated enunciations rather than as an inherent quality of the artifact. Like the wallet, this case represents a dialectic nature of the object between autonomous being and passive tool, as some sounds are closely associated with the user's and others with the tape's agency.

"Assembly Line"
Philipp Lehmann, Julian Kraan, and Didier Bertschinger

In this prototype of a futuristic assembly line for car manufacturing, robotic arms are remotely controlled through touch screens. The manipulative gestures on the touch screens are mediated by a multilayered sonic environment. Abstracted synthesized representations, for instance of the engine, represent different components of the car. This provides the basis for sounds that represent changes in precision, compatibility, and overall quality by modulating consonance and pureness in the synthesized timbres. The sounds are constantly modulated until a closure is reached, for instance by resolving dissonance. The scenario focused on conveying the metatopics of transition and metamorphosis in the car's complex composition. The disembodied interaction via the touch screen is reembodied through the sonic representations and becomes more than mere feedback because of the rich and complex nature of the resulting sounds. The overall result is a multilayered, atmospheric sound composition with a musical quality, turning assembly line work into an artistic performance.

18.3   Discussion and Findings

The cases described, together with results from previous work, make it possible to develop elements of a heuristic framework for designing sounds for interactive commodities. In the following sections an overview of the core points of this framework is provided, which certainly is not complete but can serve as a starting point for design practice and further investigations.

18.3.1   Classes of Narrative Metatopics

The following classes of narrative metatopics emerged from the cases as the most relevant. Please note that the terms are derived from the folksonomy of the workshop participants and may not be in congruence with taxonomies from established scientific fields.

• *Artifact-user relationship* circles around the narrative of suitability and appropriateness of an artifact (and its characteristics) to a user. The perspective can be either on the artifact or on the user. The enunciation of this relationship can follow a notion of *acceptance* (foregrounding psychological enunciations) or *compatibility* (foregrounding traces of material processes).

• *(In)correct use* relates to the way an artifact is handled by a user.

• *(In)appropriate use* relates to the use of the artifact in a specific sociocultural context.

• *Quality of energy*   Energy was perceived as the force that initiates, maintains, or ends processes associated with interactive commodities. Apart from materiality, energetic quality is a fundamental aesthetic element of schizophonic artifacts. Energy, manifested through sound, can transcend its physical framework and be used in higher-level semantics, metaphorically or symbolically, and even for expressing a positive, supportive or negative, dangerous or adversarial quality of an artifact. Moreover, an artifact's (metaphorical) life cycle can be expressed through "energetic" sounds.

• *Source of control or energy*   An artifact, in particular a networked one, may be controlled or "possessed" by a known or obscure agent. This relates to the perception of an energy source, which may be explicitly placed beyond the artifact, possibly by sounds that are not interpreted as coming from the artifact itself.

• *Animated artifact*   Here an ensouling or subjectivization of the device is expressed sonically. This also involves transitions—dead matter becomes "alive," and vice versa.

• *Atmospheric artifact*   "Atmosphere," as it was used in the discourses during the workshops, had two meanings. First, it refers to artifacts that express a mood through sound. Second, it can be defined as constituting the "in-between" between environmental qualities and human sensibilities [2]. Atmospheric sound in particular leads us outside ourselves, between our body and the sound's origin. Böhme's description [2] seems to suit the implicit understanding of the artifacts in question. When speaking about this quality, people were alluding to the effect of the machine's presence, mediated by sound, to the human's sensibilities.

In most of these metatopics, an essential semantic trajectory is established between the two poles of the *location of agency* in the experience of interactions: *artifact* or *user*. The sonic mediation of this location of agency can be ambiguous and may even shift during the course of an interaction.

18.3.2   Situational Categories and Related Heuristics

The cases described have been grouped according to the sociocultural situations they relate to, as this strongly influences how sounds are designed and interpreted. The categorical headings given here represent only the topics that emerged in the context of the workshops described and are neither definite nor complete nor mutually exclusive. But they are well aligned with established categorizations (e.g., "Smart Home," "Wearable Computing," or "Computer-Supported Collaborative Work").[5]

Private Spaces

Private spaces, in particular domestic environments, provide good opportunities to use sounds that reflect personal image. Sounds here become elements of personal experiences and narratives [14]. Moreover, a poetic or atmospheric quality can be important. In the case of the dangerous microwave, the same atmospheric quality expresses a disruption of homely safety. Another aspect that emerged prominently is the reflection of (moral) values, as in the fridge's sonic reflection of the food selection.

The need to sonically accommodate both private and social situations was identified. In the private setting the prototypes tend to foreground companionship, using gentle sounds and smooth transitions. This changes to some extent in a more sociable setting. In general, a tentative design heuristic could be that the more an artifact serves to promote a self-image explicitly, the more its sound should call for attention and stand out from the soundscape. Also, if value judgments are expressed, the sound may have to be designed so that the quality of being a statement is understood only by the concerned person. However, as people visiting somebody at home are usually either family or friends, a noticeable expression of personality may still be of interest.

All these points translate well to other private situations that are not covered by the reported cases, such as the bathroom (as a place for cleansing of both body and soul, as it were), systems for environment-friendly energy management (where in a social setting moral values may be more relevant), or even personal vehicles.

Wearables and "Pocketables"

Wearable technology shares the aspect of being a tool for personal expression with the artifacts for private spaces, with the difference that it is much more difficult to control who actually will hear the sonic expressions. Therefore, expressive quality is meant to be directed at the user exclusively, in such a way that it is clear enough for the user to notice but will not be heard by bystanders. The "angry wallet" uses this quality to emphasize negative feelings such as guilt. Both the "angry wallet" and the "moody hat" reveal the uncanny potential of intimate yet autonomous, or even "possessed," high tech. In the case of the wallet, however, the object oscillates between a very personal artifact (as a container of *your own* money and all wishes and fears associated with it) and a neutral functionality (as a container of *some* money).

Professional Tools for Professionals

Tools are professional in the sense that they convey professionalism at all stages of interaction and also in the sonic aesthetics they employ. This allows or even demands the construction of more complex sonic semantics. The assembly line prototype showed that the combination of abstracted everyday sounds with musical design strategies, guided by musical theory, is a powerful design strategy here. Musical systems allow the construction of complex artificial sonic structures. For instance, dissonance, cadence, rhythmic patterns, and the interplay of voices (e.g., counterpoint) can be used to express states, relationships, dynamics, and qualities. The tools are thus also *for* professionals because they require a certain level of mastery and a familiarity with a specific code. The "ancient tape measure" example supports this insight ex negativo.

18.3.3 Semantic Differentials of a Situational Heuristic

Several trajectories across situational categories that define the relationship between interactive commodities and their circumstances have been identified. In the following, their poles are described, which can be used for the conception of appropriate sound design strategies. Usually several aspects will overlap and characterizations will shift during the process of interaction.

Private-Public

As sound pervades space, this is a very important aspect. Obviously, in a private situation the design seems to be much freer. However at the same time the sound will also stand out more, and its occurrence may be more significant, which may require a finer tuning of repetitions (variations), complexity, and elaboration. The cases described show this clearly, also in the attempt to create different sounds for more social private situations. In public spaces, on the other hand, it is important to make sure that the sound(s) will be heard by those concerned. This depends less on volume than on the plausibility of elements of the acoustic ecology. Subtle modifications allow the sounds to be picked out of the soundscape by those who know the sonic pattern of the modification. The more private an interaction is, the more also the aspect of ergo-audition can be played out (see section 18.3.5). In public this requires that the sound be produced within a suitable acoustic community. It is interesting to note that the intersection of "wearable computing" and "public space" combines requirements from both private and public settings.

Tool-Assistant

From the point of view of task-oriented interactions, the artifact as "mere tool" always remains objectified and by definition is not autonomous. Sounds here are a manifestation of the work done *with* the artifact rather than produced by the artifact itself. The sound design thus relates to the qualities of the work rather than to those of the artifact, unless the latter is "authentic." In this case, the sounds need to be based on the given sonic properties of the artifact.

The "assistant," on the other hand, is a subjectified artifact that may act independently and whose sound design aims at working out an autonomous character. Note that these poles correspond to the level of autonomy as well (see section 18.3.1). In terms of task-oriented activities, an artifact-as-assistant can be positively characterized as "supporter" or "companion." An interesting variation of these basic types could be the artifact as slave, where "toolness" and a higher level of autonomy are combined.

Casual-Professional

The casual pole stands for an incidental quality and a low level of attention during interaction. This reduces the need for a detailed elaboration of the relationship between action/gesture and the sound. Also, the complexity of the sounds can be rather limited to the minimum that makes the sound interesting in a given situation. A casual interaction also implies that the attention may shift to other artifacts in the process, which suggests that the sounds be fine-tuned to fit into the acoustic ecology, allowing a smooth shift of focus to and away from it. The opposite is the case for the professional artifact, where a sound can be as complex as required for the interaction, and its relationship with control gestures, functions, and procedures is essential. In a professional environment the soundscape is defined by the tools used and certainly can claim attention to a certain degree.

18.3.4 Narrativity and Enunciation

To operate with the concept of narrativity is a helpful strategy for sound design for interactive commodities because it supports a focus on the interpretation of the artifact's agency and the meaning of events and processes. However, as the cases here have shown, it is in some way also misleading, as interactive technology is not narrative in the way, for example, that films are. The study of computer games provides some valuable hints because in many ways interactive games incorporate both narrative and performative aspects [13]. Further work has to clarify what the impact of this ambivalence is on the design and experience of schizophonic interactive artifacts. One insight from the cases described here is that a narrative potential can become assimilated by the action and thus become an element of ergo-audition (see below), transcending the limits of a linear, one-directional storytelling. And what is more, the more agency remains with the artifact the more its sounds can become actual "utterances" of it.

A related quality, which was recurrent in the designs reported above, was enunciation. This term designates sensory manifestations of a narrative system, that speaks to an "audience"; that is, it is not a manifestation of a strictly diegetic event. Because an "intelligent" system can be understood as expressive to some extent, this use of sound is rather common. Now that we have said this, enunciation works only if the artifact is perceived as an abstract system rather than animated. In general, enunciations require the establishment of a metasystem of communication that can serve as source for the enunciative act.

18.3.5 Performativity and Ergo-Audition

Interaction sounds can be understood and conceived as manifestations of our agency in the world. Chion coined the word *ergo-audition* to denote this phenomenon [4]. As a result, sounds can contribute to both joyful and frustrating experiences. Ergo-audition is potentially more joyful if the action-sound relationship is not too simple and also always a bit surprising (e.g., through variations). The cases of the thermos flask and the "moody hat" confirm the design potential of ergo-audition. This is also true for the "angry wallet," albeit in a negative manner: The more carefully the user opens the wallet, the more stretched and dreadful the growl will be. In the case of interactive artifacts with shifting location of agency, it is important to note that ergo-audition will shift along with agency. In this context we can also speak of a mediated ergo-audition, in the case when a very personal artifact, such as a "companion" in the terms outlined above, creates a sound by its own initiative.

The case of the "angry wallet" also challenges the notion of a clear bipolarity of agency (either user or machine). On the one hand, it is an example of sounds expressing an autonomous character in an animated object; on the other hand, no sound is created at all without the user's manipulation of the artifact.

18.3.6 Sonic Referentiality in Schizophonic Artifacts

Certainly referential functions of sounds exist and are useful. Sonic referentiality includes all aspects of sounds as signs qua an identified source or cause. In a linguistic analogy, referentiality is the base for tropes such as metonymy, metaphor, or allegory. In particular, metaphor, the expression of a new and unfamiliar entity with the help of familiar concepts, is often discussed in sound design.

But it is problematic to reduce the possible use of sounds to metaphoric and ultimately referential functions, making causal certainty the only relevant quality criterion. Like any sensory manifestation, sound also has a phenomenal and aesthetic quality in itself, free from metaphoric and referential functions, as prominently discussed by Schaeffer [16] and Cage (quoted in Thorau [18]). The aesthetic possibilities of using sound for its own quality have been elaborated by Hug [7] and are well demonstrated in the prototypes described here, including the smart fridge, the poetic thermos flask, and the assembly line.

If referential sounds (i.e., with high causal certainty level) are used in a nonsymbolic design context (e.g., not graphical user interfaces), their integration into a physical body might conflict with the physical body or process that the sound's source represents, a consequence of the dialectic nature of schizophonic artifacts. Salient examples from the cases presented here are the "moody hat" and the "angry wallet." Thus, it is important to understand what the natural qualities and what the semiotic aspect of an interactive artifact are. Metaphors usually can be established if the sound used is clearly identified as belonging to another domain than the artifact. This requires that the type of artifact supports such an obvious aesthetic rupture. Here the typology of interactive commodities provides a heuristic.

Not all sounds that are at first understood as metaphorical are really so. Let us revisit the case of the "moody hat." It produces sounds of breaking glass when it is touched inappropriately. This certainly works metaphorically but also—and in terms of an impromptu interaction process probably more importantly—is immediately meaningful, based on our experience of a negative relationship between our touching and something breaking. Moreover, the mere sonic quality (as described, e.g., by Schaeffer's morphotypology [16]) plays an important role as well in the creation of expressive, gestural qualities. After all, if just the process of breaking needs to be conveyed metaphorically, also the sound of breaking wood could have been used, but certainly with a very different experiential result.

Finally, we have to keep in mind that "authentic" sounds may mix seamlessly with "caricatures" (e.g., in the case of the "dangerous microwave"), metaphors, and enunciations. Thus, because of the ability of sound to merge into complex amalgams, often neither the metaphorical nor the sonic quality per se is predominant.

18.3.7 Applying the Typology of Schizophonic Interactive Commodities

The typology that has been proposed and outlined above is intended to provide a base for design heuristics for interactive schizophonic commodities. In particular it helps to frame the use of narrative design strategies and metatopics in order to form an initial "design hypothesis." This, however, needs to be integrated in an interactive, participative design process to allow for adaptations and further refinement of the heuristics. It has to be noted that the typological categories often cannot be applied exclusively to an artifact but may shift. In fact, in varying stages of interaction, various types may come to the foreground. For instance, at an initial stage authenticity is a useful quality that establishes trust. Extension, substitution, and omnivalence then may become the defining quality of an artifact in the course of an interaction process. Moreover, an extension can be developed into a constituting characteristic of a new type of artifact, which is experienced as "authentic" again. After all, schizophonic interactive commodities may also be schizophrenic!

18.4 Conclusion

The workshops carried out were very beneficial for exploring possible future experiences with schizophonic interactive commodities in a structured and revisable way. The combination of narrative sound design strategies and experience prototyping using the "Wizard of Oz" method worked well and made it possible to apply and test hypothetical design heuristics.

On the basis of the analysis of the cases from the workshops, several specific contributions to a heuristic framework for designing schizophonic interactive commodities could be made. First of all, narrative metatopics have emerged from the process that can serve as starting points for an "interactive storytelling" with artifacts. Second, several continua have been formulated, grounded in the cases described, that can work as semantic differential for conception and evaluation. Furthermore, several observations were made related to the limitation of the narrative paradigm, the question of referentiality of sound, performativity and ergo-audition, and the application of the typology of schizophonic interactive commodities.

But none of these observations constitutes a definitive "guideline" or "how-to." Apart from the need for applying these tentative principles to further test and elaborate them, the experience from the workshops also has shown how important it is to integrate interpretation explicitly into the design process. Cultural codes for sonic signs for nonimmersive media have not been established, and this is an important strategy to approach hermeneutics and design heuristics for SID.

As a welcome side effect for this research through design, many creative and promising concepts for possible applications of sonic interaction design emerged. Interestingly, several projects embodied a critical approach, reflecting possible scenarios of a life with such technology. This critical approach was also helpful to stimulate response and sharpen the insights ex negativo.

One recurrent characteristic that emerged from reviewing the cases reported here was the ambivalence of schizophonic artifacts on at least two levels: first in terms of material identity and second in terms of a narrative quality. Thus, interactive commodities always potentially incorporate objectivity (as material entities) and subjectivity (as psychologized agents), as it were. In this sense, sound can be used to design "hermeneutic affordances," that is, as a way to provide the "condition of the possibility" of interpretation and understanding, merging immediate perception with higher-level semantics. Future work will need to incorporate this aspect, in particular by appropriating suitable hermeneutic and semiotic frameworks as well as performance-related theory and, last but not least, by creating more cases to study.

Notes

1. A more detailed discussion of the semantics of artifacts and their relevance for interactive commodities can be found in Hug [7].

2. A similar design-research strategy has also been used in a theatrical setting [15].

3. Professional jargon for sounds that are enhanced with other sounds and processed in order to give them a stronger profile and a specific narrative quality.

4. This term, coined by Paul Jennings, stands for a humorous theory in which inanimate objects display hostile desires toward humans, a "fact" apparent in experiences such as cars not starting when one is in a hurry or the bread always falling on the side with butter on it.

5. Toys are another relevant category and were also addressed by a case. However, such artifacts open a vast field, and their discussion would go far beyond the scope of this chapter.

References

1. Aarts, E., & Marzano, S. (Eds.). (2003). *The new everyday*. Rotterdam, The Netherlands: 010 Publishers.

2. Böhme, G. (2000). Acoustic atmospheres—a contribution to the study of ecological aesthetics. *Soundscape—The Journal of Acoustic Ecology, 1*(1), 14–18.

3. Buchenau, M., & Suri, J. F. (2000). Experience prototyping. In *Proceedings of the conference on designing interactive systems*, Brooklyn, NY: ACM Press. (pp. 424–433).

4. Chion, M. (1998). *Le son*. Paris: Editions Nathan.

5. Csikszentmihalyi, M., & Rochberg-Halton, E. (1981). *The meaning of things—Domestic symbols and the self*. Cambridge: Cambridge University Press.

6. Franinović, K., Hug, D., & Visell, Y. (2007). Sound embodied: Explorations of sonic interaction design for everyday objects in a workshop setting. In *Proceedings of the 13th international conference on auditory display*, Montreal: ICAD. (pp. 334–341).

7. Hug, D. (2008). Genie in a bottle: Object-sound reconfigurations for interactive commodities. In *Proceedings of audiomostly 2008, 3rd conference on interaction with sound*, Pitea: Audiomostly (pp. 56–63).

8. Hug, D. (2008). Towards a hermeneutics and typology of sound for interactive commodities. In *Proceedings of the CHI 2008 workshop on sonic interaction design*, Firenze: ACM Press (pp. 11–16).

9. Hug, D. (2009). Using a systematic design process to investigate narrative sound design strategies for interactive commodities. In *Proceedings of the 15th international conference on auditory display*, Copenhagen, Denmark: ICAD (pp. 1–8).

10. Kelley, J. F. (1984). An iterative design methodology for user-friendly natural language office information applications. *ACM Transactions on Office Information Systems, 2*(1), 26–41.

11. Latour, B. (2005). *Reassembling the social—An introduction to action-network-theory.* Oxford: Oxford University Press.

12. McCarthy, J., Wright, P., & Wallace, J. (2006). The experience of enchantment in human-computer interaction. *Personal and Ubiquitous Computing, 10,* 369–378.

13. Neitzel, B. (2000). *Gespielte Geschichten. Struktur- und prozessanalytische Untersuchungen der Narrativität von Videospielen.* PhD thesis, Universität Weimar.

14. Oleksik, G., et al. (2008). Sonic interventions: Understanding and extending the domestic soundscape. In *Proceedings of the 26th annual CHI conference on human factors in computing systems* (pp. 1419–1428).

15. Pauletto, S., et al. (2009). Integrating theatrical strategies into sonic interaction design. In *Proceedings of audio mostly 2009–4th conference on interaction with sound*, Glasgow: Audiomostly (pp. 77–82).

16. Schaeffer, P. (1966). *Traité des objets musicaux.* Paris: Seuil.

17. Schafer, R. M. (1977). *The soundscape: Our sonic environment and the tuning of the world* (2nd ed.). New York: Destiny Books.

18. Thorau, C. (2007). The sound itself—antimetaphorisches Hören an den Grenzen von Kunst. *Paragrana—Internationale Zeitschrift für Historische Anthropologie, 16*(2), 206–214.

19. Vastokas, J. M. (1994). Are artifacts texts? Lithuanian woven sashes as social and cosmic transactions. In S. H. Riggins (Ed.), *The socialness of things—Essays on the socio-semiotics of objects* (pp. 337–362). Berlin: Mouton de Gruyter.

Contributors

Federico Avanzini
University of Padova, Padova, Italy

Gerold Baier
Bielefeld University, Bielefeld, Germany

Stephen Barrass
University of Canberra, Canberra, Australia

Olivier Bau
Disney Research, Pittsburgh, Pennsylvania

Karin Bijsterveld
University of Maastricht, Maastricht, The Netherlands

Roberto Bresin
Kungliga Tekniska högskolan, Stockholm, Sweden

Stephen A. Brewster
University of Glasgow, Glasgow, UK

Jeremy R. Cooperstock
McGill University, Montreal, Canada

Amalia De Götzen
Aalborg University Copenhagen, Copenhagen, Denmark

Stefano delle Monache
IUAV University of Venice, Venice, Italy

Cumhur Erkut
Aalto University, Espoo, Finland

Georg Essl
University of Michigan, Ann Arbor, Michigan

Karmen Franinović
Zurich University of the Arts, Zurich, Switzerland

Bruno L. Giordano
Institute of Neuroscience and Psychology, University of Glasgow, Glasgow, UK

Thomas Hermann
Bielefeld University, Bielefeld, Germany

Daniel Hug
Zurich University of the Arts, Zurich, Switzerland

Antti Jylhä
Aalto University, Espoo, Finland

Johan Kildal
Nokia, Finland

Stefan Krebs
University of Maastricht, Maastricht, The Netherlands

Anatole Lécuyer
INRIA, Rennes, France

Wendy Mackay
University Paris Sud, France

David Merrill
Sifteo Inc., San Francisco, CA

Roderick Murray-Smith
University of Glasgow, Glasgow, UK

Sile O'Modhrain
University of Michigan, Ann Arbor, Michigan

Pietro Polotti
Conservatorio Giuseppe Tartini, Trieste, Italy

Hayes Raffle
Google, Mountain View, California

Michal Rinott
Holon Institute of Technology, Holon, Israel

Davide Rocchesso
IUAV University of Venice, Venice, Italy

Antonio Rodà
University of Padova, Padova, Italy

Christopher Salter
Concordia University, Montreal, Canada

Stefania Serafin
Aalborg University Copenhagen, Copenhagen, Denmark

Zack Settel
University of Montreal, Montreal, Canada

Simone Spagnol
IUAV University of Venice, Venice, Italy

Jean Sreng
CEA List, Fontenay-aux-Roses, France

Patrick Susini
IRCAM, Paris, France

Atau Tanaka
Newcastle University, Newcastle, UK

Yon Visell
University Paris 06, Paris, France

John Williamson
University of Glasgow, Glasgow, UK

Mike Wozniewski
McGill University, Montreal, Canada

Index